STALIN'S
LAST CRIME

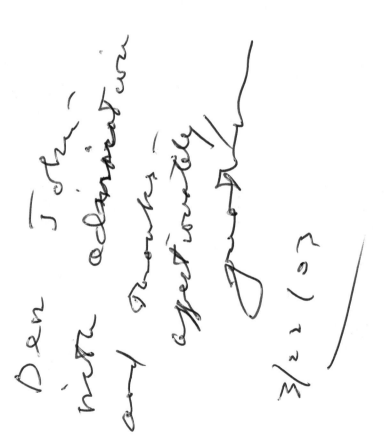

STALIN'S LAST CRIME

THE PLOT AGAINST
THE JEWISH DOCTORS
1948–1953

JONATHAN BRENT

AND

VLADIMIR P. NAUMOV

HarperCollins*Publishers*

HarperCollins books may be purchased for educational, business, or sales promotional use. For information, please write: Special Markets Department, HarperCollins Publishers Inc., 10 East 53rd Street, New York, NY 10022.

FIRST EDITION

Designed by Jackie McKee

Printed on acid-free paper

Library of Congress Cataloging-in-Publication Data

Brent, Jonathan.
Stalin's last crime : the plot against the Jewish doctors, 1948–1953/Jonathan Brent and Vladimir P. Naumov.
    p. cm.
Includes bibliographical references and index.
ISBN 0-06-019524-X
    1. Jews—Soviet Union.   2. Jews—Persecutions—Soviet Union.
3. Jewish physicians—Soviet Union.   4. Stalin, Joseph, 1879–1953—Views on Jews.   5. Soviet Union—History—1925–1953.   I. Naumov, Vladimir Pavlovich.   II. Title.

    DS135.R92B744   2003
    947.084'2—dc21   2002191930

03 04 05 06 07 NMSG/RRD   10 9 8 7 6 5 4 3 2

Doctors, doctors
Have become our sons!
A star shines over our heads
Oy!

— Anonymous Yiddish poem set to music by Dmitri Shostakovich, 1948
(not publicly performed until 1964)

CONTENTS

ACKNOWLEDGMENTS

Vladimir Naumov began working on the Doctors' Plot in 1992, having approached it through his work on the Commission for the Rehabilitation of Repressed Persons in the late 1980s. As part of the commission's work, he studied the 1952 trial of the Jewish Antifascist Committee and soon discovered that nobody had examined the details of this extraordinary case before. He read the entire case—forty-two volumes of documents including eight or nine additional volumes. He found that many of the JAC defendants had not yet been amnestied or rehabilitated. Naumov was stunned by the words of one of the defendants, Solomon Lozovsky, in which he declared his innocence and predicted that eventually the truth would become clear. Lozovsky had only one request, that when they were all eventually rehabilitated, the truth be published in the newspapers. Naumov resolved to carry out this wish.

Unlike the trial of the JAC, however, the documents concerning the Doctors' Plot were dispersed in different archives, including the KGB archive, the presidential archive, and various party archives. We thank Archive of the President of the Russian Federation, Russian State Archive of Social and Political History (RGASPI), Central Archive of the Federal Security Service (FSB), and Russian State Archive of Contemporary History (RGANI).

Our thanks go also to the many individuals without whom this book could not have been written:

Rita Yakusheva translated the English language manuscript for Vladimir Pavlovich. John Lukacs read the entire manuscript at an early stage and made valuable corrections on practically every page. His comments and criticisms have improved this book throughout. David Murphy has provided continual, unstinting help throughout the difficult process of sorting through the intelligence materials on which much of this book is based, checking every footnote and reference, and bringing invaluable materials to our attention. Holocaust historian Nechama Tec read a draft of the manuscript and caused much constructive rethinking of important aspects of the plot as they connected with Jewish history. Gary Saul Morson read with great attention and perceptiveness the entire manuscript of late draft and caught a vast number of details requiring attention; Saul greatly sustained the difficult process of revision. Sincere thanks go to Richard Pipes for reading the manuscript and offering many important suggestions for improvement. Harvey Klehr read an early draft and suggested fundamental changes in the narrative and point of view. William E. Odom read a late draft of the manuscript, offered many constructive insights, and helped support our view of the historical processes at work in this phase of Soviet history. Annabel Patterson read the manuscript in its penultimate form and urged various improvements in style and presentation. Without the continual support and encouragement of William F. Buckley Jr., the Yale University Press project of publishing Soviet documents could not have continued and this book would not have come into being.

In Moscow we must thank Nikolai Petrovich Yakovlev (who died tragically two years ago), Olga A. Varshaver, Oleg V. Naumov, and many others who have helped open up the world of Soviet archives to scholarly research. Among these we are honored to count as one who has offered continual generous support, Alexander Nikolaevich Yakovlev, the architect of perestroika under Gorbachev, and the head of the Presidential Commission on the Rehabilitation of Repressed Persons in Moscow. His dedication, passion, and sincerity have set a standard of intellectual commitment. His depth of concern for the fates so many millions suffered in a cruel and despotic system has demonstrated what the living connection between history and political action can be.

We would like to thank John G. Ryden, former director of Yale University Press, who believed in our work with the Soviet archives, and significantly helped make this book possible. John's contribution to Yale

Press over a twenty-three-year period has been rich, diverse, and large-visioned.

Dr. Lawrence S. Cohen at Yale University School of Medicine generously advised on the subjects of Timashuk's diagnosis, Zhdanov's heart condition, and the report on Stalin's illness. We are grateful for his interest, help and professional advice. Dr. Phillip S. Dickey, also at Yale University School of Medicine, read a final draft of the chapter on Stalin's death and provided sober advice on how to deal with the difficult medical material. Robert Silver, Ph.D., professor of physiology and pharmacology at Wayne State University, provided exceptional help in sorting through the medical evidence.

At Yale University Press we wish to thank Vadim Staklo, special projects manager for the Annals of Communism series over the last four years; Gretchen Rings, editorial assistant; and the entire staff of the press for putting up with what must often have seemed a fanatic preoccupation with Soviet archives.

At HarperCollins we are exceedingly grateful to Hugh Van Dusen, whose encouragement, support, and kindness have helped sustain this project. We also owe a large and happy debt to our agent John Thornton.

In the background of this book are the works of great scholars, in particular the magisterial biography of Stalin by Robert C. Tucker, whose influence can be seen and felt throughout; the groundbreaking work of Robert Conquest, who opened the door for all subsequent researchers to a serious consideration of Stalin's crimes; and the work of the numerous scholars participating in the Annals of Communism archive project of Yale University Press.

The authors dedicate this book to our families, who have sustained us in mind and body throughout the long and often complicated transcontinental collaboration.

STALIN'S
LAST CRIME

INTRODUCTION:
THE INVERTED WORLD

Like a dead spirit he stood over us.

—ALEXANDER TVARDOVSKY

Nothing impressed me so much as the doctor story," Winston Churchill wrote to President Eisenhower on April 11, 1953, a little over a month after Stalin's death, noting, "This must cut very deeply into communist discipline and structure."[1] Churchill intuitively recognized that this "story" held particular significance for Soviet and world affairs. Though it was difficult to say what that significance was, he urged the American president to take it as a sign of potential change in future U.S.–Soviet relations.

As Churchill divined, the "doctor story" was a tangled, complicated affair, deeply embedded in the structure of Kremlin leadership. It reflected many of the intentions of Stalin's foreign policy as well as the tensions within his government. Had the plot against the Jewish doctors succeeded, much subsequent world history might have been quite different. Many leading Kremlin figures would have been purged and probably shot; the security services and the military would have been decimated by purges; Soviet intellectuals and artists, particularly Jews, would have been mercilessly repressed; and the surviving remnant of Soviet and Eastern European Jewry would have been gravely (perhaps mortally) imperiled, while grievous suffering would have been inflicted on all the citizens of the Soviet Union. Another Great Terror, such as occurred in the late 1930s, was averted when Stalin suddenly died on March 5, 1953. Stalin's version of a "final solution" remained unfulfilled, and the new Soviet leaders immediately backed away from the abyss toward which Soviet society was headed. The Great Terror was a prelude to Stalin's preparations for World War II, and much new evidence shows that the Doctors' Plot would have served a similar purpose. In the thirties the enemy was Germany; after the war, it was the United States.

The doctors' story is difficult to sort out into a clear, linear unfolding of events. A great many tiny details from a great many different sources within the Soviet government and internationally contributed to its development. To see Stalin's signature on each detail he hammered into place requires the vision of the cross-eyed, left-handed craftsman in Nikolai Leskov's famous story, who engraved his initials on the head of each microscopic nail used to shoe the czar's golden flea. The doctors' story has many such nails, and this book seeks to read the inscriptions they bear. Like those of the czar's flea, its ponderous, if tiny, shoes disclose a marvel of intricate workmanship.

Unlike Leskov's left-handed craftsman, Stalin, the illegitimate son of a Georgian shoemaker, in many respects worked backward. To be sure, the investigation into the death of A. A. Zhdanov, who died in 1948, and the alleged crimes of doctor Yakov G. Etinger progressed over a period of several years, ending in 1953. But this forward progress of the investigation was something of an illusion. The investigation proceeded from the death of Etinger in March 1951 to the *Pravda* announcement of the Doctors' Plot in January 1953 as if it were uncovering facts, piecing together details, drawing conclusions, exposing crimes. In reality the crimes had been determined in advance, at least as early as the July 1951 secret letter of the Central Committee. What was "proved" in 1953 had been stated as fact and established from the beginning. A good example of this general principle of working from conclusion to fact is the January 1953 article in *Pravda* revealing the sensational confessions of various doctors, even though the doctors had not yet confessed to these crimes; another example is the evolving, if posthumous, role of Politburo member A. A. Kuznetsov throughout the investigation; a third example is the invisible but all important shift in the 1951 secret letter from a case against one man, Dr. Etinger, to a "conspiratorial *group*." While it appears that the evolution of Kuznetsov's role, and the change from an individual crime to a group conspiracy may have occurred as the investigation gathered more facts, the truth is that Kuznetsov's role and the mass nature of the conspiracy were essential to the plot *from the outset*. As a result, the story we tell must zigzag back and forth from 1948 to 1953 so that details occurring at an earlier stage can be understood in their final context. Narratives move linearly in time; Stalin's plots did not.

Stalin's Doctors' Plot came to worldwide attention in January 1953, two months before his death. It has been called "the provocation of the century"; it has also been described as the irrational product of the aging dictator's diseased mind. Though Stalin died before he realized his intentions, the archival record allows us to reconstruct his purpose with some confidence. This record shows that until February 17, 1953,[2] thirteen days before his stroke, Stalin retained a vigorous, purposive mind; his ambitions were clear; his hold on power assured and complete. Nothing in the archival record suggests that Stalin's last days as leader of the Soviet Union were not consistent with his first.

From the distance of fifty years it is now possible to see more clearly what Churchill only guessed, that the story of the doctors was a state-sponsored conspiracy of mass proportions in the service of Stalin's political vision. It grew together gradually over a period of several years. Though sketched out in July 1951, it achieved its final form only in the late fall of 1952. Announced to the world on January 13, 1953, it came to an abrupt halt with Stalin's death in March. Those who followed him were quick to obliterate all trace of it.

The *dyelo vrachey* (case of the doctors), as it was called by the Soviet government, was alleged at the time to be a widespread conspiracy in the Soviet medical profession organized by Jewish physicians against Kremlin leaders. A. A. Zhdanov was one victim; A. S. Shcherbakov was another. Many other prominent names were added including Georgi Dimitrov, former head of the Comintern; several Soviet generals; Grigory Malenkov, a powerful Politburo member; and important foreign communists as well. Jewish doctors were accused of either murdering these leaders or planning their murders in league with American intelligence and a corrupt Ministry of state security (MGB). Hundreds of doctors were arrested over a period of five months, beginning in October 1952 and ending in February 1953. The consequences of the plot did not stop there, however. Fantastic rumors circulated that Jewish doctors were poisoning Russian children, injecting them with diphtheria, and killing newborn infants in maternity hospitals.[3] Terrible clouds gathered on the horizon of Soviet society. After Stalin's death, in March 1953, the core group of thirty-seven doctors and their wives was released from prison. More releases followed in a general amnesty; apparent calm returned to daily life. Of the thirty-seven, only seventeen were Jews. Of the original group of six doctors accused of murdering A. A. Zhdanov in 1948, none was Jewish with the exception of the EKG technician, Sophia Karpai. This fact is one of the plot's deepest tangles.

Today it is generally thought that the Doctors' Plot instead of being a conspiracy of doctors against the government was a conspiracy by the government against the Jewish doctors. The documents assembled in this book tell a different story. They show us that the "case of the doctors" was actually a conspiracy of the government, in the person of Stalin, against itself. Had it succeeded, its rabid, anti-Semitic character would have had devastating consequences for the trapped Soviet Jewish population, but it had far wider implications, well beyond those of the 1952 trial of the Jewish Antifascist Committee or the ugly suppression of Jewish rights that occurred during the so-called anticosmopolitan campaign that had begun in 1947–1948.

To understand how this could be so, it is necessary to reflect on the long history of the relationship between medicine and political power in the Bolshevik state, starting with the death of Lenin. Lenin was not murdered or alleged to have been murdered, but his illness gave Stalin and other Bolsheviks the chance to isolate him physically and deprive him of authority in the last days of his life. Trotsky's absence from Moscow at the time of Lenin's death significantly weakened his stature in the Soviet government and allowed Stalin to seize center stage at a crucial moment. Some fifteen years later, at the 1938 show trial of Nikolai Bukharin and other Kremlin leaders, a trio of doctors, Lev Levin, Dimitry Pletnev, and Ignaty Kazakov, were accused of having murdered prominent Soviet figures including Maxim Gorky; his son; V. V. Kuyibishev, a high-ranking Bolshevik who died in 1935; and V. R. Menzhinsky, former head of state security, who died in 1934. The then security chief, Genrikh Yagoda, was accused of having instigated the murders to overthrow the government. Of the three doctors, one, Levin, was Jewish. The charge proved effective then, and in 1953 Stalin used it once more against his enemies.

It is paradoxical that the Bolsheviks' intention to build a utopian society based on radical equality produced so many enemies so quickly. Part of this had to do with their need to turn the assumptions of Western liberal democracy upside down, thereby concentrating all power in the state. Only this could guarantee such equality. The result was that from a political standpoint, the individual counted for nothing. It was an equality based on an absence rather than a presence of rights. The world-famous Russian bass F. I. Chaliapin recounted in his memoir that in 1917 a Bolshevik named Rakhya explained to him that the first thing the new communist government must do when it came to power was to "cut off" talented people—the intellectuals and artists. Chaliapin was dumbfounded, but the explanation was simple: In the new state "no one must have an advantage over another. Talent destroys equality."[4] The first thing Chaliapin did after the revolution was to flee for France.

Standing at the apex of the state, Stalin had absolute power. He achieved this not because absolute power was conferred on him by the state, but because he succeeded in finding means to delegitimize the state itself. The Doctors' Plot became his most powerful weapon in the last years of his life in pursuing this end; it starkly demonstrates that Stalin's power did not derive from the state and its institutions but from the underlying system that allowed him to manipulate them. This book is a study in the exercise of this enormous power.

Terror and naked force became the system's principal means of achieving legitimacy in a world in which equality was based on mass disenfranchisement. The resulting inversion of the Western ideal of the social contract caused radical distrust to prevail throughout the social-political-human world that Lenin founded and Stalin developed. Tracing the evolution of the Doctors' Plot takes us into the inner life of this inverted world where most previous norms of political, religious, and social life had been discredited. As the case of the quack botanist T. D. Lysenko demonstrates, rational thought itself was subject to the dictates of the state.

Coinciding with the early phase of the Cold War, Stalin's conspiracy against the Jewish doctors reflected both the general external and internal conditions of the Soviet Union at the time. As such the plot has significance far beyond specific Kremlin rivalries, Stalin's personal anti-Semitism, or the malevolence of state security underlings like M. D. Ryumin, whom some credit with having masterminded it. The Doctors' Plot was the natural outgrowth of the bureaucratic, political, psychological, and moral structure of Stalin's system of government. Examining it reveals much about this system and the men who perpetuated it.

There have been few firsthand accounts of the Doctors' Plot, either by victims or by victimizers. Khrushchev's memoirs shed little direct light on it; most of the information Khrushchev conveyed was fabricated or incomplete. No other member of the Politburo wrote about it at all. Molotov, who was not directly involved except as a potential victim, provided a sketchy and largely misleading picture in his published conversations with Felix Chuev. Bits and pieces have come to us in the diary entries of V. A. Malyshev, who recorded parts of Stalin's important speech at the December 1, 1952, meeting of the Presidium of the Central Committee. Outsiders to the plot, such as Pavel Sudoplatov, Peter Deriabin, Zinovii Sheinis, Dimitri Volkogonov, Arkady Vaksberg, and Roy Medvedev, have fallen prey to a plenitude of misinformation or partial truths. The general defect of their retellings, however, is that each is fragmented and reflects a specific set of interests, whether the fate of Soviet Jews or intrigues in the Ministry of State Security. None has grasped the plot in its totality, largely because the relevant state and party documents have remained top secret for over fifty years. These are the documents on which this book is based.

The doctors themselves left no record at all. It has been said that the principals who survived met at the house of a colleague after they had been released from prison and took a solemn vow never to discuss their ordeal. For good reason. No one knew which way the political winds would blow after Stalin's death. Beyond this, no one knew what to say. Dr. Vladimir N. Vinogradov, Stalin's so-called personal physician, had been arrested in November 1952 and was tortured into making false confessions. He put the matter succinctly when his interrogators asked why he did not respond to a certain question. "I find myself in a tragic position," he told them, "and have nothing to say." Vinogradov and his colleagues, who had treated A. A. Zhdanov in 1948, all found themselves in this situation. Each was innocent and each was guilty. Though Lidia F. Timashuk received the Order of Lenin in January 1953 for her role in "uncovering" the saboteur doctors, there were no heroes in the Doctors' Plot. Her award was revoked by the new leaders soon after Stalin's death. She spent the rest of her life attempting to reclaim it.

The plot began haphazardly, without plan or clear intention on Stalin's part. The plan came to him over time. From A. A. Zhdanov's death in 1948 to the announcement in *Pravda* in January 1953, Stalin seized upon fortuitous events, manipulated some, and fabricated still others out of whole cloth. He employed all the means at his disposal—the Ministry of State Security, the Central Committee, the media, private channels both within and around the government—to achieve his ends. He remained the *khozyain* (boss) to the end.

Just at the point at which he thought he had achieved his goal, he died. The conspiracy that had consumed almost all of Stalin's energy and attention in the last year and a half of his life died with him and was immediately repudiated by his followers. One of Beria's first official acts, on March 13, 1953, as the new minister of state security (now called the Ministry of Internal Affairs, or MVD), was to order an intensive inquiry into the *delu arestovannykh vrachey* (case of the arrested doctors). On March 31 a decree would be issued terminating the case against the doctors and freeing them; on April 6, one month after Stalin's death, an article would appear in *Pravda* exposing the entire plot as the invention of "criminally adventurist" MGB officers and reaffirming that "the inviolability of the person" was guaranteed under Soviet law. Those held responsible for the plot were arrested; most were shot and others were dismissed from service. The speed with which Beria, Khrushchev, and their close colleagues acted suggests that Stalin's death may have been more than providential.

The origin of the Doctors' Plot was not its beginning. It had its origins in two apparently unrelated streams of events. The first was Stalin's anti-Semitic, anticosmopolitan campaign that began in 1947 and culminated with the execution of the members of the Jewish Antifascist Committee in August 1952. The second was the death of Andrei Zhdanov, a powerful Politburo member, in 1948. The Doctors' Plot did not begin until Stalin connected these tributaries into a single, united river that, as Beria put it in 1953, would "swallow up" the security services and threaten to engulf the entire nation. It is possible to date this point of unification to July 1951, when the Central Committee, following Stalin's lead, issued the *zakrytoe pismo* (secret letter) that outlined the eventual features of the plot. Stalin privately must have reached this point much sooner, but when remains a matter of speculation.

One MGB officer subsequently testified that significant orders given to the interrogators of the doctors were oral, not written.[5] Much was not written down and must be reconstructed from sometimes misleading or contradictory documents and testimony. Understanding Zhdanov's death poses special difficulties in this respect. Significant gaps exist both in the medical record and in the record of Stalin's decisions. No order depriving Zhdanov of his responsibilities in the Central Committee exists; no directive from Stalin ordering his death will ever be found; no record of the conversations among the doctors in Valdai, the government health resort where Zhdanov died, exists, except as they have been reconstructed by witnesses who often had an ax to grind. Nevertheless, a wide and varied array of materials substantiates the view that Stalin wished to punish Zhdanov severely in July 1948, even to the point of physical elimination. The doctors treating him in Valdai gradually became aware of Stalin's wish, and the care they provided Zhdanov strongly suggests that they knew what role they were being asked to perform.

Lidia Timashuk, an ordinary doctor in the Kremlin Hospital who was also an MGB operative, wrote a letter two days before Zhdanov's death informing the authorities that the doctors had not understood the gravity of Zhdanov's condition. The letter was duly passed to Stalin, and he ignored it. Perhaps it told him nothing he did not already know. After Zhdanov's death the doctors were not punished; instead, Timashuk was demoted. Three years later, in 1951, their positions were reversed. The doctors by now had come under suspicion and were gradually being arrested. Eventually they would be formally charged with having murdered Zhdanov, and Timashuk's letter would be used against them.

The plot, however, could not be realized until the accused doctors, none of whom was Jewish except the EKG technician who had preceded Timashuk in Valdai and had left three weeks before Zhdanov's death, could be linked to a specifically Jewish conspiracy. The continuing, vicious anti-cosmopolitan campaign provided a socially accepted basis for a mass action against the Jews, but a point of ignition was necessary.

It seems likely that with the arrest of Dr. Yakov Etinger, Stalin soon recognized his opportunity. Etinger, who was Jewish and widely regarded as one of the leading Soviet diagnosticians of his time, was arrested on November 20, 1950, on charges that he had uttered anti-Soviet thoughts to family members, friends, and colleagues. His arrest was part of the action taken against the Jewish Antifascist Committee, with which Etinger had had connections. Whether the result of the scheming of M. D. Ryumin, the MGB officer who interrogated him, or because Stalin had already implanted the notion in Ryumin's mind, Etinger was alleged to have referred to the death of A. S. Shcherbakov, a member of the Central Committee who had died in 1945, as a case of medical "wrecking."[6] The circumstances surrounding Ryumin's interrogation of Etinger are murky. Despite numerous official and unofficial interrogation sessions, only five protocols of interrogation were ever written down. Etinger's supposed confessions were never recorded. He died in March 1951, returning to his cell from an interrogation. Ryumin denounced the then minister of state security, Victor Abakumov, in July 1951 for having "smoothed over" or "slurred over" the Etinger case, concealing the elderly doctor's explosive confessions. Abakumov was removed; Ryumin was elevated to deputy minister of state security; the next year and a half was spent in feverish activity by the MGB to knit the plot together using Etinger's unrecorded confessions and Timashuk's letter as the foundation.

The plot grew far beyond the boundaries of an anti-Semitic action. It encompassed the security services and led to widespread purges. Kremlin leaders like Molotov, Mikoyan, and Voroshilov were denounced as spies. Citizens' committees were formed to identify and denounce Jews and other dubious individuals. Scientists, doctors, and intellectuals were arrested or came under increasing suspicion. As newly discovered documents show, in the months preceding Stalin's death four new, large concentration camps were put under construction.

Many of the prisoners fell into "a state of extreme physical collapse, deep moral depression," as one document put it. Many lost what this same document termed, rather evocatively, "their human aspect." The loss of a "human aspect" was not confined to the doctors. Built on radical distrust and the continual threat of state violence, Soviet society represented a limit to what civilization could endure. It jeopardized the "human aspect" of an entire people and posed a threat to civilization across the globe.

Its complex interweaving of personal motivation, political machination, and international intrigue makes the Doctors' Plot Stalin's last great criminal conspiracy. It is the work of his old age, revealing Stalin's mastery of all the resources at his disposal, his final masterpiece of deception.

A word about the documents in this book. Some have never been seen before; many have never been published before; none has ever been integrated into the large historical perspective we have sought to establish in this book. Some of the most important new documents are the July 11, 1951, *zakrytoye pismo* (secret letter) of the Central Committee; Beria's debriefings of S. D. Ignatiev and S. A. Goglidze after Stalin's death; the confessions of M. D. Ryumin and his July 2, 1951, letter to Stalin; the February 1953 order to begin construction of new concentration camps for "especially dangerous" criminals; Yuri Zhdanov's letter to Stalin; the criminal indictment of Abakumov with Stalin's handwritten marginal comments; and the unpublished, undated letter of Jewish intellectuals castigating their compatriots for anti-Soviet sentiments and behavior in supporting Israel. The bizarre case of I. I. Varfolomeyev and the "Plan of the Internal Blow" has never before been told. Perhaps the most suggestive new document is the medical report on Stalin's death. This report demonstrates conclusively that Khrushchev knowingly misrepresented Stalin's death and that all previous accounts have been incomplete, faulty, or fabricated. The implications of this for understanding the end of the Vohzd (Leader) and his regime are great. These documents have been stored for over fifty years in the archives of the KGB (now renamed FSB), the presidential archive, and other former party and state archives. We have been privileged to make use of them in this book through the good offices of those high officials in present-day Russia who wish the truth to be told about their history.

ONE

THE UNTIMELY DEATH OF COMRADE ZHDANOV

MOSCOW, SEPTEMBER 6, 1948

———

You went your glorious way, comrade Zhdanov
Leaving eternal footsteps behind

—ALEKSANDER ZHAROV, *PRAVDA*, SEPT. 1, 1948

With what biting sarcasm he threatened all the old enemies, all those
who would poison young Soviet literature with their pernicious
poison! With what annihilating scorn and hatred he exposed
the mercenary bourgeois literature of the West today, that
corrupts the minds of readers, that aids reaction, that served
fascism once in no small way and now serves the lackeyism of the
imperialists—the instigators of a new war!

—MIKHAIL SHOLOKHOV, *PRAVDA*, SEPT. 2, 1948

The workers of the entire world mourn the untimely
death of com. A. A. Zhdanov

—*IZVESTIA*, SEPT. 3, 1948

On August 31, 1948, at about 4 o'clock in the afternoon, Andrei Aleksandrovich Zhdanov, a powerful member of Stalin's Politburo, "unexpectedly" died in Valdai, a health resort for members of the Soviet political elite, about two and a half hours by car to the north and west of Moscow, on the road to Leningrad. Zhdanov was fifty-two years old. His death provoked a public outpouring of grief throughout the Communist world. Condolences were published over many days. Mao Zedong from China and Georgi Dimitrov, the new president of Bulgaria, expressed their sorrow, as did the leaders of Communist parties of Great Britain, France, and Austria, as well as all the Soviet satellites. His body was borne on a gun carriage through Red Square to the Great Hall that housed the Politburo, followed by a procession of dignitaries and mourners, with Stalin in the lead. Eulogies by Molotov and other Politburo leaders were declaimed in Red Square to a sea of mourners. Poems were dedicated to his eternal memory. Mikhail Sholokhov and other noted writers, along with members of the cultural elite, referred to the "great heart" and the "crystal bright mind" of this "true son of the Motherland and the party." It was said at the time that no Kremlin leader had been buried with such public attention since Sergei Kirov, who died in 1934.

A. A. Zhdanov had been the former, brutal boss of the Leningrad party and the architect of Soviet postwar ideology and cultural policy. It was he who gave the keynote speech at the 1934 International Writers' Congress, where he had invoked Stalin's dictum that writers were the "engineers of human souls." In 1946 Zhdanov had issued the infamous "report" on literature and ideology that included, among other things, harsh criticism of the poetry of A. A. Akhmatova, and led to her banishment from the Writers' Union. Akhmatova, Zhdanov wrote, was "A nun or a whore—or rather both a nun and a whore who combines harlotry with prayer." He described her poetry as "utterly remote from the people. . . . What can there be in common between this poetry and the interests of our people and State? Nothing whatsoever."[1] Another time he had rebuked Shostakovich for not writing music the average Soviet worker could hum. Zhdanov's report unleashed what came to be termed the "Zhdanovshchina," which terrorized Soviet arts and letters for a decade until the Khrushchev "Thaw."

After the Second World War Zhdanov had been represented as a hero of the siege of Leningrad; he had given his life to the "interests of the State." At the time of his death he was considered by many to have been the most powerful member of the Soviet government after Stalin. He had defended Stalin and Stalinism all his life and had earned the respect of the people as well as Stalin's confidence. In September 1947 Stalin entrusted Zhdanov, not Molotov, with delivering the keynote speech at the Szklarska Poreba conference in Poland that set out the basic terms of Stalin's Cold War vision to leaders of the satellite countries and foreign Communist parties. Zhdanov depicted the postwar world as "irrevocably divided into two hostile camps."[2] American expansionism, he charged, was comparable to that of the fascist states of the 1930s.

More than just professional or political sympathy drew Stalin and Zhdanov together. Stalin was said to have personally preferred him because Zhdanov was educated and more literate than most of the other Politburo leaders. Zhdanov enjoyed playing the piano at Stalin's dacha and discussing literature with Gorky. In the spring of 1949 his son, Yuri, married Stalin's daughter, Svetlana Alliluyeva, something that was possibly more a sign of political realities than of personal choice.[3]

But in early July 1948 Zhdanov had fainted on the way to his office in Old Square near the Kremlin, returning from a Politburo meeting held in Stalin's "nearby" dacha the Blizhnyaya. Though just over fifty, he had had a bad heart for some time, the result of years of overwork, drinking, and hypertension. Earlier that spring Stalin had given him other reasons to be anxious. In April, only nine months since Zhdanov's historic speech at Szklarska Poreba, Stalin had reacted very negatively to the role Zhdanov's son, Yuri, had played in discussing the botanist T. D. Lysenko at a meeting held under the auspices of the Central Committee, saying that those responsible for this provocation should be punished in "exemplary" fashion. Although not formally dismissed from the posts of Secretary of Ideology of the Communist Party and head of the Leningrad party, Zhdanov knew his position in the party was now at risk and his life might also be in jeopardy. Stalin's suggestion that he recover in Valdai had carried with it the same solicitude that had accompanied the demotion of other Kremlin leaders since the thirties. The head of the Kremlin Hospital signed the medical certificate authorizing Zhdanov's leave and he was sent to Valdai on July 13. "Strict bed rest" was ordered for him to safeguard his health, but its other purpose, as Zhdanov might have guessed, would have been to keep him out of active political life.

On September 6, 1948, one week after Zhdanov "went his glorious way," an emergency session of experts was convened in the Kremlin Hospital to investigate whether the doctors who treated him in Valdai had misdiagnosed Zhdanov's illness and had provided criminally negligent treatment. The Kremlin Hospital doctor responsible for the last EKGs of Zhdanov's heart had sent a secret letter accusing the doctors of this negligence to Lieutenant General Nikolai Vlasik, the Head of the Kremlin

Bodyguards, known as the *Okhrana*. In her letter of August 29, 1948—two days before Zhdanov's death—Lidia F. Timashuk asserted that the attending doctors had underestimated the "unquestionably grave condition of comrade Zhdanov."[4] Soon after his death this evaluation turned into an accusation of outright murder.

For some fifty years this accusation has been viewed as the essential element in the instigation of the Doctors' Plot, and for fifty years, ever since the accusation against the doctors was dismissed by the Soviet government after Stalin's death, it has been widely assumed to have been a false charge, stemming from a premeditated plan to launch a conspiracy against the doctors. Therefore, the question of whether Timashuk's accusation was false and whether in 1948 it was part of a premeditated plan deserve careful scrutiny. Only by examining the details of the session of September 6 and the events immediately leading up to it will the full complexity of this essential component of the Doctors' Plot be understood.

The special September 6 session took place in the office of Dr. P. I. Yegorov, head of the so-called *Lechsanupra* (*lechebno-sanitarno-upravleniye*) of the Kremlin, the complex of medical units, known as the Kremlin Hospital. He himself was one of those Timashuk had accused of misdiagnosing Zhdanov and providing "incorrect treatment." All the doctors summoned to examine the charges of medical malfeasance were those alleged to have been complicit. No other participants, except Timashuk, attended this extraordinary session.

Even though Timashuk's initial letter had not raised the accusation of murder (Zhdanov was still alive when she sent it on August 29), all concerned instantly realized the menacing implications of her charge. Before August 1948 Timashuk had been a "rank-and-file doctor at the Kremlin Hospital,"[5] managing the electrocardiograph unit, with little authority to dispute the findings of the distinguished professors and doctors who had tended Zhdanov and made the original diagnosis of his illness. Timashuk's secret letter to Vlasik included the transcript of Zhdanov's EKG examination of August 28, on the basis of which she disputed the conclusions and directives of her superiors. She warned of a "fatal outcome" if Zhdanov did not receive a strict bed rest regimen that her diagnosis of a myocardial infarct (heart attack) would have indicated for him.

I consider that the consultants and physician doctor MAIOROV underestimate the unquestionably grave condition of com. ZHDANOV, permitting him to get up from bed, stroll about in the park, visit the cinema, and so forth, that this provoked the second attack [of August 29] and worsened the indications of the EKG of August 28, and in the future this regimen may lead to a fateful outcome.[6]

As the head of the Kremlin security guards, Vlasik was responsible for ensuring the physical safety of all leading party and Politburo members. He was the right individual to be informed about this matter, and Timashuk expected him to show her letter to the Politburo, if not directly to Stalin himself. This, she assumed, would bring the responsible parties to justice so that correct medical treatment could be provided.

She could hardly have expected what happened next. Her letter of denunciation had been expeditiously delivered through covert and secure channels, but she soon suspected that it had nonetheless fallen almost immediately into the hands of Dr. Yegorov himself, chief among those she had accused. From the Politburo, from the Ministry of State Security (MGB), from Stalin she heard nothing. Nor would she for several years to come.

What went wrong? On September 4, 1948, two days before the special session held in the Kremlin Hospital, Dr. Yegorov summoned Timashuk to his office. Timashuk recalled that she had scarcely "crossed the threshold of [Yegorov's] office, when Yegorov began to scold me. 'What have I ever done to you?' " he demanded. " 'Why did you write about me? What have you got to complain of?' " When Timashuk began to defend herself, saying that she had never written any such document, Yegorov exploded: " 'What do you mean you didn't write it? You wrote about us in a statement saying that we had incorrectly treated A. A. Zhdanov. And this statement . . . was given to me because they believe me and not some sort of Timashuk.' "[7]

Timashuk tried to deflect Yegorov's anger by keeping silent, but he continued his reproach. According to Timashuk, he ran around the office, pounded the table with his fists. Afraid that the accusation had left the lim-

its of the Kremlin Hospital, he shouted, " 'Why didn't you send the statement to me? If you didn't agree with us, you could have turned to Yefim
Ivanovich Smirnov [minister of public health of the USSR].' " She tried in
vain to deny that she had written any such document. " 'How dare you
deny that you wrote this statement about me?!' " Yegorov countered.
" 'I've even got it. I've read it myself. You mentioned my name twice in it,
saying that we had incorrectly treated A. A. Zhdanov!!' "[8]

Timashuk's behavior demonstrates much self-confidence. Yegorov was
not, after all, just another colleague. As head of the Kremlin Hospital, he
was her boss. He could unilaterally demote, reassign, or fire her—or con-
ceivably worse. Furthermore, Yegorov held the military rank of major general, and, as Timashuk surely knew, had close administrative ties to important governmental networks, including the MGB (the KGB of the time),
an organization that wielded ubiquitous, incalculable, and inexplicable
power in the world of Stalin's Russia. Although the details of this
September 1948 encounter come from Timashuk's own testimony and
serve her own interests, other sources confirm her remarkable poise under
Yegorov's attacks and her confident insistence on the correctness of her
diagnosis. Only much later—in fact not until the Doctors' Plot was officially repudiated by Soviet authorities following Stalin's death—did it
emerge that Timashuk herself was a covert agent of the MGB, one among
many, working in the Kremlin Hospital system.

Though Timashuk seems to have been unshaken by the fact that her
letter had come into Yegorov's possession, the meaning of this inevitably
lodged in her mind. Did it signal a possible deeper betrayal of her personally but also of the security institutions in which she and the state had
placed their trust? A letter from her to Lieutenant General Vlasik should
have gone straight to the Politburo; to the minister of state security, V. S.
Abakumov; and if necessary to Stalin. Appropriate action should have
been taken. The safety of the state was in question. Her urgent warning
should have elicited an instant response. She remained upset, and four
days after Zhdanov's death, on September 4, 1948,[9] wrote another covert
letter, this time to Suranov, her handler in the MGB. She complained
again about the medical mistreatment of Zhdanov. On September 15[10]
she wrote the first of two letters to A. A. Kuznetsov, a Leningrad protégé
of Zhdanov, then a secretary of the Central Committee and head of the

Directorate of Cadres, who was entrusted with supervising the organs of state security.

The assassination of state and political leaders in the Soviet Union had a long, dark history. The assassination of Sergei Kirov in 1934 had ignited the Great Terror. Had Zhdanov also fallen victim to a criminal conspiracy of immense proportions? Both Kirov and Zhdanov had been leaders of the Leningrad party. Both had been powerful Politburo members. Both had been close to Stalin. Rumors persisted, after almost twenty years, of an alleged medical conspiracy in the deaths of Maxim Gorky and V. V. Kuyibishev. Could forces again be at work linking high political conspiracy with medical malpractice? Could the leaders of the Soviet Union simply ignore Timashuk's account of the machinations of doctors who willfully disregarded the objective results of her EKG examinations and had thereby murdered another Kremlin leader? This troubled Timashuk. She could not understand why nothing had been done.[11]

Timashuk's behavior set the first small piece of the machinery of conspiracy into motion. On the one hand, she acted out of what she thought were the interests of the state; on the other, she acted out of her own self-interests. If she did not denounce the doctors, she herself was in danger of being denounced. The Kremlin Hospital system, like every government agency in the Soviet Union, was filled with informants such as Timashuk who routinely supplied the security service with details about what they saw or heard. Being part of this system, Timashuk knew that she also was under surveillance and would be brought to account if she did not denounce her supervisors in Valdai. With or against her will, the system of covert surveillance compelled her to act as she did.

Unexpectedly, however, she now found herself, not the doctors, on the defensive. *She* was the one being unmasked and interrogated by a livid Dr. Yegorov. "They believe me and not some sort of Timashuk," he had taunted. Yet his mocking tone disclosed *his* panic and insecurity. Yegorov, too, knew how the system worked; now he had discovered he had somehow miscalculated and had been denounced by "some sort of Timashuk." What neither Yegorov nor Timashuk could know was who else might be involved and how.

All those present at the extraordinary September 6 session in Yegorov's Kremlin Hospital office were leading figures of Soviet medicine: the consulting professors of medicine V. N. Vinogradov (who had accompanied Stalin to the Tehran Conference in 1943 as his personal physician[12]) and V. Kh. Vasilenko, both of whom had been with Zhdanov during his final illness; Zhdanov's attending physician, G. I. Maiorov; and the Kremlin pathologist who performed the autopsy, A. N. Fedorov. The ostensible purpose of the session was to examine Timashuk's statement, show Timashuk her error, and persuade or force her to retract her charges. It probably had a second purpose that was not so obvious. The doctors needed a document that would formally exonerate them by establishing an authoritative account of the "correctness" of the treatment given Andrei Zhdanov, in the event such a document ever proved necessary. The official stenogram of the meeting, dated September 6, served this purpose. As the head of a significant division of the Soviet government's bureaucracy, Yegorov had ample reason to fear what he and his colleagues would suffer if they were not successful. His peremptory and exasperated manner betrayed his apprehension.

The meeting began calmly enough. Yegorov called the session to order and stated the key question under review: the correctness of the initial diagnosis made of Zhdanov by Vinogradov, Vasilenko, Maiorov, and himself. He noted that he considered "it necessary yet again to discuss the established diagnosis in connection with the statement of doctor TIMASHUK, known to all present here, that the diagnosis of the consulting physicians consisting of academic VINOGRADOV, professor VASILENKO, the attending physician MAIOROV, was incorrect."[13] Yegorov argued plausibly that it was on the basis of no more than "a single study of an EKG" that Timashuk determined that Zhdanov had suffered a heart attack and that in doing so she had completely ignored all the other clinical facts of his illness. He pointed out that by taking no notice of the clinical data, it was Timashuk, not Yegorov and his colleagues, who had been negligent. As final proof, Yegorov stated that the autopsy, performed by Fedorov, confirmed the original diagnosis established by the team of doctors and consultants: no myocardial infarct.

Zhdanov suffered the first of two severe attacks of a cardiac nature on July 23, 1948. It had been treated in the standard way with strict bed rest and various medications. After a couple of weeks he began to feel better and this regimen was abandoned; Zhdanov was then permitted to get out of bed, walk around, and go to the cinema.

At seven-thirty in the morning of August 28, Timashuk was summoned to Valdai. Why she was summoned to Valdai on that date has never been explained. From July 23 to that time, the doctors' notes did not record a significant alteration in his condition. No EKG had been taken since August 7. Timashuk flew there from Moscow with Yegorov, Vinogradov, and Vasilenko. Dr. Maiorov, Zhdanov's personal physician, was already at Valdai, having arrived with Zhdanov in July.[14] Timashuk took an electrocardiogram of Zhdanov's heart at about twelve noon, and she became convinced that Zhdanov had recently suffered a myocardial infarct, though she could not say just when. She immediately brought this information to the attention of Maiorov, Vinogradov, Vasilenko, and Yegorov. Maiorov thought this was a "mistaken diagnosis and could not agree with it. His view was that Zhdanov did not experience an infarct, but rather had suffered a 'functional disturbance,' sclerosis and hypertension."[15] She was told to alter her conclusion, a demand with which she did not comply. Timashuk returned to Moscow with Yegorov and Vinogradov that evening.

On the following day, August 29, the date of her first letter to the authorities, Zhdanov suffered his second severe attack in Valdai; again Timashuk was called from Moscow, where she found Vinogradov, Yegorov, Vasilenko, and Maiorov in attendance. Yegorov and Vinogradov instructed her to delay taking another EKG until the following day, August 30. She could not understand this and interpreted it as another indication of inadequate medical attention. She found the doctors' refusal to entertain the idea of a heart attack or take appropriate measures both inexplicable and inexcusable, and she expressed her anxiety to the chief of Zhdanov's bodyguards, A. M. Belov. She told him that she had reason to suspect the doctors of mistreating Zhdanov. "Belov proposed that I write a statement about this to the MGB USSR, directed to General VLASIK [sic]. I did this. . . . I thought that my statement would be taken seriously because it had to do with the incorrect treatment of com. ZHDANOV, A. A., that threatened his death. My surprise was great when I discovered that the statement, given to BELOV, whom I trusted . . . turned up in the hands of the one I accused of criminal treatment of com. ZHDANOV . . . that is, YEGOROV."[16]

Timashuk stated that the doctors had instructed her to rewrite the conclusion of her medical report "not indicating a 'myocardial infarct' "[17] or in any way mentioning the possibility of one, insisting that she should merely observe that Zhdanov's medical situation required careful attention. When the doctors learned later on August 29 that she had not rewritten her conclusion, Yegorov himself confronted her at dinner. She was evasive, saying that she had no official form on which to write anything. They gave her a form. But instead of writing the conclusion the doctors wanted, Timashuk simply noted, "See the previous electrocardiograms" performed by Dr. Sophia Karpai in July and August. Timashuk believed that these substantiated her diagnosis.[18]

Timashuk's August 29 letter, addressed to Vlasik, was sent at once through covert security channels "in a wax-sealed packet with the heading 'top secret, deliver immediately' " written on it.[19] It was duly received and was at once transmitted to the proper authorities. The grammatical mistakes, cross-outs, and the nervous, almost illegible hand with which it was written suggest the stress Timashuk felt, knowing what was at stake.[20]

The EKG taken on August 30 convinced Timashuk even more that Zhdanov had suffered a heart attack. Once again the doctors categorically instructed her to write nothing about this in her report. The following day Zhdanov died. Agitated by the situation in which she found herself, Timashuk began to interview the nursing staff. What she discovered appalled her still more. The nurses told her that, instead of strict bed rest, Zhdanov had been allowed to go for walks in the park, take massages, go to the movies, and get up from his bed to use the lavatory. Her August 29 letter to Vlasik emphasized the fatal possibilities of such treatment.[21] Two days later Timashuk was certain that the "fatal outcome" she had predicted had resulted from negligent treatment. In her mind the charge had now become murder.

The hastily prepared autopsy, performed in Zhdanov's bathtub in Valdai on the evening of August 31, with the staff masseuse, nurse Turkina, taking notes, showed otherwise, but Timashuk could not be quieted or convinced. No one in a position of high authority seemed interested, and no one responded. She found herself in a situation in which the certainties of party discipline and her medical training seemed to have vanished. Timashuk sensed collusion, and in her mind the September 6 special session proved it.

After his introductory remarks at this session, Yegorov asked Dr. Fedorov, the pathologist, to read out the official diagnosis based on the autopsy report:

> Hypertension. Widespread arteriosclerosis. Arteriosclerosis of the aorta and coronary vessels with the primary lesion of the left branch and its offshoots. Angina. Cardiosclerosis . . . Progressive failure of the coronary circulation with the primary lesion of the anterior wall of the left ventricle and the intraventricular partition . . . Hypertension and enlargement of the heart, primarily on the left side. Failure of the mitral valve. Remaining manifestations after micro-strokes. Moderate stage of pneumonia-sclerosis in the left, lower lobe, after pneumonia . . . [22]

Fedorov summarized his findings: "The death of comrade ZHDANOV, A. A., followed from the paralysis of a morbidly changing heart that was the consequence of sharp arteriosclerosis of the coronary vessels in combination with general arteriosclerosis. As a result of heart failure there was a sharp attack of emphysema."[23] No heart attack.

Yegorov was pleased: "We may conclude that we have almost complete congruence of the clinical and the pathological-anatomical diagnosis. . . . What else is there to say?"[24]

To his dismay, Timashuk, who had been silent up to this point, raised again the evidence from the electrocardiogram. Timashuk insisted on what she saw set out before her in black and white. *She* did not make the diagnosis of a heart attack, the electrocardiogram did: "I did not write it," she told the doctors almost poetically, "the heart wrote it, and I set it onto paper."[25] The state now found the literary products of Zhdanov's heart as useless to its impenetrable interests as they had once found Akhmatova's.

Timashuk was quite familiar with the normal electrocardiograph of Zhdanov's heart because she had been assigned to him as far back as 1941, with minor interruptions. She told the doctors that she remembered them well and that when she saw the EKG of August 28 she instantly recognized "a colossal difference between this last electrocardiogram and the previous ones." She told the doctors on September 6 that

Maiorov had rebuffed her when she had asked about the diagnosis and only reluctantly informed her that the diagnosis was "cardiosclerosis and hypertension."

"I have evidence of something worse," Timashuk warned. "I have evidence of an infarct in the area of the left ventricle and the ventricular partition." Maiorov would not listen. "Nonsense," he said. "Nothing more than a functional change. No sort of organic change. You've spoken only of functional changes, lesions, and not of organic changes." Timashuk persisted. "I have evidence," she said, "of an infarct on the electrocardiogram; the electrocardiogram has never let me down. I have always had confirmation of what I have reported."[26]

At the Kremlin session, the doctors did not deal directly with the essential question of the treatment; for the most part, they attempted to prove the correctness of the diagnosis by attacking Timashuk's basic assumption that Zhdanov had suffered a recent infarct. If Yegorov could show that Timashuk's diagnosis was wrong, then he could argue a different treatment was necessary. When he asked whether on August 28, 1948, Timashuk had evidence of a "fresh infarct," she replied, "No."[27] If the EKG gave no indication of a fresh infarct, Yegorov asked, why should Zhdanov have been treated for one? How was it possible to treat an infarct a year or so after it might have occurred?

Timashuk wavered, conceding that she couldn't date the occurrence of Zhdanov's last heart attack, but insisted that it was "recent" (*svezhii*)—perhaps no more than a month old. Therefore, she argued, there was cause for continued concern. His heart functions were deteriorating. Measures should have been taken in light of this. Instead, the doctors were contemplating sending him on car trips to Moscow in two weeks. This made no sense, Timashuk argued. Could they not see this?

TIMASHUK: *I did not think it was an old infarct. I thought it was a recent infarct because at the end of 1946 there was no sign of an infarct and in 1947 there had been no infarct.*

YEGOROV: *If there had been an infarct, although two months previously, how would you treat the patient from the point of view of prolonging his life?*

TIMASHUK: *On the electrocardiogram there was evidence not only of one infarct but of one in both the ventricle and in the partition and it's entirely possible that he repeatedly had several attacks. Consequently there was not one infarct, but two or three, when I looked at the electrocardiogram. Therefore, I would have kept the patient in bed for perhaps more than two months.*

I asked that the old electrocardiograms be given to me, but I was told that they weren't available. They were in Moscow. I asked to be given the electrocardiogram that was made after the first attack. The patient, apparently, had more than one infarct. . . .

YEGOROV: *You did not answer my question. Even if there had been an infarct several months ago, what would you have done concerning the necessary therapy, particularly as regards [the patient's] movements?*

TIMASHUK: *I would have held the patient to a regimen of the strictest bed rest from 4 months to 1/2 a year, as we do for such patients in the hospital!*

(Outcries of objections erupt from those present)[28]

Timashuk returned again and again to the question of treatment, while Yegorov kept pressing the nature of the diagnosis. Timashuk's obtuseness appears to have taken Yegorov by surprise. Her insistence on the EKG stood in their way. They never challenged her reading of the EKG itself, nor did they invite another EKG technician to dispute it, which might have immediately settled the issue.

YEGOROV: *Do you consider it legitimate, from the standpoint of understanding the illness, for the diagnosis to be based only on an auxiliary medical investigation without your knowing or wanting to know the whole complex of the clinical course of the illness?*

TIMASHUK: *I provided a conclusion on the basis of the electrocardiogram.*

YEGOROV: *Is it possible to establish a diagnosis, without looking at the patient and not considering anything besides the electrocardiogram?*

TIMASHUK: *In particular circumstances it is possible. Seeing the electrocardiogram I made the diagnosis . . .* (They interrupt)

YEGOROV: *We know what an infarct is without you—on the basis of an attack of stenocardia. . . .*
Allow me to tell you that if you had the interests of the patient at heart there would have been no divergence between us. You would have come to me, to professor VASILENKO, to academician VINOGRADOV, to doctor MAIOROV, and if you had come to us with your view, there would have been no difference of opinion. But you allowed yourself to write an official statement accusing us of incorrect therapy. [29]

At this point another important theme in what will become the "case of the doctors" invisibly entered the dispute. The question of whether Yegorov actually read Timashuk's letter eventually became crucial in the case the government built, but the testimony concerning this point proved highly contradictory. On September 4 Yegorov had revealed to Timashuk that he had read her letter to Vlasik, but his responses at the September 6 meeting put this into question. Now when Timashuk stated that she did not accuse the doctors of incorrectly treating Zhdanov, Yegorov ironically responded, "I'm glad you declare that you have not accused us of any-thing—of incorrect therapy. I'm glad you have declared this." Timashuk responded, "I wrote a report regarding only my electrocardiographic diagnosis and that I would have demanded the strictest bed rest regimen."

Yegorov dismissed this subtle difference. "Incorrect treatment is what was written there." But Timashuk flatly denied the charge, "No, that isn't so," she said. "I didn't write that the treatment was incorrect. I do not know how they treated him. I based my conclusion on the electrocardiogram and said that for such a patient the treatment—an even more stringent regimen. Not this regimen that permitted him to stroll around!"[30]

From Timashuk's August 29 letter to Vlasik we know that initially she had *not* accused the doctors of "incorrect [*nepravilnoe*] treatment"—at least in so many words. She wrote: "I think that the consultants and the physician Dr. Maiorov underestimate the unquestionably grave condition of comrade ZHDANOV, permitting him to get up from bed, stroll about in the park, visit the cinema, and so forth . . ." In the letter the question of incorrect treatment was implied but not explicitly stated.

Yegorov claimed to have read her statement. How or why did he get this detail wrong? And why would he have committed this mistake when he knew that the proceedings were being recorded for future use? Though small, this is an important detail because it suggests that Yegorov knew more about her accusation against the doctors than was contained in her letter and might well have signaled a breach in security protocol. The fact that Yegorov could know that Timashuk had accused him and his colleagues of "incorrect treatment" suggests that he might have had a source of information within the intelligence community in direct contact with the case. Yegorov knew that the accusation had subtly but decisively changed into something much sharper than what was contained in the original letter to Vlasik.

In testimony given in 1952 Timashuk threw light on this question. She recounted that she had indeed been concerned about the treatment Zhdanov was receiving:

> I spoke of this because of the regimen of treatment that had been prescribed for comrade ZHDANOV, A. A. This regimen was *absolutely incorrect (nepravilnii)* for a patient with an infarct. For such a patient there should be a strict bed rest regimen. [Emphasis added.][31]

Timashuk eventually made an even more damning assertion:

> Knowing to what fateful consequences the *criminal treatment* of comrade ZHDANOV, A. A., by YEGOROV, VINOGRADOV, VASILENKO and MAIOROV might lead, on August 29, 1948 I informed BELOV, an MGB worker dedicated to ZHDANOV, A. A., stationed in Valdai.
>
> Hearing what I said, BELOV suggested that I write a statement about this and send it to General VLASIK in the MGB. I did this. Indicating in the statement that YEGOROV, VINOGRADOV,

VASILENKO, and MAIOROV were incorrectly [*nepravilno*] treating comrade ZHDANOV, A. A. and that this might conclude with his death, I gave the statement to BELOV.

I thought that my statement would be seriously received because I said that comrade ZHDANOV, A. A., was incorrectly [*nepravilno*] treated and was threatened by death. Imagine my surprise when I discovered that the statement given to BELOV, in whom I believed as a man guarding comrade ZHDANOV, A. A., turned up in the hands of the one I accused of criminal treatment of comrade ZHDANOV, A. A., that is, YEGOROV. [Emphasis added.][32]

Timashuk had good reason to trust Aleksander M. Belov, who had been Zhdanov's bodyguard since 1935 and did not depart from the Zhdanov family circle until February 1949 when he received another assignment from the MGB.[33] Though he claimed to suffer from a poor memory, Belov recalled the following:

After A. A. ZHDANOV's second heart attack, which occurred on August 29, 1948, doctor TIMASHUK—whose [first] name and patronymic I can't remember—came to Valdai along with several other doctors. . . . After TIMASHUK took the electrocardiogram and analyzed the condition of A. A. ZHDANOV, she wrote out a report addressed to me. . . .

TIMASHUK informed me that in her opinion, the diagnosis of the illness of A. A. ZHDANOV by the attending physicians was incorrect. . . .

I can't now remember what was said concretely about the diagnosis. However, as I remember the statement, in the diagnosis established by TIMASHUK, the condition of A. A. ZHDANOV was considerably more grave than that found by doctor MAIOROV and the professors.[34]

Belov added an important detail. After receiving Timashuk's statement, he immediately informed Zhdanov's wife, who was with her husband in Valdai, of the conflicting diagnoses and the gravity of her husband's condition. Zinaida Aleksandrovna was not just Zhdanov's wife. Obstinate, pushy, and bold, she did not hesitate to address Stalin directly and had her own channels to the Politburo. Even her husband feared her tongue. She "immediately called together all the doctors treating A. A. ZHDANOV who were in Valdai."[35] Belov did not know what happened at this meeting, but it must have become clear to all that the stakes were being raised. On the same day Belov dispatched Timashuk's statement by airplane to Vlasik in Moscow. What became of Timashuk's statement, Belov did not know. In his handwritten confession dated October 18, 1951, Belov said that he never passed this statement on to Dr. Yegorov. Yegorov himself later confirmed this.[36]

However, in February 1953, a month before Stalin's death, Yegorov finally put the blame on Belov: "BELOV told me that TIMASHUK had sent a statement to the Minister of State Security [Abakumov] or to the head of the Chief Directorate of Administration of the Guards of the MGB [Vlasik] . . . in which she wrote that A. A. ZHDANOV had had a recent myocardial infarct and that the treatment he was receiving was incorrect. At the same time, BELOV allowed me to read the statement of TIMASHUK. . . . the statement of TIMASHUK was with me for only a half an hour. After I became acquainted with its contents, I returned it to BELOV."[37]

Yegorov was a well-educated, intelligent doctor who accused Timashuk unequivocally of having written to Vlasik that he and other doctors were guilty of giving Zhdanov "incorrect treatment," and further that this "incorrect treatment" could well have resulted in Zhdanov's death. Yet, as Timashuk herself asserted at the special session held in Yegorov's office, she had *at that time* written nothing of the kind.

Both Timashuk and Yegorov now found themselves in that twilight world of *Macbeth* in which "Nothing is but what is not." On paper to Vlasik, Timashuk did *not* accuse the doctors; in conversation with others she did. Timashuk might have written a second letter to Vlasik making the explicit accusation of "incorrect treatment," but no evidence supports it. More probably, Timashuk's claim of incorrect treatment was so deeply a part of her thinking and had been so widely spread by word of mouth throughout the staff at Valdai that both Timashuk and Yegorov confused the written with the oral statements he might have heard while treating Zhdanov. This is more plausible in light of Belov's admission that he had shared Timashuk's concern over the diagnosis with Zhdanov's wife, and Zhdanov's wife then called all the doctors together to discuss the question.

It is also possible that Timashuk's September 4 letter to her MGB handler Suranov had been divulged, though probably also not shown, to Yegorov. Timashuk's frustration over not having a response from the authorities had grown to a high pitch by the time she wrote to Suranov on September 4. That is the letter in which her suspicions would have turned to angry accusations and her implications became assertions. Being informed of this letter, even without reading it, Yegorov would have known precisely the accusation against him and the other doctors. Only Vlasik, in whose directorate Suranov worked, could have made this information available to him. Vlasik's possible role in this conspiracy shows how it was expanding in unforeseeable ways and in effect had turned the security services against itself. Dark questions filled everyone's mind on September 6. Something had begun and no one knew where it might lead.

When Yegorov said to Timashuk that "they" believed him and "not some sort of Timashuk," he probably meant, at least in part, that Vlasik had revealed the content of Timashuk's letters to him. Vlasik's role in the initial stages of this conspiracy became the subject of intensive MGB scrutiny in the fall and winter of 1952, when it was revealed that Yegorov and Vlasik frequently drank together at Vlasik's dacha and that from time to time Vlasik would share sensitive MGB information about Lechsanupra operatives, such as Timashuk, with Yegorov.

In 1948, however, the situation was very murky. Yegorov may or may not have actually read Timashuk's letter; he knew more or less what Timashuk was saying to those around her in Valdai, but he did not know what would come of this. Had he actually read Timashuk's letter to Vlasik, Yegorov would have known that Timashuk had, as she told him, not used the words "incorrect treatment." Timashuk, however, had no hesitation in using this language with her MGB contacts and eventual interrogators, and probably with members of the Valdai medical staff as well.

The autopsy report read out by Fedorov at the special session of September 6 should have closed the case if Timashuk had been concerned only with those matters outlined in her letter to Vlasik. The case was not closed. Yegorov was agitated at the September 6 meeting because he knew that something more than a simple medical diagnosis was at issue. But what this was no one could tell. The scope of the Timashuk problem became evident to Yegorov as he began to see the net of interrelations of which Timashuk's accusation was only a part. The letter had gone to Abakumov and to Stalin. No doubt Poskrebyshev, Stalin's secretary, knew of it. Another letter had gone internally to the security organs through Suranov. Vlasik was sufficiently concerned to provide reassurance. But clearly the reassurance was not sufficient to prevent Yegorov from calling the special September 6 session. Information was being leaked from various sources, and the leadership was silent.

Yegorov could not be certain what was going on, though he knew well enough from Vlasik about the system of covert surveillance and informing in the Kremlin Hospital. When he was appointed the head of the hospital system, Vlasik and Yegorov reviewed the files of some two hundred workers.[38] Yegorov knew that the Zhdanov case had now entered into the vast informant web in which Timashuk revealed her conviction that the doctors' behavior was *criminal*, not merely incorrect.

"They believe me and not some sort of Timashuk" was Yegorov's way of signaling that he had reason to think he knew *what* they believed. Or at least he wanted Timashuk to think this was the case, perhaps to intimidate her so that she would desist in her efforts at undermining him, perhaps to send this information back to her contacts in the MGB apparatus. Yegorov's "they" would, in the end, have included Stalin himself.

Why Yegorov might have felt sure, but also unsure, of his ground lies at the heart of the narrative that follows. What Stalin believed or did not believe, what he intended or did not intend has mystified journalists and scholars until now, largely because he rarely wrote anything down, kept no diary, and destroyed nearly everyone around him who knew too much.

Stalin almost never issued a direct order. Rather, he would approve or have others approve orders prepared by others. He worked through proxies much of the time and created a system in which loyalty and intuitive knowledge of his wishes counted for more than any other qualities. Loyalty to Stalin meant knowing Stalin's enemies. Stalin instilled this system throughout Soviet government and society. It became known as "vigilance." Although much in Khrushchev's memoirs is simply untrue, fabricated, or incomplete, his description of "vigilance" is compelling.

Stalin had instilled in the consciousness of us all the suspicion that we were surrounded by enemies and that we should try to find an unexposed traitor or saboteur in everyone. Stalin called this "vigilance" and used to say that if a report was ten percent true, we should regard the entire report as fact . . .

Stalin's version of vigilance turned our world into an insane asylum in which everyone was encouraged to search for nonexistent facts about everyone else. Son was turned against father, father against son, and comrade against comrade. This was called "the class approach."[39]

Poskrebyshev, who had been Stalin's secretary since 1929, and Vlasik were closest to Stalin in his inner circle. Before them the Vozhd often spoke freely with the intent of having what he said conveyed to others, and the test of political loyalty often consisted in whether Stalin's wishes were properly interpreted. Stalin seldom committed his deeper political strategies to writing. His thinking continually adjusted itself to circumstances. He could wait for years, as he did with Bukharin, before eliminating a rival. He took advantage of circumstances and often changed his views in apparently inconsistent fashion. Rather than being guided by ideological commitment, Stalin's political tactics depended on circumstances. He was a master of waiting, of moving on different tracks until he thought the moment was ripe.

Although ruthless and cruel, Stalin was rarely precipitous. He could keep his ruthlessness in check, his brutality in abeyance. He was complex, wary, dark, often inconsistent, cruel, changeable, and cautious. We can often only trace out what those around him did within the system in which he held absolute power. All governmental and state structures were ultimately contingent on his will, down to such details as the travel plans of all members of the Politburo and Central Committee. For anyone to travel outside of Moscow, Stalin's approval was required in advance. Those who showed the slightest sign of independence or autonomy were punished, sometimes with banishment, often with worse.

Stalin's will shaped a world. And only through looking at the totality of this world can his will reveal itself. He worked largely through indirection, holding many moves in suspension for a time, waiting for the right moment to seize his opportunity while covering his intentions with his rhetoric. He outmaneuvered his rivals and enemies not because he was more intelligent or more charismatic, but because he was more cautious and more cruel, less constrained by those principles out of which his ideology was constructed and in accordance with which so many genuinely dedicated Communists attempted to live. "Why did we prevail over Trotsky and the rest?" Stalin asked rhetorically in 1937 in a famous toast. "Trotsky, as we know, was the most popular man in our country after Lenin. . . . We were little known, I myself . . . We were field workers in Lenin's time, his colleagues. But the middle cadres supported us, explained our positions to the masses. Meanwhile Trotsky completely ignored those cadres."[40] Controlling the middle cadres required a deep, instinctive pragmatism that was also a kind of genius.

Yegorov, though not the most ideologically dedicated or astute Communist, was among those middle cadres. He understood the system and had been the beneficiary of it. Born in 1899 into the peasant village of Rudnits in the Leningrad district, Yegorov had worked his way up from his peasant origins to become a major general in the Army and in 1947 the head of the Kremlin Hospital system. He was a typical Soviet bureaucrat, of medium height, with a short neck, gray eyes, thick lips and a large nose. He evidently enjoyed his position of power. He seems to have been a basically lazy man and often treated those beneath him in the Kremlin Hospital crudely and with contempt while fawning over those above him. While treating the gravely ill Zhdanov in Valdai, Yegorov felt compelled to return to Moscow the evening of August 28 to have dinner with Smirnov, the minister of public health. Timashuk took this as another sign of Yegorov's indifference to Zhdanov. It also revealed his obsequious, calculating nature.

> How inconvenient! Today I must have Yefim Ivanovich SMIRNOV [minister of public health] and his wife at my place, and despite everything I must return home. Otherwise, it would be very uncomfortable for me in front of SMIRNOV.[41]

He often drank a good deal with his friend Vlasik. In addition, he would from time to time meet with Vlasik to vet the MGB files of incoming medical personnel. He often spoke with the minister of state security,

Abakumov, by telephone to answer questions concerning the health of Kremlin leaders.[42] Yegorov was well aware of the operative political structures in his medical facility, though he didn't necessarily know the identity of the operatives who routinely informed the MGB of the affairs of the Kremlin Hospital. It would have come as no surprise to him that Timashuk was one of them.

Knowing through *his* sources that the accusation of criminal misconduct—not merely that of a mistaken diagnosis—was the heart of Timashuk's denunciation, Yegorov stated overtly at the September 6 session what Timashuk had communicated covertly. "Incorrect treatment is what was written there," Yegorov insisted. By means of this signal, Yegorov communicated that he, too, was part of the larger machinery of Soviet intelligence that had now been set in motion.

Deep and opposed interests clashed during this September 6 session. By attempting to drive Timashuk out of the larger political picture, Yegorov indirectly revealed the larger invisible battle over covert influence, privilege, and power that was not confined to the Kremlin Hospital system. Plot fit inside plot in the intricate underworld of Stalin's universe, as figure fits inside figure in the traditional Russian Matrioshka dolls. At this moment in the Kremlin Hospital, one of the tiniest figures is disclosed. What enveloped it and what further plots and subplots it contained is the subject of the rest of this book.

Yegorov relentlessly pursued Timashuk at the September 6 special session, and what emerged further suggested that Belov was telling the truth when he said that he had not shown Yegorov Timashuk's statement.

YEGOROV: *Why did you emphasize that I and academic VINO-GRADOV forbade taking the electrocardiogram on Sunday and Monday?*

TIMASHUK: *I was impatient to take the electrocardiogram because when there is an attack we are in a hurry to determine the picture of the heart functioning, and [the delay] seemed strange to me.*

YEGOROV: *Were you forbidden?*

TIMASHUK: *No. There was no prohibition. It was only that it wasn't suggested that I take the electrocardiogram on August 29, but that I take it the following day.*

Yegorov seemed satisfied with Timashuk's response. In fact, what Timashuk had written to Vlasik was "but *by the order* of academician VINOGRADOV and Prof. YEGOROV the EKG was not taken on 8/29, the day of the heart attack, but was scheduled 8/30."(Emphasis added.)[43] She did not write that she had merely been urged to delay the EKG for one or *two* days—Sunday and Monday—as Yegorov stated. Rather, it was "*by the order*" of Professor Vinogradov and Yegorov that she delayed, but only until Sunday. The fact that Timashuk's own memory may have proved faulty suggests that she might have written one thing and orally reported another so frequently that she mixed them up. Or that she wrote one thing to Vlasik and another to Suranov. But if Yegorov had actually read her statement to Vlasik, and had a copy of the letter, he would not have been satisfied with her answer. This fact strengthens the view that he was informed of the totality of Timashuk's charges only after the letter had been received by Vlasik and after Timashuk had discussed the matter extensively with others. This further confirms the suspicion that it was not Belov who informed Yegorov of the seriousness and import of its contents, but someone at a later date and higher up—probably Vlasik, or someone even higher up in the echelon of Soviet intelligence or the party.

The remainder of the September 6 proceeding shows Yegorov in a state of near panic as he tried to force Timashuk to keep silent. Although he felt the support of someone high up in the organs of state security, he knew enough to realize that he did not know how far he could trust it. Timashuk was also not without support. Why else would this obscure nobody possess the temerity and stubbornness to continue to challenge him, even after he had revealed in what amounts to a threat that he, too, had highly placed sources of information?

Yegorov somewhat feebly argued that it was the doctors' concern for Zhdanov's well-being that caused them to wish to delay the EKG until the following day so as not to tax him too severely.[44] He then asked, "Why did you state that I personally ordered you to alter the diagnosis?"[45]

"We were intending to depart [from Valdai]," Timashuk recounted, and she had given her conclusion to Maiorov, stating that "there [were] signs of an infarct in the anterior wall of the left ventricle and in the intraventricular partition. He looked it over and said: 'She has written: "myocardial infarct." ' " Here, rather dramatically, she turned to face Yegorov: "And you . . . said that it was impossible to write a diagnosis without the agreement of the attending physicians and a clinical diagnosis."

Yegorov did not dispute this. Instead he asked, "Do you now think that your step was reasonable, rational, leading to the good, in the interest of the patient and his relatives? For your information," he continued, "you poisoned the spirit of the relatives. Answer me: Do you think your statement was reasonable, rational, leading to the good of the patient?"

Timashuk remained consistent, saying that the electrocardiogram told the truth. But she added: "I would have acted further, out of the anxiety that the patient walked around." Yegorov became angry: "You are incorrigible. I cannot let you to work here any longer." The stenogram quotes Yegorov as summing up as follows:

> The diagnosis is always established in combination with the clinical facts. For your information, we invited the leading specialists of the Soviet nation—academician ZELENIN, Professor NEZLIN, Professor ETINGER. They evaluated this electrocardiogram differently on the basis of the clinical data. Is it possible that you really believe yourself to be the Alpha and the Omega . . . ? I can no longer permit you to work in the *Lechsanupra* system, because you create discord and dangerous confusion in the treatment of each individual . . . You are not even interested in x-rays. Clinical facts don't interest you . . .[46]

"I am interested," Timashuk responded, ignoring his threat to fire her. She did not otherwise react to Yegorov's attack. Although not indifferent to being dismissed, as her subsequent behavior makes clear, she, too, may have felt herself adequately protected. Two days later she was in fact demoted to a peripheral job in the system.

"Permit me to bring to your attention," Yegorov continued, "what the professor-specialists said about the diagnosis of the illness of Andrei Aleksandrovich ZHDANOV."

Yegorov then read out to Timashuk a stenogram of a consultation held in Moscow on the question of the diagnosis of Zhdanov's illness on August 31, the day Zhdanov died. The meeting began at three o'clock in the afternoon and at its conclusion at 4:10 P.M. it was announced that Zhdanov had died. Vinogradov flew back to Moscow from Valdai the morning of August 31 in order to participate. Yegorov also returned to Moscow on the thirty-first but did not attend the session. The stated purpose of Vinogradov's mission was to hear the views of leading Soviet specialists to find a way of prolonging Zhdanov's life, based on an analysis of all the physiological and clinical data at their disposal. In other words, the stated purpose was to find a way of saving Zhdanov's life, not simply confirming the correctness of the diagnosis. In fact, as the stenogram of the August 31 meeting shows, the real objective was to secure agreement about the diagnosis.

Yegorov read sections from this stenogram, probably such as the following:

PROFESSOR ETINGER: *There are no indications of a coronary thrombosis. What is present is a grave form of chronic, coronary failure in combination with attacks of cardiac asthma.*

PROFESSOR NEZLIN: *I concur with Professor Etinger that against the background of hypertension, the patient had severe coronary arteriosclerosis and above all suffered lack of nourishment to the anterior wall. . . . Facts pointing to a thrombosis do not exist either clinically or from the electrocardiograph; there is no evidence of a major infarct.*

ACADEMICIAN ZELENIN: *True, there is the circumstance that there were two attacks of asthma, but they were not accompanied by the clinical picture of a myocardial infarct. This forces us to think that there had been no infarct. I suggest that there had not been a major, massive infarct, but rather something similar . . .*

Dr. Markov, the chairman of the meeting in the absence of Yegorov, summed up the findings somewhat oddly:

> It is not necessary to decide on the prognosis. What is important is to establish precisely the *correctness* of the diagnosis and conduct of the treatment. You understand the gravity of the situation and the condition [of the patient]. We invite you to check the diagnosis yourself using your own means. You confirm the stated diagnosis— that the diagnosis is following correct paths and in addition you confirm that the applied therapeutic measures are correct. In this regard we can feel moral satisfaction; the situation is being handled in an entirely responsible manner . . . [Emphasis added][47]

The words "correct" or "correctness" appear three times in Markov's brief conclusion. Though the correctness of the diagnosis *and* the therapy were the declared concerns of those present, in fact they spent almost the entire session confirming the diagnosis. The question of what concretely could be done to save Zhdanov's life never came up. Different therapeutic strategies were not proposed or debated. No discussion of the sort took place. Perhaps nothing could have saved Zhdanov's life, but the intent of this August 31 meeting in Moscow appears to have been to achieve a unanimous medical opinion to deflect Timashuk's criticism, rather than to deal with the concrete facts of his illness. No one present knew that Zhdanov would die that day. It is doubtful that Timashuk would have shared the doctors' "moral satisfaction."

The timing of this August 31 session is important. What provoked it? In Valdai the medical evidence had been treated unambiguously up to this point by the doctors. They showed no hesitation or doubt about either the treatment or the diagnosis. The evidence of the EKG did not convince them, and Yegorov explained plausibly that by itself the EKG could not guide a complete diagnosis. If so, why hold the August 31 session in Moscow?

Timashuk had rendered her opinion originally on August 28. If Yegorov had doubts about the diagnosis, why did he wait until August 31 to convene this consultation? The doctors could have and should have acted sooner, particularly when dealing with a life-threatening situation of a high government leader who suffered a significant cardiac event again on August 29. But Timashuk did not write her letter until August 29 and it was not received in Moscow until August 30. The doctors sought additional advice only at the point at which they knew that their treatment and diagnosis had been impugned, not at the point at which Zhdanov's condition had significantly worsened. Timashuk's letter had been forwarded to Moscow on August 29, and on August 30 it would have been read and its impact, if any, would have been assessed. Meanwhile, Zhdanov's condition became so critical on August 29 that the doctors ordered Timashuk to wait to take the EKG until August 30, allegedly so as not to disturb his rest. Or did they insist she wait so that no additional facts would disturb their diagnosis and force them to consider altering their treatment?

We can see how Vinogradov shaped the opinions of the consultants on August 31 toward this end from the outset. He began the session by telling them, "Having carefully discussed the matter, and not having any other facts, we [Yegorov, Vasilenko, Maiorov and himself] decided that there could hardly have been an infarct in this case."[48] Timashuk's EKG of August 28 was never discussed on August 31; her objections to the treatment were never raised; in the end everyone agreed with Vinogradov.[49]

Vinogradov felt he could count on Etinger, Zelenin, and Nezlin to provide the confirmation he needed:

> I said nothing directly to ZELENIN, ETINGER and NEZLIN, but I conducted the consultation in such a way that it was clear to them what conclusion I wanted them to come to.
>
> I had known ZELENIN for ten years. He is a professor of the old, pre-Revolutionary school strictly observing the rule: "Do no harm to another," and I was certain that if he understood my difficult position he would always offer to help. And so it happened. ZELENIN provided a fuzzy conclusion that subsequently allowed me to conclude that the consultation found no evidence of a myocardial infarct.
>
> ETINGER was also a person close to me. My relationship with him allowed me to hope that he would not let me down; and Nezlin, his student, always followed in the footsteps of his teacher. To put it succinctly, after I had stated significantly at the beginning

of the consultation that in my opinion the patient had not suffered
an infarct, all three—ZELENIN, ETINGER and NEZLIN—con-
curred with my point of view.[50]

Vinogradov's recollection of this meeting might be suspect if what he
told his interrogators did not coincide so closely with the stenogram of the
August 1948 meeting and with later statements given in March 1953, after
Stalin's death. There can be no doubt but that he was more concerned
with demonstrating that he and his colleagues in Valdai were correct than
with finding ways to prolong Zhdanov's life.

Despite the tendentious and collusive nature of the August 31 consultation, the doctors' case at the September 6 meeting had many plausible
elements. Timashuk was made to appear overzealous, stubbornly clinging,
like an inexperienced student, to an important but erroneous thesis after
having mastered only a part of the material, and unwilling to give it up
despite new evidence or wider analysis. By the end of the September 6 session, Timashuk even admitted that she had no evidence of a "recent" heart
attack, and she conceded that any final diagnosis had to be based on clinical as well as physiological data.

Yegorov summarized the September 6 session, saying, "I called this session for you, doctor TIMASHUK, to consider, to correct your views (and
not for something else), to listen to what learned people say . . . How you
might wish to see another treatment of the question."

Timashuk remained unmoved. "I am in full agreement," she said.
"This does not diverge from the electrocardiogram . . . after a month to let
the patient walk around, you might just as well have let him walk around
after his July attack!"[51] It appears that Vinogradov purposely concealed
from his colleagues on August 31 the full extent of the movement
Zhdanov was allowed. We know that Zhdanov was permitted to stroll
around in the park and go to movies, but on August 31 he stated merely
that

> about August 20 he was able to get up a bit from his bed, walk to
> the bathroom, but no more and until recently he was able to walk,
> but was taken out onto the verandah in a wheel chair. He moved
> about little and felt satisfactory.[52]

On September 6 Vinogradov stated concisely that further "discussion is out of place. Doctor TIMASHUK put forward one diagnosis. To our great regret, this diagnosis was not confirmed. We came to a resolution of the question and to a complete confirmation of all of our convictions by means of a grave catastrophe [Zhdanov's death]. And now citizen TIMASHUK is attempting to escape from an inescapable position—nothing more."[53] Escaping the inescapable was what Vinogradov and the other doctors were also attempting to do. Yegorov challenged Timashuk directly: "Be honest: is it correct, is it legitimate to behave as you are: to put forward a diagnosis and to construct an argument about the therapy and the diagnosis without being interested in the clinical facts?"[54]

They endlessly circled the issue. The doctors, insisting on the clinical data, the opinion of the experts, the fact that Timashuk herself could not tell when Zhdanov might have suffered the infarct she claimed to have detected on August 28; Timashuk, insisting on the facts of the EKG and what standard treatment of such a case should have been. They debated the nature of an attack of cardiac anemia, the levels of leukocytes in Zhdanov's blood, his temperature, and the ROE readouts. Yegorov then concluded that Timashuk's answers demonstrated "medical ignorance. . . . Our diagnosis was significantly more serious [than Timashuk's] . . . (hypertension, widespread arteriosclerosis, arteriosclerosis of the aorta and the coronary vessels, angina cardiosclerosis, remaining manifestations of previous infarcts . . .)—with a myocardial infarct people recover and work another ten years. But our diagnosis was significantly more serious."[55] Yegorov argued that it was possible to achieve an EKG resembling the one Timashuk took of Zhdanov on August 28 yet be led to a diagnosis other than that of a heart attack.

Yegorov's summary, like the consultation held in Moscow on August 31, evaded the crucial issue of what treatment the doctors prescribed for Zhdanov. Yegorov did not answer the question why, if the diagnosis of the doctors was even more serious than Timashuk's, they nevertheless did not follow the medical practice of the day and establish absolute bed rest for a prolonged period of time, but rather allowed Zhdanov to walk in the park, go to the movies, and receive leg massages.

Professor Vasilenko's attempt at doing so was full of contradictions. Vasilenko noted that two days after Zhdanov's first attack on July 23, 1948, the doctors were presented with a "picture of extreme, sharp left ventricular failure. We held a consultation and discussed the question: did we have before us a massive, morbid part of the heart or a temporary ischaemia with subsequent sites of degenerative muscle tissue. In the first 24 hours this question was not conclusively decided. And only in the following days when the blood circulation regained its previous pressure and went up to 90–100, when there was an absence of leukocytes, ROE was around 10–12 an hour, when the temperature was normal in the following days were we able to come to the conclusion that we had a patient with hypertension, coronary sclerosis, and damaged myocardium caused by hypertension."[56]

Vasilenko was unequivocal: "The situation was thus . . . the patient had a grave lesion of the heart. The question existed whether this was massive necrosis, deadness of the myocardium. The condition of the patient was so severe that it was entirely different from, even more severe than simply a myocardial infarct. A myocardial infarct occurs and scars over."[57] Restating Yegorov's point that the doctors recognized an even graver illness than the one diagnosed by Timashuk may have been an important tactical point— for the record—because Timashuk had specifically charged that the doctors "underestimate[d] the unquestionably grave condition of com. ZHDANOV."[58] Or it may have been a statement of fact.

Timashuk saw it as the former. But that is not the only possible interpretation. Dr. Lawrence Cohen, a cardiology specialist at Yale University Medical School, after recently reviewing all the medical material, including the doctors' daily reports of Zhdanov's illness, concluded that Vasilenko's assessment may well have been medically responsible. There was an attack, Cohen concluded, "related to pulmonary edema, filling up of the lungs with fluid," but Vasilenko's conclusion that there wasn't a heart attack, per se, was based on the absence of leukocytes. "When there's a heart attack," Cohen explained, "the white blood cell count—the leukocytes—usually rise, and that did not happen." Vasilenko may therefore have been expressing the doctors' initial doubt as to whether the pulmonary edema was due to a heart attack or was just an expression of longstanding heart disease related to hypertension, with the loss of a lot of muscle and scarring. This situation in turn may explain why the doctors did not feel complete bed rest was necessary and why very mild walking might have been appropriate, and why they projected two weeks down the road that Zhdanov could have gone to Moscow.[59]

We shall return to Dr. Cohen's assessment of the case in the next chapter. For the present we have a very contradictory situation in which the medical diagnosis is corroborated but colored by the fact of collusion among the consultants on August 31, real negligence in the medical care provided in Valdai, the unexplained nonresponse of higher authorities, and the strong-armed bullying of Timashuk on September 6 that suggested a cover-up. If the medical situation is relatively straightforward, the political situation is not.

The picture presented by Vasilenko provided enough credible detail to allow Dr. Cohen to say that it was indeed a defensible interpretation of the available medical data. If this were so, then Timashuk's accusation was indeed false. It was, however, fatefully prescient.

It is the second part of Vasilenko's statement that inspires doubt. Vasilenko acknowledged that the picture of Zhdanov's condition was "progressive" and life threatening. After the July 23 attack Zhdanov was given "strict bed rest," although the clinical facts, Vasilenko stated, "demanded in the first place that he not lay horizontally." This was so because "the stagnation in the lungs with the presence of wheezing provoked fears about the return of pneumonia; in the second place, a prone condition would increase the flow of blood to the heart."[60] Therefore, after three weeks, or in mid-August, the doctors struggled with whether the patient "could put his feet down, whether he could be raised up from the bed in order to allow blood to circulate to the extremities."[61] According to Vasilenko, the doctors were continually worried that Zhdanov would suffer an attack of emphysema, which in their view "could have been fatal and was much more threatening to him than to some other patient with a myocardial infarct."[62] In other words, they recognized an even greater degree of immediate danger from absolute bed rest, because this posed the danger of a potentially life-threatening attack of emphysema and a return of pneumonia. Further deterioration of the heart, Vasilenko suggested, posed a lesser danger than the possible emphysema.

Vasilenko concluded by telling Timashuk that the doctors had an overview of Zhdanov's entire condition:

All of us are of different ages and of different schools, but we all came to the same conclusion that he was a very gravely sick patient and that he could not make a single movement without our knowing how it would tell on his circulation.

Your anxiety was great, but you should have shared your anxiety with the attending doctors. If you had shared it with the doctors your anxiety would have become even greater because you would have known the extent of the severity of the illness and therefore you would have said both to the relatives and to comrade BELOV that at any moment there might be a catastrophe and that it might come in a period of his best general state and the longest bed rest.

We always spoke about the progressive failure of the coronary circulation.

If you take the whole picture into account, the picture becomes so grave that the electrocardiogram can be seen as only a small particle from which to judge the gravity of the illness. [Emphasis added.][63]

Vasilenko had two points. The first is that the doctors saw the whole picture rather than simply the detail of the EKG; the second is that the situation was even graver than Timashuk had imagined.

Vasilenko's statement ignored the fact that Timashuk had, in fact, immediately gone to the doctors with her anxiety. Instead of the enlightenment Vasilenko suggested they would have provided, she was met with contempt and her worries were rebuffed. The gravity of the doctors' concerns was not reflected in Yegorov's departure for a dinner party on August 28, in the negligent nursing care, or in the fact that Yegorov himself did not attend the August 31 consultation in Moscow. The question is not whether Zhdanov should have been allowed to get out of bed or have his legs massaged, but rather the degree of concern the doctors actually demonstrated for his well-being.

At the September 6 session Yegorov did not follow up on Vasilenko's observations as he had with Fedorov's autopsy report. He could have emphasized that the doctors saw the situation as even bleaker than Timashuk imagined. But there was a problem in his doing this. Yegorov must have known that to emphasize this point might expose the doctors to another, even greater risk: Such concern was not evident in what they said or did in Valdai. Yegorov dropped the matter.

He simply asked Timashuk to recount what she had written to Vlasik. This should have been unnecessary, if Yegorov had actually had a copy of the letter and had shared its contents with the other doctors as he claimed to have done. Apparently he had not done so and some doubt remained.

Dr. Maiorov asked whether Timashuk had spoken "with the personnel and to the relatives" about her doubts.[64] On the one hand, the personnel were the nursing staff; on the other, they were the security staff stationed in the house of which everyone was aware. Timashuk now divulged the details that caused her as much alarm as the EKG. Her account of this corresponded with her subsequent testimony to the MGB and with that of the nurses at Valdai who were eventually interrogated. The doctors did not dispute it.

Timashuk said that family members had assured her, repeating (she surmised) what the doctors had told them, that Zhdanov had not suffered a heart attack, and that he had actually begun to feel better, that he walked around and even went to the movies.[65] After taking the electrocardiogram, Timashuk asked one of the nurses whether Zhdanov was, indeed, allowed to get out of bed and walk around.

> "Yes," [the nurse] answered. "He walks around and the last time did not feel well." "What was the matter?" [Timashuk] asked. "Cyanosis appeared. His lips turned blue." "But why didn't you at once say something about this to Gavriil Ivanovich [Maiorov]?" I asked. She answered: "Gavriil Ivanovich was fishing. I said that it was necessary to send for him, but Andrei Aleksandrovich [Zhdanov] did not permit this, saying: 'it's not necessary to bother him. Let him relax.' "[66]

One of the highest-ranking members of the Politburo, the head of the Leningrad party, and a personal friend of Stalin should have received a different kind of treatment. His personal physician should not have gone fishing. Yegorov should not have been called away to a dinner party in Moscow on the evening of August 28. There were other disturbing details.

Timashuk revealed that one of the nurses had fallen asleep during her watch, and that this had forced Zhdanov to get out of bed during the night to close his window.

A nurse told me, in great confidence, not to say anything to Piotr Ivanovich [Professor YEGOROV]. I promised not to say anything and asked what the problem was. She said that it concerned the fact that Valentina Nikolaevna slept during her watch, and Andrei Aleksandrovich called for her 6 times. In the morning he said to her: "I tried to wake you six times, Valentina Nikolaevna, but you did not get up, and I was cold. Apparently, you were asleep." He himself got up, closed the window, and used the chamber pot. When the nurse became very upset about this, he said: "Let this be a secret, don't be upset. Besides you and me, no one will know about it." But the nurse in a moment of openness told the other nurses about this, and one nurse told me.[67]

Maiorov responded simply: "I knew nothing of this." He never denied, however, that Zhdanov had suffered the exhaustion the nurse described or that he had in fact gone fishing on more than one occasion while in Valdai.

The other doctors said nothing. None of this detail was in Timashuk's letter to Vlasik. Apparently Yegorov thought the best strategy was to say as little about such matters as possible, and he made no attempt to correlate their understanding of the gravity of Zhdanov's illness with Maiorov's fishing trips, the negligent nursing care, or his own apparent unconcern.

Yegorov changed the subject. "We convened this session," he stated, "in order to illuminate your error for you." It was not convened so that Timashuk could illuminate their errors to them. No further discussion was necessary. Timashuk, however, remained unsatisfied. "You said that there had been excellent nursing care," she scolded. "But here a very gravely sick patient was forced to get up!"

"What do you yourself make of this?" Yegorov shot back.

"For myself," Timashuk replied, "I make the following: Between the Electrocardiogram, the clinical and pathology-anatomical facts there is no divergence."[68] What she implied was that her diagnosis agreed with the clinical picture as presented to her by the nurses and the EKGs she had taken. Therefore, the treatment Zhdanov received should have reflected heightened concern and greater attentiveness to his needs.

With this, Yegorov called the meeting to an end. "We may conclude," he said, little knowing the extent to which this "conclusion" would be no conclusion at all.

Doctor TIMASHUK, exceptionally partial to the electrocardiogram, not understanding it well, not taking into account the clinical facts, made a mistaken diagnosis. In addition, on the basis of this discussion, we are convinced that she is an insufficiently trained doctor; she does not strictly attend to the difference in the diagnosis between a stenocardiac attack and a myocardial infarct. Probably, we are all agreed that in the future, comrade TIMASHUK must not act this way again, deviating from the interests of the patient. Wherever you may work, you must not be so categorical, stubborn and obstinate as to disregard the clinical data.

If this is how you think, why do we have professors and doctors? Just take the electrocardiogram and prepare the diagnosis. But this will never be the case, and, if in the future you proceed this way, you will do harm to our valuable patients. Is this true, Vladimir Nikitovich [Vinogradov]?

"This is true," Vinogradov replied.[69]

There are many competing truths in this narrative—the electrocardiograms, the autopsy, the clinical "facts," the doctors' behavior, the covert surveillance. Perhaps most important was the truth of the interests of the party to which Zhdanov himself insisted true Soviet art must adhere. These are not simple things to disentangle, and one EKG or medical evaluation cannot, as Vasilenko said, answer the question of what was true and what was false in the doctors' treatment of Andrei Zhdanov.

The only clear fact is that Timashuk's letter and the truth of the EKG did not yet serve the "higher" truth of Stalin's purposes. They were ignored by everyone who saw them. "They believe me and not some sort of Timashuk," Yegorov said, and for the moment he was right.

But Timashuk did not accept this conclusion, insisting that "this question . . . be decided by a third party, and not only between our two sides— me and you."

"That is your affair," Yegorov told her. "You may apply to the Ministry of Public Health. I hope that you yourself will make the organizational conclusions. Allow me to conclude this session."[70] The session was over but the history of the "case of the doctors" was just beginning.

According to Timashuk, Yegorov in fact had used more threatening words than were recorded in the stenogram: "you are a dangerous person," she reported him as saying, "who has no place in Kremlin Hospital. I'll fire you."[71] This he did the next day.

Yegorov had been condescending and rude from the beginning, and his closing rebuke to Timashuk had something of Sarastro's dramatic repudiation of the Queen of the Night in *The Magic Flute*, whom he brands *ein stolzes Weib!* (an arrogant woman) and banishes forever from the kingdom, as "The sun's rays drive out the night,/destroy the ill-gotten power of the dissemblers!"[72] Unmasked as a dissembler, her conclusions overthrown, this latter-day *stolzes Weib* was summarily dismissed on September 7 by Yegorov as manager of the electrocardiographic unit of the Kremlin Hospital and thrown into the outer darkness of a considerably inferior position in the second polyclinic of the system. Dangerous, stupid, ill-educated, incendiary—in Yegorov's view, Timashuk was an individual to be treated with contempt. How much personal animus on Timashuk's part went into her pursuit of the doctors henceforth is impossible to know, but certainly she must have loathed Yegorov and those with whom she thought he had conspired, particularly after her demotion.

Knowing that she would get no justice from Yegorov or his immediate associates, Timashuk next wrote a letter to A. A. Kuznetsov, the Central Committee secretary charged with security affairs. She restated her opinion about the diagnosis of Zhdanov's illness but added one important detail that perhaps reflected her personal feelings and also suggested the underlying web of connections that would guide the slow course of this case into the full-blown scenario of the Doctors' Plot. In her letter to Kuznetsov, dated September 15, 1948,[73] a week after the session with Yegorov, she emphasized what earlier she had claimed was never part of her accusation against the doctors: "the treatment and regimen given A. A. Zhdanov *were incorrect*; that is to say, the disease of a myocardial infarct demands strict bed rest for the course of several months. In fact, he was allowed to move around (strolls in the park twice per day, going to the cinema, and further physical exertions)." (Emphasis added.) She continued: "Rudely, improperly, without any lawful grounds, Professor Yegorov on 8 September has thrown me out of the Kremlin Hospital to an affiliated polyclinic for the ostensible purpose of improving the work being done there."[74] Timashuk received no reply. Eventually she would write another letter to Kuznetsov in early 1949. This one, too, went unanswered.

Nothing happened.

The "they" referred to by Yegorov remained unresponsive. Did Kuznetsov act alone? Did Vlasik act alone? In late 1952 Kuznetsov, who had been shot in 1950 as part of the so-called Leningrad affair, was posthumously accused of not having brought Timashuk's letters to the attention of the Politburo or Stalin, and thus having participated in a conspiracy against the government. Vlasik was indicted in December 1952 and removed from office for, among other offenses, not having properly verified Timashuk's statement about the treatment given to Zhdanov. He, too, was accused of not having passed the letter on to Stalin.

By 1953 Timashuk's situation dramatically changed. On January 20 Timashuk was summoned to the Kremlin to meet personally with Poskrebyshev. From Poskrebyshev she was taken to see Grigory Malenkov, a powerful member of the Politburo, and was told that Stalin himself had begun to occupy himself with her case. Malenkov praised Timashuk's "patriotism." He said that her letters to the government unmasked the "criminal activity of the professor-doctors."[75]

The full power of the newly aroused Soviet state was then turned against these "criminals in white coats," as they came to be called in the press. Mass arrests began; large-scale concentration camps were built supposedly to be filled with Jews deported from Soviet cities; a vicious anti-Semitic campaign already under way found new life and allegedly produced evidence of international espionage, betrayal, terrifying plots to overthrow the government and poison the nation's health. A great public trial was rumored to have been planned for the end of March 1953. But on March 5 Stalin died. On March 31, 1953, the Ministry of Internal Affairs (which had replaced the MGB) recommended that the arrested doctors be exonerated. And on April 6, 1953, *Pravda* printed an article accusing the security services in general and the former minister of state security, S. D. Ignatiev, of dereliction, and an end was put to the organized madness threatening to overwhelm Soviet society.

In the wake of the nightmare of denunciations, accusations, interrogations, arrests, and torture, it has been assumed for nearly fifty years that the 1953 recipient of the Order of Lenin for patriotic service to the Soviet Union, Lidia Timashuk, had been little more than a small-minded nobody irrationally bent on the vilification of her superiors, deserving the contempt Yegorov showed her. The doctors, Stalin's victims, were prima facie innocent. Furthermore, Beria recommended their release, and they were pardoned in full. Although Beria was eventually shot, his recommendation concerning the Doctors' Plot was never questioned by his successors. Along with Beria, those responsible for inciting the plot were thrown in jail, and most of them were subsequently shot.[76]

But were all the doctors who sat in judgment of Timashuk on September 6, 1948, *completely* innocent? This is the question the new documentation and circumstantial evidence provokes. It is the question forced upon us in particular by Vinogradov's confessions and subsequent statements after Stalin's death.

Finally released from nightmarish interrogation and brutal torture, Vinogradov had no need to fear further retribution or punishment. In these circumstances, he made the following admission:

As is evident from the stenogram of [the August 31consultation], conducted by Professor A. M. MARKOV, who replaced the absent head of the Lechsanupra [Yegorov], professors V. F. ZELENIN, V. Ye. NEZLIN and Ya. G. ETINGER acknowledged that the electrocardiograms taken of A. A. ZHDANOV in July and August 1948 did not display an absolutely typical form of a severe myocardial infarct and might represent a condition of chronic alteration of the heart. *Nevertheless, it is necessary to acknowledge that the autopsy of A. A. ZHDANOV, who died on 31 August, disclosed that he had suffered a recent myocardial infarct. Therefore the rejection of this by me, professors V. Kh. VASILENKO, and P. I. YEGOROV and doctors G. I. MAIOROV and*

S. E. KARPAI was a mistake on our part. I must categorically declare that we had no evil plan in making the diagnosis or carrying out the treatment. [Emphasis added.][77]

Could the best minds of Soviet medicine—Zelenin, Nezlin, Etinger, Vinogradov, Yegorov, Vasilenko, Maiorov, and Karpai—have been wrong and Timashuk with her EKG been right?

UNDERLYING THE OFFICIAL RESPONSE TO THE TIMASHUK letter was a many-layered network of communication among the security services, the Politburo, and the Kremlin Hospital leadership. For four years Timashuk's letter remained officially "unread," until inexplicably in August 1952 the MGB "found" it in its archive and decided to investigate.

The dark silence becomes only greater when we read the following newly discovered document:

Top Secret
TO COMRADE STALIN, I. V.

I present to You herewith a statement by the manager of the electrocardiographic office of the Kremlin Hospital—doctor TIMASHUK, M. F. [sic], in connection with the health of comrade ZHDANOV, A. A.

As is evident from the statement by TIMASHUK, the latter insists on her conclusion that comrade ZHDANOV has suffered a myocardial infarct in the region of the anterior wall of the left ventricle and the intraventricular partition, at the same time that the head of the *Sanupra* of the Kremlin YEGOROV and academician VINOGRADOV suggested to her to alter her conclusion so as not to indicate a myocardial infarct.

Attachment: Statement of com. TIMASHUK and the electrocardiogram of comrade ZHDANOV.

ABAKUMOV
30 August 1948

Below the date is Stalin's signature indicating that he personally received the material. Below his signature in large script is Stalin's handwritten comment: "Into the archive."

IN STALIN'S WORLD OF 1948 TIMASHUK WAS WRONG AND THE doctors were right. The ambiguity of the medical situation reflected the protean political realities. All that would eventually change, but for the time being the question of right and wrong, truth and falsity was signaled by the very silence Timashuk could not comprehend. It is otherwise impossible to explain the fact that Belov, Zhdanov's bodyguard, was not questioned about the events in Valdai until *nearly four years afterward*. The fact that Vlasik conducted no inquiry into Timashuk's charges also cannot be explained, except to assume that he was involved in a conspiracy (for which there is no evidence) or that he received no order to do so. Kuznetsov was eventually accused of hiding the Timashuk letter as part of his plan to undermine the Soviet government, but in 1948 he was in no position to undertake an investigation not authorized by Stalin. Vlasik was eventually accused of criminal negligence, though Stalin knew well enough that it was he who sent Timashuk's August 29 letter "into the archive." The pervasive contradictions cannot be dismissed as mere human error. Yegorov's "they" consisted in such indirect reassurances from Vlasik and the knowledge that Timashuk's letter had been buried with Zhdanov.

Denunciations of Timashuk's type were rarely simply ignored. Hers was not a random letter from an unknown individual. It was from a vetted MGB agent working in the Kremlin Hospital. In addition, Timashuk's letter was handled through the proper reporting structure. Timashuk herself admitted that on September 4, unhappy with the inaction of both the government and the security services, she wrote a "detailed letter to SURANOV, Georgii Nikolaevich, a worker in the MGB USSR, concerning the incorrect treatment of com. ZHDANOV, A. A."[78] Suranov, her MGB handler, passed the information up the chain of command. Nothing came of that, either.

Nothing came of Timashuk's efforts at alerting the government just as Soviet intelligence could not spur Stalin into action against Hitler on the eve of the June 22, 1941, invasion. *Stalin did not wish to act.* He had his own reasons.

Stalin received this material, read it, and may have decided to trust the judgment of the head of the Kremlin Hospital. But that is unlikely. Stalin trusted no one. Nor does the possibility that Stalin trusted Yegorov explain the silence of Kuznetsov, the inaction of Vlasik, or the fact that Stalin's knowledge of the matter was not disclosed for fifty years after the affair terminated. Nor does it explain Timashuk's stubbornness.

It is inconceivable that Vlasik, the Head of the Kremlin Guards, would not have checked with Stalin either directly or indirectly in a matter of such importance. Vlasik knew that the letter was received by Abakumov, the minister of state security, and he probably also knew what had become of it; that is, the letter had been sent "into the archive," which was Stalin's formula for disposing of matters of no importance to him. Had Stalin been assured that Yegorov was to be trusted, Timashuk could have been persuaded to discontinue her protests. But she persisted.

Vlasik recalled the following at his interrogation in January 1953.

VLASIK: *The statement of doctor TIMASHUK was presented to
me . . . on August 30 or 31 of 1948. Exactly when is difficult for
me to remember. . . .*

    *Now I remember that I had the statement of doctor TIMASHUK
of August 29, 1948, by August 30 or 31. I did not myself read the
statement, but on the same day that they brought the statement to
me, I sent it on to ABAKUMOV. He also did not read it, but it
remained with him and he gave me no instructions to verify it; he
promised to inform the party of the statement.*

INTERROGATOR: *You yourself were responsible for verifying the
statement. Why didn't you do this?*

VLASIK: *I never received instructions from ABAKUMOV concerning
the necessity to verify the statement, and the head of the
Lechsanupra YEGOROV said that the statement, verified by him,
was not confirmed. He allegedly informed the Central Party about*

this. All of this quieted my fears, and I didn't take any measures myself to verify the statement of Doctor TIMASHUK. I declare this to be criminal negligence and I acknowledge myself guilty of it.[79]

The trouble with Vlasik's account is that we know from the Abakumov memorandum to Stalin that both Abakumov and Stalin read Timashuk's statement. It is likely that Vlasik read it as well. Like all well-trained security operatives, Vlasik lied to protect those above him in the system. Though now in prison, he was protecting Stalin himself.

Vlasik knew immediately that Abakumov had passed the letter to Stalin. He also knew that if Abakumov gave him no order to take further action, it was because Stalin wished it so. Stalin acted through Abakumov, who acted through Vlasik, who acted through Yegorov. Had Abakumov not shown the statement to Stalin, we might suspect Abakumov of working behind Stalin's back. But he dutifully presented the statement to Stalin on the same day he received it, proving that this was not the case.

It may appear that the security bureaucracy had broken down. It had not. Rather, the system whereby Stalin manipulated that bureaucracy was at work. All evidence suggests that Belov was not the source of the leak to Yegorov, and the chain of command from Belov to Vlasik to Abakumov to Stalin remained intact. Why Stalin chose to remain silent will be the subject of Chapter Two.

It is highly unlikely, almost inconceivable, that the doctors—the most noted in their profession in Russia—could *all* have made a mistaken diagnosis ignoring the results of a standard diagnostic test: Timashuk's EKG. Vinogradov's statement that Zelenin, Nezlin, and Etinger all came around at the August 31 consultation as a favor to him is also not entirely credible. Each of them would have understood the grave risk in doing this since the patient was Andrei Zhdanov. These were men who had come through the Terror of the 1930s and had witnessed the ruthlessness of Stalin's system. Could they really expect that once Timashuk's denunciation was received by the authorities, the security services would not pursue them to the ends of the earth? What brought the three specialists to concur with Vinogradov might well have been a conviction, no doubt indirectly conveyed, that Yegorov's invisible support within the government would support them as well.

The mortal contest between Yegorov and Timashuk was not limited to a disagreement over a medical diagnosis. It was a symptom, rather, of a vastly complex political sickness that threatened to overwhelm Soviet society. Behind Yegorov was a network of political and intelligence interests, reaching to the very top of Soviet leadership. Timashuk's boldness at the September 6 special session in Yegorov's office appears to have been inspired by a set of similar, though opposing forces. Both were working in the dark of Stalin's Russia.

By September 1948 A. A. Zhdanov was dead, and an intense conflict, not yet a war, had begun in the Soviet security service between those at the top—Abakumov, Vlasik, and their immediate subordinates who had direct access to Dr. Yegorov—and as yet unknown mid-level operatives with whom the likes of Dr. Timashuk had dealings. This conflict originated in the results of an EKG administered by Timashuk to Zhdanov and appeared to have begun by itself, without influence from either above or below. But was this the case? This is the question we must take up in the next chapter.

TWO

STALIN'S SILENCE

VALDAI, AUGUST 1948

---

*Glory to the great Stalin, the leader of the people and
the Coryphaeus of vanguard science!*
—T. D. LYSENKO, AUGUST 7,1948

A fact is the most stubborn thing in the world.
—MIKHAIL BULGAKOV, *THE MASTER AND MARGARITA.*

What was the meaning of Stalin's "into the archive"? Why did he do nothing about the health of one of his most trusted, valued, and talented comrades? Stalin's silence is at the heart of this story. "After Kirov, Stalin loved Zhdanov best," Molotov recollected,[1] noting, "only Zhdanov received from Stalin the same kind of treatment that Kirov enjoyed."[2] To understand what that treatment was we must return to an examination of what transpired in Valdai in the summer of 1948 and in Moscow the previous spring.

Most recently the Doctors' Plot has been described as "the apotheosis of the huge postwar purge . . . fabricated at Stalin's will [that] was preceded by a propagandistic attack initiated in 1949 against so-called stateless cosmopolitans . . ."[3] According to Louis Rappoport, an Israeli journalist, Timashuk was a virulent anti-Semite (a charge for which there is no proof whatever), "seething with ambition and hatred of the Jews."[4] She concocted the plot out of whole cloth.

"The Doctors' Plot was the logical culmination of Stalin's entire illogical system. It defies common sense—all one can do is attempt to analyze the series of events that led up to it," Yakov Rapoport, the only arrested doctor to have produced a memoir, wrote in 1988.[5] Though largely correct, this view suggests that the Doctors' Plot grew out of an irrational system and therefore there is no use trying to understand it. The truth of the Doctors' Plot may be quite different. It may have grown out of logical motives and political structures that are all the more dangerous for being so.

Yakov Rapoport's view ignores the fact that the doctors themselves never said that Timashuk's EKG did not show *something* indicative of a gravely serious cardiac condition. On the contrary, Vasilenko stated, "If you take the whole picture into account, the picture becomes so grave that the electrocardiogram can be seen as only a small particle from which to judge the gravity of the illness." At the September 6 session Yegorov argued (1) that Timashuk's EKG results had been produced by a condition different from that of a heart attack; and (2) that the EKG had to be viewed in the context of the totality of the clinical facts at the doctors' disposal before a proper diagnosis could be made.

Timashuk's allegation became central to the government's case against the doctors; its presumed incorrectness has been central ever since. Two factors were essential in establishing the conspiracy—the rightness or wrongness of this diagnosis, and the disposition of those in power. Before we look at that disposition, we must examine the diagnosis more carefully.

The truth of the diagnosis may not be as clear as either Timashuk or Yegorov, Stalin or Khrushchev thought. Vinogradov eventually acknowledged that the autopsy showed that Zhdanov had suffered a heart attack.[6] Yegorov admitted this only after intensive interrogation and torture. Vinogradov never retracted his confession. Yegorov did.

The murder of A. A. ZHDANOV was supposed to have been consciously conducted by me with a group of doctors (VINOGRADOV, VASILENKO, MAIOROV, KARPAI) using insidious methods. Supposedly, we intentionally did not recognize the myocardial infarct, and supposedly we permitted him more forceful physical effort (premature sitting up in bed, premature getting up, walking around the room, to the bathroom, etc.) at an earlier date than was appropriate in the circumstances.

The truthful illumination of this question has great significance not only for my fate, not only for the prestige of Soviet doctors, but also for historical truth.[7]

Yegorov attempted to portray the "historical truth" as it unfolded in Valdai and afterward, denying the existence of a heart attack and any plot to murder Zhdanov.

> In relation to A. A. ZHDANOV or any other leader of the Soviet state and the Communist Party of the Soviet Union, I personally never inspired any kind of enemy plans or had any thought to do any kind of conscious harm to their health, let alone to cut short their lives. I never gave anyone an evil order of this kind—neither to the doctors nor to the professors. . . . It never happened.[8]

What did happen? Both Vinogradov's admission and Yegorov's denial may be consistent with the facts. Returning to the assessment by Dr. Cohen cited earlier, we find a medically ambiguous picture. Cohen thought the electrocardiogram of August 29 was "very abnormal, representing clearly what is called a left bundle branch block," a so-called conduction abnormality. It represented "severe underlying heart disease, sometimes due to a myocardial infarction, sometimes due to longstanding hypertension," sometimes to other causes. It showed a "primary disease of the heart muscle" but in itself does not necessarily confirm that a heart attack has occurred. Therefore, in Cohen's view, Timashuk's EKG was more or less accurately interpreted by Vasilenko. The problem is that Cohen could not rule a heart attack out as a cause. Since it can't be ruled out, the doctors should have investigated further and taken necessary precautions. Cohen thought that Timashuk made a claim "beyond her abilities," because nothing in the electrocardiogram proved the left bundle branch block was caused by a heart attack.

The question remains whether, under those circumstances, the care the doctors in Valdai provided was responsible. This depends, in Dr. Cohen's view, on the history of Zhdanov's illness, in particular whether his state on August 28, when Timashuk took the EKG, was in some way a signal of a changed situation. The key, then, to understanding the situation in Valdai is not whether the EKG showed a heart attack, but whether the doctors provided "intensive observation and more intensive medical therapy," in Cohen's words, and gave due attention to the *full* medical picture, as Vasilenko insisted they had. The full medical picture included the ambiguity of the data before them in light of Zhdanov's progressively deteriorating condition.

The truth of the diagnosis was less important than the effort of the doctors to achieve certainty. Until August 31 when Vinogradov went to Moscow, they made no such effort and showed only contempt for Timashuk's concerns. If the doctors had even some reason to think that he had had a heart attack, Cohen thought, their "actions would be somewhat extraordinary."[9]

Timashuk's diagnosis was wrong, yet the doctors did not do everything to achieve the kind of clarity they needed; Timashuk may have been right that the doctors had underestimated the gravity of Zhdanov's illness. By the middle of August, Zhdanov had begun developing new symptoms, despite periodic improvements: cyanosis, bloody phlegm, difficulty breathing. Nevertheless, these symptoms did not spur the doctors to look more closely into Zhdanov's condition. Their notes from August 18 to August 30 suggest that Zhdanov's health had somewhat stabilized.[10] But as we will see, this impression was misleading.

August 18: "The day passed with [the patient in] a good general state until 18:00."

August 19: "He slept well at night. . . . Good general state."

August 23: "During the day he was twice out on the verandah. General state in the course of the day was satisfactory. No organic change from the morning."

August 26: "General state during the day was good."

August 27: "The day passed satisfactorily. He went to the cinema in the evening. . . . Prescriptions the same."

August 28: "General state is satisfactory. Sleep was variable. . . . There is light fatigue during movement. . . . Shortness of breath is not upsetting. . . . Recommendations: Cut off the capsule with *etrimin*, continue the digitalis, slowly increase movement; from 9/1 allow trips in the car. The rest the same. 9/9 permit the question about a visit to Moscow."

The following day Zhdanov had a powerful attack of some kind. The day after that he died.

No doctor notes are recorded for August 29, the day Zhdanov suffered a major new attack and Timashuk was recalled to Valdai for the second EKG. Although Vasilenko asserted at the September 6 session that the doctors recognized his very grave condition, evidence of this is found only in their notes of August 30, the day after Timashuk wrote her letter:

> Last evening was quiet. He slept well. Complaints about great weakness, difficulty breathing, he is troubled by a cough with definite bloody phlegm. He continues to look pale with a shading of cyanosis. Sweating. Breathing −20′. Significant amount of wheezing of the right and lower left, continuing to the middle of the shoulder . . .
>
> Conclusion: Suffering from high blood pressure, general arteriosclerosis with the primary lesion of the vessels of the heart and brain, the patient in the past was observed to have a micro-stroke of the brain, attacks of stenocardia and cardiac asthma over the course of the last month and a half, against the background of chronic failure of the left side of the heart. The patient has had extremely grave repetition of attacks of sharp weakening of the left part of the heart with the appearance of cardiac asthma and emphysema.
>
> Taking into account the clinical picture and the facts of the EKG investigation, it is necessary to acknowledge the presence of progressive failure of the coronary circulation with the primary lesion of the anterior wall of the left ventricle and the intraventricular partition with the presence of sites of degenerative muscle tissue.
>
> Prescription: 1) as before, absolute quiet . . .

Vinogradov, Yegorov, Vasilenko, and Maiorov signed this report. It has several problems. The most important is that while he may have slept well during the evening of August 29, earlier that day he had been so gravely ill that the doctors recalled Timashuk from Moscow to Valdai and then forbade her from taking the EKG until the following day in order not to disturb him. Written after the doctors had become aware of the contents of Timashuk's letter, these notes confirmed Timashuk's evaluation and recommended exactly what Timashuk had insisted upon twenty-four hours earlier—"absolute quiet." In these notes the doctors expressed for the first time the sense of urgency Cohen thought necessary, as they attempted to review the *total* medical situation, by "taking into account the clinical picture *and the facts of the EKG investigation . . .*" (Emphasis added). The doctors acknowledged the findings of Timashuk's EKG for the first time on the same day Stalin received Timashuk's letter. Their recommendation of "absolute quiet" is consistent with Timashuk's own conclusion that Zhdanov required "a strict regime of bed rest."

> I consider that the consultants and physician doctor Maiorov underestimate the unquestionably grave condition of com. ZHDANOV, permitting him to get up from bed, stroll around in the park, go to the movies and so on. . . . I . . . insist on the observation of a strict regime of bed rest for com. ZHDANOV.

The credibility of Zhdanov's medical care up to that point is called further into question by the doctors' admission that over "the last month and a half," Zhdanov had suffered from "attacks of stenocardia and cardiac asthma . . . against the background of chronic failure of the left side of the heart." These symptoms were nowhere remarked in the doctors' notes from August 18 to August 30. More had been going on in Zhdanov's medical condition than had been recorded.

According to Timashuk, when she first brought her EKG results to Yegorov, he responded: "There was no infarct. This electrocardiogram is just like the previous ones."[11] According to Dr. Cohen, Yegorov was right that the EKG did not disclose an infarct, but the EKG was demonstrably *not* the same as EKGs taken in previous years, or even in July by Sophia Karpai, who had left Valdai on August 7. Even following the episode of August 29, as Timashuk reported, the doctors refused to allow her to mention the possibility of a heart attack in her conclusion; nor did they acknowledge that absolute bed rest was required, even though that is precisely what they officially recommend in their notes of August 30: "po-prezhnemu absolyutnyi pokoi . . ." ["*As before* absolute quiet . . ." (Emphasis added)].[12] This was signed by all four of the attending doctors, yet on August 28 they recommended continued activity for Zhdanov and contemplated sending him to Moscow by car, even though they had for unexplained reasons taken the precaution of summoning Timashuk to Valdai.

When Timashuk learned from the nursing staff what Zhdanov's symptoms had been, she recommended absolute bed rest. But when she raised this with Yegorov in Valdai on August 29 he rebuffed her crudely: "'So in your opinion, the patient's shit should also stay in the bed?'"[13]

The phrase "as before" in the doctors' notes of August 30 suggests that the doctors had prescribed this regime up to that point. This had not been the case. Following Zhdanov's attack on the evening of July 23–24, bed rest was prescribed but it was abandoned after about two weeks, and Zhdanov was allowed to move around freely.

Vinogradov did not disclose the extent of Zhdanov's physical activity to the consulting doctors during the August 31 session in Moscow, even though he disclosed Zhdanov's symptoms.

> ZHDANOV was asthmatic, suffering from angina, was treated with glucose, with *strofant*. Shortness of breath. ZHDANOV often attempted to hide his condition from the doctors who nevertheless insisted that he take a vacation from work. On July 13, 1948, ZHDANOV left Moscow, his general state was so-so. On the night of July 23-24 an unpleasant incident occurred. *There was a business call on the telephone.* He became very upset and was not able to finish the conversation because of a severe attack of cardiac asthma. He remained motionless, breath was gurgling, phlegm from the lungs containing some blood, and so forth. [Emphasis added.][14]

Vinogradov understood the grave condition of the patient, something affirmed by all those present at the council session. In his March 27, 1953, statement to Beria, Vinogradov wrote that the results of the EKG showed a "left bundle branch block" in accord with Dr. Cohen—but with an important discrepancy. Vinogradov stated that this blockage "disappeared" in the later stages of Zhdanov's illness and was no longer detectable in the EKG: "If at the beginning [in July], as far as I remember, there was a blockage of the left leg of the bundle of Gisa, then in the subsequent EKGs it disappeared."[15] This is only partly true. On July 31 Sophia Karpai, who had first detected this blockage on July 25, found that it had vanished.[16] Timashuk's EKG of August 28 showed that it had returned.

Etinger's comments at the August 31 session reflected his sense of Zhdanov's critical condition. "The patient had a severe lesion of the coronary vessels owing to hypertension . . . Definite indications of a coronary thrombosis are not present. There is a severe form of chronic coronary failure that coincides with cardiac asthma. Manifestations of heart failure have grown up. In connection with *a grave form of chronic coronary failure* the formation of small centers of necrosis is possible (emphasis added)."[17] Zelenin appeared to concur, saying, "Undoubtedly this patient is an old hypertensive, with progressive hypertension. . . . This circumstance—that there were two attacks—is not associated with the clinical picture of an infarct, and I think that there was no massive infarct. . . . And from the point of view of prognosis each successive attack would be severe and even threatening."[18]

Vinogradov eventually admitted that "ZELENIN gave a garbled conclusion that subsequently allowed me to say that the consultation did not find a myocardial infarct in A. A. ZHDANOV."[19] This was, in fact, what happened when on August 31, Vinogradov summed up the opinion of the consulting committee:

So, this is the conclusion that was formed at the time . . . there is no objection to this diagnosis? (All the professors agree.) I would like to turn to you with the question: After the first attack [in July], he was subjected to strict quiet and was kept to his bed. He was not permitted to get up, he used a bedpan. Only after approximately two weeks did he get up a bit. He was permitted to use the toilet; he was given massages on his hands and legs. They did not touch his spine. We permitted him to get up, make a few steps in front of the bed and at the maximum he twice went about 50 meters around the verandah. I want to ask you to turn your competence to this question: Was this incorrect?[20]

"No," Professor Etinger immediately declared. "It was correct."[21] Vinogradov did not mention the lengthy strolls in the park prior to August 28. Nor did he mention that Zhdanov was allowed to go to the movies, something that would have required a different kind of sustained physical and emotional exertion.

The August 31 session concluded at 4:10 pm, and at that moment the doctors were informed that comrade Zhdanov was dead.[22] The degree of the doctors' panic is betrayed in Yegorov's subsequent behavior. Though Zhdanov was in critical condition in Valdai, both Vinogradov and Yegorov had traveled to Moscow on August 31. Only Vinogradov attended the special session. Yegorov's whereabouts were never explained. He was never asked in interrogations or subsequent debriefings why he left Valdai at that supremely critical moment. Timashuk also had returned to Moscow. The Valdai medical team had been cut by 60 percent on the day the doctors' notes state that Zhdanov had suffered an "extremely grave repetition of attacks of sharp weakening of the left part of the heart with the appearance of cardiac asthma and emphysema" and had complained of "great weakness, difficulty breathing." Yegorov's trip to Moscow was probably motivated by political, not medical concern.

We know that Yegorov was immediately informed of Zhdanov's death and at once called Poskrebyshev to arrange for the autopsy to be conducted in Valdai. Poskrebyshev approved this. Two hours later, at 6 P.M., Yegorov prepared to fly back to Valdai with the pathologist Fedorov. They arrived at approximately 10 P.M. Instead of taking the body back to Moscow, they performed the autopsy in Valdai, in the bathroom of Zhdanov's dacha.[23] This was done in the presence of Politburo members A. A. Kuznetsov and N. A. Voznesensky, close colleagues of Zhdanov from Leningrad (the same Kuznetsov to whom Timashuk would write in September), and P. S. Popkov, a secretary of the Leningrad party, who also immediately flew in from Moscow upon news of Zhdanov's death.

Poskrebyshev's approval for the autopsy implied Stalin's consent for what several people, including Timashuk, thought was a very unusual step. With the exception of Fedorov, the rest of the staff was untrained in the necessary procedures. Standard practice approved for leading government figures dictated that two MGB officers who *did not know each other* should have been present, in order to ensure the objectivity of the process. Why this protocol was not followed remains one of the dark mysteries surrounding Zhdanov's "untimely death."

The haste with which it was done struck the duty nurse Aleksandra Panina as unusual. She "found it personally strange that the autopsy was not done in Moscow where the body of A. A. Zhdanov could have been quickly transported, but in Valdai. It seemed to me that there was some kind of a hurry in the conduct of the autopsy. Attention should also be paid to the fact that the protocol of the autopsy was entrusted to the masseuse Turkina, and not to a doctor, as in my view it should have been done."[24] It is not known whether Poskrebyshev or Yegorov proposed conducting the autopsy in Valdai, but Poskrebyshev's consent represented collusion at the highest level for a non-standard medical operation performed on a Politburo member.[25]

Zhdanov's doctors never exhibited the sense of urgency while he lived that they showed now upon his death. Such collusion limited the participating medical personnel to those already present, thus excluding Timashuk and any other possible low-level Kremlin Hospital informants. Information could be strictly controlled. Even Vinogradov, it appears, did not return to Valdai for the autopsy. Kuznetsov, Voznesensky, and Popkov were present only with the direct consent of Stalin, who approved all travel plans for members of the Central Committee.

Voznesensky had been dispatched to Valdai *before* Zhdanov died, arriving ahead of Yegorov.[26] What Voznesensky's business was we do not know, but it may have been connected to Stalin's decision to hand over some of Zhdanov's former duties to Malenkov, Zhdanov's chief rival on the Politburo. This prospect had greatly alarmed Zhdanov when it was conveyed to him by Shepilov in July and was the cause of his first attack. It was the objective sign that his power was rapidly diminishing as member of Stalin's inner circle, and Zhdanov knew that this was the first step to political expulsion, if not physical elimination. Though not as close to Zhdanov as Kuznetsov, Voznesensky was nevertheless a Leningrad colleague. Was this the final step in the "exemplary" punishment Stalin meted out to the one guilty of the Lysenko fiasco? Had Stalin purposefully enlisted one of Zhdanov's Leningrad associates to convey the punishing news of his expulsion from power, turning, as Khrushchev put it, "father against son, and comrade against comrade" in a final twist of the knife? Or was Voznesensky's task the crueler one of trying to calm Zhdanov's fears just before the ax was going to fall? All three colleagues would be shot in October 1950 as part of the "Leningrad affair."

In a sense, Timashuk's EKG had been a red herring from the beginning. It neither proved nor disproved the doctors' medical evaluation of

Zhdanov. It neither confirmed nor disconfirmed their guilt. Rather, their guilt or innocence can be measured by their guilty or innocent behavior in the days leading up to and following Zhdanov's death. Although no document exists, or ever existed, that could decide the matter, the doctors demonstrated uncharacteristic lack of concern from August 7, 1948, when Karpai was dismissed, to August 29, when Zhdanov suffered his second major attack. Vinogradov's assertion that the "left bundle branch blockage" indicated by the July EKGs had disappeared by mid-August is contradicted by Timashuk's August 28 EKG that showed precisely a "left bundle branch blockage."

In addition, the doctors' notes from this period suggest that they did not fully record the progressive alteration in Zhdanov's medical condition or the treatment they provided. From August 7, when Karpai left Valdai, to August 30, the doctors appear to have been consciously manipulating the official medical record so that Zhdanov's death could eventually be presented as "sudden" and "unexpected." Unfortunately for the doctors, Timashuk showed up and wrote the letter to Vlasik. This changed the situation considerably and put the doctors on alert. On August 31, the day after the letter would have been read in Moscow, they acted to protect themselves by convening the Moscow meeting with Etinger, Zelenin, and Nezlin. Again, unfortunately for the doctors, Zhdanov died that day, lending dramatic credibility to Timashuk's accusation. If the doctors were doing little to ensure that Zhdanov lived, they would not have wished him to die so close to Timashuk's denunciation. This created the possibility of many unforeseeable consequences. These, however, would not materialize for several years to come.

The doctors were helped to believe that they had acted appropriately by the Moscow authorities, who paid no attention to Timashuk's initial or subsequent letters, and by the fact that Stalin continued to employ Vinogradov as his personal physician until the fall of 1952, when Vinogradov and the other doctors were arrested and formally charged with Zhdanov's murder. The supportive nature of Yegorov's relations with the central authorities is further attested by information he gave at his final recorded interrogation on February 7, 1953, which his interrogators passed over in silence. There is little wonder in this. Stalin would not have been eager for the line of inquiry suggested in Yegorov's statement to have been developed any further, suggesting as it did his own direct participation.

The interrogator intensively questioned Yegorov about how he obtained Timashuk's August 29 letter to Vlasik and with whom he had shared it. Yegorov told them (or was made to tell them) that Belov, Zhdanov's bodyguard, passed it to him. He held it for half an hour, read it thoroughly, then returned it to Belov. From there, Yegorov stated, he had no idea where it was sent.

> I do not know precisely where this document was sent by BELOV, but soon, *actually while A. A. ZHDANOV was still alive*, ABAKUMOV called me and asked—whether I knew of the statement of TIMA-SHUK. I told ABAKUMOV that I had read this statement in Valdai. Then ABAKUMOV informed me that he had sent this statement to the head of the Soviet government. [Emphasis added.][27]

While denying he knew where Timashuk's letter was sent, Yegorov revealed that Abakumov informed him that it was sent to Stalin. Yegorov could draw two important conclusions: (1) Stalin had read the letter; and (2) he did not act on it. Furthermore, Yegorov might logically have surmised that Stalin himself had instructed Abakumov to call him. Timashuk was not alone in registering the silence of the Kremlin. What Timashuk saw as betrayal, Yegorov counted as affirmation.

We do not know whether Yegorov received the call from Abakumov on the thirtieth or the thirty-first, but it is likely that it was the thirty-first, perhaps as a response to an urgent inquiry of Yegorov's while he was in Moscow that day. It would have been the catalyst for Yegorov's subsequent actions—the autopsy in Valdai and the assurance he felt that the authorities believed him and not "some sort of Timashuk."

What happened in Valdai did so with Stalin's knowledge and tacit approval. At the same time, Zhdanov's unexpected death on August 31 raised the problem that Timashuk's continued efforts to alert the authorities about criminal mistreatment might eventually initiate an inconvenient inquiry. Therefore Yegorov called, or was instructed to call, the meeting on September 6 to silence Timashuk.

Yegorov's interrogator did not pursue this line of questioning about the head of the Soviet government because the investigation had to conceal, not disclose, the extent of Stalin's prior knowledge. By introducing this information into the proceeding, Yegorov may vainly have hoped to persuade the MGB that he was innocent because Stalin knew of Timashuk's letter and did nothing about it. The strategy failed. In 1952 Yegorov was savagely beaten, and now faced the same silence Timashuk had in 1948.

"These doctors," Khrushchev wrote, "could only have been the best and most trusted of their profession. Only men well known and much respected in the Soviet medical world had been enlisted to work in the Kremlin Hospital."[28] To understand why the doctors—the cream of the Soviet medical profession—were listening more attentively to Stalin's silence than to Timashuk's diagnosis, we must look at the complicated situation then existing in Stalin's relations to his closest comrades in the Politburo in the postwar period.

Although it was widely known, as Molotov put it, that "After Kirov, Stalin liked Zhdanov best," the relationship between Stalin and Zhdanov was far more complex than either the Soviet public or foreign governments might have understood at the time. To grasp why Stalin might have turned on Zhdanov and written "into the archive" at the bottom of Abakumov's August 30, 1948, memo to which Timashuk's letter was attached, it is necessary to examine how their relationship had evolved in the year and a half preceding Zhdanov's death.

At the end of the war Stalin had two fundamental objectives. The first was to define the Soviet Union's postwar relationship to the West, in particular to the United States. The second was to redefine his relationship with the Central Committee and the Soviet government so as to retain supreme control over the country. All historians and memoirists of this period have noted Stalin's strained relations with nearly all his old colleagues after the war.[29] His relationship with Zhdanov during this time reflected this.

Having suffered some kind of major physical collapse—either a heart attack or a stroke—immediately after the war, which caused him to recuperate for long periods of time in his residence in Sochi, Stalin had to demonstrate that he was still the "master of the house," able to control foreign policy, domestic policy, the security services, and the military, as well as Soviet agriculture. The victory over Hitler had had a paradoxical result. Stalin knew that it was by virtue of becoming a world power that his people had become irretrievably part of the larger world in a way unprecedented in Russian history. During the war large numbers of Soviet peoples had direct experience of the West—with American and British soldiers, with the former democracies of Poland and Czechoslovakia, and with the elite cultures of Hungary and Germany. After the war Soviet soldiers stationed in Germany, Austria, Czechoslovakia, Poland, and Hungary soaked in the material and spiritual culture of the West.

This contact posed a significant threat to Stalin's hold on total power, particularly as famine raged in the Ukraine in 1946, where, according to Khrushchev, there were widespread chaos, unimaginable suffering, and more than one report of cannibalism.[30] The demonstrably worse living conditions in the USSR were borne in on every Soviet citizen who had any contact with the West.[31] The effect was often devastating.

Those who had risen to governmental positions of great eminence before and during the war now found themselves being pushed aside. It was in this context that Zhdanov rose to the zenith of his power in September 1947, when he delivered the important Szklarska Poreba speech in Poland, outlining the hard-line position the Soviet bloc countries of Eastern Europe should take in the coming postwar conflict with the West. Normally, Molotov, who in early 1947 had been foreign minister and the Politburo member in charge of foreign affairs, would have delivered such a speech. He did not. Stalin no longer trusted him or approved of what he perceived to be his tendency to think that the USSR and the USA could come to terms with each other.[32] Zhdanov's preferment was a signal not only to Molotov. Stalin wished to demonstrate to all the members of the so-called big four of the Politburo—Molotov, Beria, Mikoyan, and Malenkov—that they were not in control of the government. He and only he was in control. An example of the division between Stalin and the inner circle occurred when *Pravda* published a glowing speech by Churchill addressed to Stalin on November 9, 1945. Churchill spoke admiringly of the Soviet leader, describing him as a genuinely great man and the father of his country. Molotov, Beria, and the others saw nothing wrong in this speech and allowed it to be published in full. On November 10 Stalin wrote to them from Sochi:

I consider the publication of Churchill's speech, full of praise for Russia and Stalin, a mistake. This praise is necessary for Churchill in order to quiet his dirty conscience and mask his enemy relationship to the USSR and in particular to mask the fact that Churchill and his disciples in the Labor Party are the organizers of the Anglo-American-French bloc against the USSR . . . By publishing this speech, we are helping these gentlemen. . . . We now have many responsible workers who have gone into foolish raptures from the praise on the part of Churchill, Truman and Birnsof, and on the other hand fall into depression from any unfavorable word these gentlemen utter. I think these moods are dangerous because they develop obsequiousness in us before foreign figures. We must carry out a harsh struggle against obsequiousness before foreigners. If in the future we will publish similar speeches, we will instill such obsequiousness and groveling. I'm not even speaking of the fact that Soviet leaders don't need such praise from foreign leaders. As for me personally, such praise only jars on me.[33]

There were other warning signals. Molotov had foolishly written to Stalin in Sochi that "he thought it necessary to be more liberal toward foreign correspondents and allow the foreign correspondents to operate without strict controls."[34] Soon news services around the world, including Reuters, the *New York Times*, and the *Daily Herald* were carrying stories about a relaxation of censorship in the Soviet Union under Molotov. All this could only have provoked Stalin's fury.

Zhdanov had remained untainted by these developments with the "big four" because he was something of an outsider, having lived in Leningrad, not Moscow. As a consequence he did not enjoy close relations with the others. He did not know their families intimately, eat or drink with them. This enhanced his prestige with Stalin. Although he had been a member of the Politburo since the 1930s, it was not until the spring of 1946 that Zhdanov became a member of this informal inner circle that effectively ruled the country.

Although formally retaining their high government positions, the other members of Stalin's inner circle could sense they were losing favor. Beria had been removed in 1945 as head of state security and put in charge of the atomic bomb project.[35] This was not technically a demotion, but it removed him from direct access to the reins of *political* power. Malenkov, who had been in charge of the party apparatus and unofficially Stalin's deputy on party matters, was demoted in the spring of 1946 and made deputy chairman of the Council of Ministers, though formally remaining a member of the inner circle. At that time he was put in charge of agriculture and was sent to Siberia to oversee grain procurement. Somewhat later Molotov "at his own request" was released from his position of overseeing foreign policy. Stalin had specific reasons to be dissatisfied with Molotov, who had officially been his first deputy. Because of Molotov's comments to journalists in 1946 concerning possible rapprochement with the West and the liberalizing of the Soviet media, many foreign observers, much to Stalin's displeasure, deduced that Molotov would replace Stalin after he retired. Nothing could have irritated Stalin more than speculation about his retirement and a policy of appeasement or accommodation with the West.[36]

When Zhdanov became a member of what had by now become the "big six" (Beria, Molotov, Mikoyan, Malenkov, Zhdanov, and Stalin), he took over the party apparatus and cadres that had previously been overseen by Malenkov. Zhdanov's powers were in reality even wider because he also oversaw ideological matters and foreign policy. In effect he took over some of Malenkov's as well as Molotov's functions. As Molotov fell into disfavor in 1947, Zhdanov's power grew, to the point that it was he and not Molotov[37] whom Stalin appointed to articulate publicly his postwar vision.

In his unpublished diary Anastas Mikoyan provides a picture of Stalin's behavior at this time toward many of his former colleagues.

In December 1948, when Stalin returned to Moscow from vacation, I went to see him at the *Blizhnyaya* dacha together with other members of the Politburo. Poskrebyshev was there. Though rarely at *Blizhnyaya,* in connection with the arrival of Stalin from his vacation, he was present. Apparently his presence was agreed to beforehand with Stalin and had a definite aim (as I afterwards realized). A typically pleasant conversation took place for such an occasion. I sat next to Stalin, but at the corner of the table. Poskrebyshev sat on the other side of the table, the last after the other members of the Politburo. Kaganovich sat next to him.

Suddenly in the middle of the meal (it was evening) Poskrebyshev got up from his place and says: "Comrade Stalin, while you
have been relaxing in the South, Molotov and Mikoyan in Moscow
have prepared a conspiracy against you."

This was so improbable that I shouted out: "Bastard!" I picked
up the chair to throw it at him.

Beria, sitting next to me, grabbed me by the hand, and Stalin
said: "Why do you shout here, you're my guest, he's a CC member."

I said: "It's impossible to listen to anything like this that is not so
and could not be so."

I sat down, everything was electric. Stalin spoke calmly without
agitation. Apparently, all of this was discussed beforehand in the
South, and Poskrebyshev fulfilled Stalin's assignment. Before now
there had never been anything like this, and at the moment there
was no cause for this outburst.

Molotov turned pale, but didn't say a word, he sat like a statue.
The others also didn't say a word.

Stalin gradually turned the conversation to another theme. But
this incident made such an impression on us that our visit was much
shorter than usual and we soon left . . .

Apparently, Stalin wanted to test us through Poskrebyshev to see
how we would react.[38]

At a closed session, following the Nineteenth Party Congress in
October 1952, Stalin "remembered" these words of Poskrebyshev and
denounced both Mikoyan and Molotov, just as he was later to "discover"
Timashuk's letter to Vlasik.[39]

The case of Molotov has special relevance partly because his wife was
Jewish—as were the wives of several other Soviet leaders[40]—and had
enjoyed a close relationship with Stalin's second wife, Nadezhda Alliluyeva
(1901–1932), the mother of Svetlana; partly also because it further illustrates
Stalin's method of working against former colleagues. On January 21, 1949,
Stalin had ordered Molotov's wife, Polina Zhemchuzhina, formerly Pearl
Karpovskaya, arrested.[41] Molotov claimed that Zhdanov was responsible for
"laying out" the case against her in 1948, though there is nothing to substantiate this, and other more plausible explanations can be found.[42] Though

the case against Zhemchuzhina had been prepared in May 1948, Stalin did not move against her for another nine months—another example of the caution with which he often proceeded against his "enemies."

According to Molotov, Zhemchuzhina, a fierce Stalinist even in prison, had been accused of no small crime—planning an attack on Stalin's life,[43] but it had been her association with Solomon Mikhoels, Golda Meir, and other so-called Jewish bourgeois nationalists that had in fact led to her arrest. Formerly the deputy minister of the fishing industry, Zhemchuzhina was deported from Moscow to a camp in Kazakstan, where she remained until after Stalin's death. But she was not Stalin's main target. Working through Zhemchuzhina, Stalin was closing the net around Molotov, whom he would not publicly denounce until 1952.

What happened to Andrei Zhdanov between September 1947 and July 1948 dramatizes a course that was diffferent in its details but fundamentally the same. As early as 1942, during the siege of Leningrad, signs of possible future trouble between Zhdanov and Stalin emerged. Stalin reprimanded Zhdanov twice during the siege for acting without permission. The first incident occurred when Zhdanov authorized a committee for the defense of the city without approval from the Central Committee, and at the same time allowed military units that went to the front to select their own leaders. In the second incident Zhdanov did not inform Stalin promptly enough of the military situation on the Leningrad military front in 1942, provoking the angry telegram, "Where do you think you are—in the middle of the Pacific Ocean? . . . Show the contents of this conversation to Kuznetsov."[44] Stalin recognized the potential for unacceptable independence in Zhdanov. In the spring of 1948 Stalin found his old concern about Zhdanov confirmed again.

At that time, the fates of Andrei Zhdanov and the Jews of the Soviet Union became improbably conjoined with the quack botanist T. D. Lysenko. Lysenko, who ruled the world of Soviet agriculture for some thirty-five years, was to Stalin something of what Rasputin was to Nicholas II. He was a man who promised a miracle. Lysenko achieved the pinnacle of his power in July 1948.

Lysenko came to prominence because he had claimed to have perfected a new strain of wheat that could alleviate the recurrent agricultural shortages that had bedeviled Soviet agriculture since the revolution. After the war, starvation ravaged the European part of the country, particularly the Ukraine in 1946-1947. In practically every city there were daily queues to buy bread. Khrushchev wrote that "normal" grain production did not return to the Ukraine until 1949.[45]

Stalin actively worked to advance Lysenko, an adherent of the so-called Michurin school of genetics, through the ranks of Soviet science. Honoring Stalin on his seventieth birthday, Poskrebyshev observed, "Stalin not only helped the Michurin school of geneticists to smash that of Weismann and Morgan, but had also shown how advanced scientific methods could be applied in practice," by introducing eucalyptus trees on the Black Sea and melons in Moscow.[46]

At the core of Lysenko's scientific "theory" was a crude neo-Lamarckism, the key idea of which was that acquired characteristics could be inherited. The session of the All Union Academy of Agricultural Sciences in July-August 1948, where he promulgated this theory, was perhaps the gravest assault on scientific thinking and rationality since the trial of Galileo. In the words of the historian, Zhores Medvedev:

> The famous August, 1948, session . . . was for a long time sorrowfully designated the historical one. It did, indeed, become an event never to be forgotten in the history of science and mankind. It will remain in the annals of human history as an example not only of the senseless destruction of theoretical and practical achievement in biology, but also of the arbitrary and outrageous violation of scientists' convictions.[47]

Adam Ulam described Lysenko in these terms.

> We have already encountered the notorious charlatan before the war when his at first genuine and then faked experiments with developing new strains of wheat earned him the support of the regime and its protection against an enraged scientific community. But the pinnacle of his fame, fraud and influence was to be reached in the 1940s, when he became the virtual Stalin of Soviet biology. His was truly a false idea whose time had come. Lysenkoism

swamped Soviet biological sciences and threatened to invade other disciplines.[48]

The scene of this triumph took place July 28–August 7, 1948. It began with the customary invocation of the party, Lenin, and Stalin by Lysenko.

Our Academy, with its work for the good of our kolkhozes and sovkhozes, for the good of our Motherland, and bearing the great name of V. I. Lenin, must justify the high degree of faith, concern and attention that it has been shown by our party, our Government, and personally by comrade Stalin (*applause*).[49]

Justifying Stalin's faith took precedence over all else.

The penultimate statement at the conference was a personal letter to Stalin read by Lysenko.

Dear Josef Vissarionovich!

The participants of the session of the All-Union academy of agricultural science in the name of V. I. Lenin—academics, agronomists, cattle breeders, biologists, machine operators, organizers of the socialist agricultural industry send You their sincere Bolshevik greeting and very best wishes.

He concluded with the words: "Glory to the great Stalin, the Leader of the people and the Coryphaeus of vanguard science." The stenogram noted: "(Tumultuous, long, undiminishing applause, leading to an ovation. Everyone stands.)"[50]

But Lysenko was not done. In his "Concluding Word," he took his support "for the good of our Motherland" a step further to demonstrate to the remaining skeptics in the audience that his views were the only acceptable ones.

Comrades! Before going on to the concluding word, I consider it my duty to state the following.

In one of the notes they ask me what kind of relationship exists between my report and the Central Committee of the Party. I answer: The Central Committee of the Party reviewed my report and approved it. (*Tumultuous applause, leading to an ovation. Everyone stands.*)

The significance of these statements cannot be underestimated. Lysenko dedicated his work to Stalin, the "Coryphaeus of vanguard science." Stalin stood behind Lysenko. The entire Central Committee stood behind Stalin. It was a closed system from which there was no exit. The Central Committee had read Lysenko's report on the biological sciences *beforehand*, and approved it. The option of open discussion and revision was ruled out from the start. Those who questioned Lysenko were therefore in grave danger. His speech was calculated not to persuade but to crush the opposition. All in that room knew it, and the few voices of opposition provided servile testimonies of allegiance by the conference's end.

With his political base secured, Lysenko proceeded to summarize the essential tenets of his biological theory.

Now I will go on to an accounting of several of the conclusions of our sessions.

Speaking here, the proponents of the so-called chromosomal theory of inheritance denied that they were Weismanists.[51] . . . At the same time . . . it was clearly shown that Weismanism and the chromosomal theory of inheritance—are one and the same. Foreign Mendelists-Morganists[52] do not hide this in any way. . . . Weismanism (and this is idealism in biology) offers a distinct depiction of inheritance, recognizing the division of the living body into two principally different essences: the ordinary living body, allegedly not possessing the power of inheritance but subject to change and transformation, that is, development; and the specific, inheritable substance, allegedly not dependent on the living body and not subject to development in connection with the conditions of life of the ordinary body, called the soma. This is inarguable. No sort of attempt by its proponents and by its proponents at this session in defense of the chromosomal theory of inheritance, attempting to give their theory a materialistic appearance, could change the character of this theory which is idealistic in its very essence. (*Applause.*)

The Michurinist direction in biology, because it is also materialistic, does not distinguish the property of inheritance from the living body and the condition of its life. Without inheritance there is no living body, without the living body there is no inheritance. The living body and its conditions of life—are indivisible. It is correct to deprive the organism of its conditions of life only when it becomes dead. . . .

The principal significance of our divergence from the Weismanists also produces the divergence over the great history of the question of inheritance of acquired characteristics by plants and animals. Michurinists proceed from the possibility and necessity of the inheritance of acquired characteristics. . . . The Morganists, in which group were some at the present session, are not able to understand this position, not having broken completely with their Weismanist imaginings.

Up to now it has not been clear to some here that inheritance is constituent not only of the chromosomes but other parts of the living body. . . .

Experiments in vegetable hybridization incontrovertibly show that the entire living body possesses inheritability, every cell, every part of the body, and not only the chromosomes. Inheritance, it is true, is distinguished by a specific type of exchange of substance. If you are able to change the type of exchange of the substance of the living body, you change inheritance.

. . . I promised academician Zhukovsky that I would show him vegetable hybrids and here at this session I have had the pleasure of showing them to him.

In this case, in a kind of grafting procedure a sort of potato-tomato was conceived; that is, a fruit with serrated leaves, as is usual with tomatoes, but similar to that of a potato. The fruit of this sort— are red and oblong.

Another having been part of the grafting process with tomatoes has the leaves such as are usually to be seen on the tomato plant— serrated; the ripe fruit is not red but white, yellowish.[53]

Many of the finest minds of Soviet science were beaten into submission with Lysenko's oblong, red potatoes and his yellowish white tomatoes—which subsequently were shown to be fakes. Those who protested on the basis of empirical data were deemed reactionary, idealist, bourgeois enemies of the people; most were summarily dismissed from their positions and sent to the gulag. Hundreds of biologists, physicists, philosophers of science, and agricultural specialists suffered repression because of attempted opposition to Lysenko's nonsensical formulations.[54] Hundreds of others fell silent.

Zhdanov's connection to Lysenko was sketched out by Aisik Moiseevich Krongauz, an otherwise obscure man, with no connection to Lysenko, the world of science, or Andrei Zhdanov. Krongauz, a Jew, had been arrested sometime in either 1949 or 1950 as part of the massive "anticosmopolitan" campaign. By July 1951 he found himself in a deportation prison somewhere in either Novosibirsk or Irkutsk—he could not recall exactly where or when. However, in this deportation prison, he met the son of the famous doctor Yakov G. Etinger, who had also been recently arrested on similar grounds. Krongauz related the following story:

> Etinger [the son] told me that the son of one of the leaders of the VKP(b) [the All-Soviet Communist Party] allegedly prepared a speech against academician LYSENKO with the help of his father.
>
> This became known to one of the leaders of the party and the Soviet government, which summoned the father and had a stern conversation with him. This conversation acted so negatively [on the father] that soon after that he died.[55]

Krongauz apparently knew nothing of the Lysenko conference or the events that had taken place in Valdai in July–August 1948. He did not seem to know the identity of the individual who prepared the speech against Lysenko. Nevertheless, he recognized something important in the death of a government leader that was precipitated by Stalin's anger, and he wrote about this in a letter he tried to send out of the country through the Israeli mission. By any standards this was important news. Krongauz's letter was intercepted by the authorities. When pressed, Krongauz admitted he knew nothing more about any of this and had put in the letter only what Dr.

Etinger had told his son whom he met in the camp. This "leading government figure," whom Krongauz associated with a criticism of Lysenko, could only have been Andrei Zhdanov.

The fact that Krongauz knew this story proves that information had leaked out of the confines of Valdai and had become known to Etinger, who had not been in Valdai at the time of Zhdanov's illness, but had been a member of the consulting group with whom Vinogradov met on August 31, 1948. He had been present when Zhdanov's death was announced, and it may be that Vinogradov mentioned this to Etinger and others at that time.

The connection with Lysenko was fateful for Andrei Zhdanov. On April 12, 1948, his son, Yuri, who had recently been appointed the head of the science section of the Central Committee, discussed Lysenko's theories, if not Lysenko himself, at a lecture on the situation in the biological sciences for regional party committees. Yuri's lecture precipitated an inquiry authorized by Stalin and an emergency session of the Politburo. The events of this meeting have been described by Vladimir Karpov in his memoir of Dimitry Shepilov in *I primknuvshii k nim SHEPILOV*, published in Moscow, 1998.[56]

In 1948 Shepilov worked under M. A. Suslov, the head of the agit-prop section of the Central Committee. Suslov worked under Andrei Zhdanov, the head of ideology. Soon after his new appointment, Yuri became aware of the controversy over Lysenko's theories, which had many opponents in Soviet science. The Central Committee had received their letters over the years but until now had never formally responded. Yuri therefore proposed a meeting on the subject to Shepilov, who thought it a good idea and brought it to Suslov. Suslov approved it, and the meeting went forward under the auspices of the Central Committee. Yuri's presentation, entitled "Points of contention in contemporary Darwinism," though mildly critical, was, according to Shepilov, immensely tactful and was greeted "with great sympathy" by all those present.[57] He did not mention Lysenko by name. However, either Lysenko himself or some of his sympathizers soon learned of the session, understood its potentially damaging consequences, and immediately reported it to Stalin.

On the following day Suslov, Shepilov, Yuri Zhdanov, and his father were summoned to a meeting of the Politburo in Stalin's Kremlin office. Stalin was frowning. He did not sit at the table but paced the perimeter of the room. A stenogram of the session was in his hand. Stalin spoke in a "hollow, quiet voice," just above a whisper.[58] No one dared speak louder.

"Has everyone read the report of Zhdanov, the young Zhdanov?" Stalin asked.

"We've all read it."

"This is unheard of. They presented a report by the young Zhdanov without the knowledge of the Central Committee at a meeting of lecturers. They really gave it to Lysenko. But on what grounds? Who authorized this?"

No one spoke. Suslov should have answered, but he didn't. The silence persisted. Then another unheard-of thing happened. Shepilov rose from his chair.

"I authorized it, comrade Stalin," he said.

Stalin glared.

"On what grounds? Don't you know that all of our agriculture hinges on Lysenko?"

"Comrade Stalin," Shepilov said, "they have reported falsely to you about Lysenko's work. Leading scholars and agronomists have come to me. . . . These scholars have not been able to name a single new sort [of plant] actually produced by Lysenko. Not a single, actual leading scholar occupied with the agriculture of our country. I am prepared to accept any punishment. But I earnestly ask you to appoint a special commission to examine the work of Lysenko. Without the highest commission of the CC no one would be able to decide this matter correctly."

Stalin said nothing. According to Shepilov, he turned white. He had been standing across from Shepilov but now went to his chair at the table, took out some tobacco, and lit his pipe.

"No," Stalin finally intoned. "No one could do that. We must appoint a special commission of the CC to examine the matter. We must punish the guilty in exemplary fashion."[59] As the meeting concluded, Stalin ominously observed that "it was necessary to question the father and not the children."[60] The father in this case was Andrei Zhdanov, who fainted on the way to his office in Old Square three months later, following another session with Stalin at which he was advised to improve his health at Valdai.

Predictably, there were no immediate consequences. Shepilov remained head of one of the sectors of agit-prop, and at the order of

Malenkov, who was in charge of agricultural matters for the Central Committee, he prepared the Lysenko conference held in July–August 1948. The ax fell only later in August, at the time Zhdanov was in Valdai, when Shepilov found himself out of a job.[61] For the moment Andrei Zhdanov remained reprimanded but untouched in his position on the Politburo, but in early July he found himself isolated in Valdai. He knew he had fallen into extreme displeasure. It was he who would be punished "in exemplary fashion."

All who were connected to the Lysenko affair knew their careers, if not their lives, were now in jeopardy. All except Yuri Zhdanov himself. In the spring of 1949 he married Svetlana Alliluyeva, and by all accounts remained a favorite of Stalin's until the dictator's death. Soon after Stalin's death, they were divorced.[62]

Though crucial, the Lysenko affair by itself would probably not have been decisive. Stalin had invested much of his own prestige in advancing Zhdanov above the other members of the "inner circle." The slowness and stealth of his actions toward Zhdanov amply testify to his need to protect this prestige. Nevertheless, as Dimitri Volkogonov observed, "Stalin could not forgive independent, free thinking."[63] This was the decisive element. Those who would not back Lysenko could not be trusted. Stalin was higher than rationality, and the party stood above any mathematical or scientific proof. Rationality and empirical science, along with literature, art, and music had to serve "the interests of the state." So did Andrei Zhdanov. Not only Soviet agriculture was at stake. Together with Stalin's previous criticisms of Zhdanov for alleged "independence from the Central Committee," the Lysenko affair finally pushed Zhdanov into the same circle of those Stalin no longer trusted, which included Molotov, Voroshilov, Mikoyan, and Beria.

In Stalin's mind, Yuri Zhdanov's criticism of Lysenko, however mild and indirect, challenged his authority. The lecture had been approved by Shepilov and Suslov, but Andrei Zhdanov was the head of ideology under whom they both worked. Though he probably knew nothing of it, he became the target. He, not Yuri, posed the threat. According to Krongauz, Dr. Etinger had said that the elder Zhdanov had helped prepare his son's speech criticizing Lysenko. It is doubtful that Zhdanov would have taken such a step even if he had known of the meeting in advance. Although

Krongauz's statement was undoubtedly erroneous on this point, it strongly suggests that this was the view held by the doctors in Valdai. If it was their view, it must have been the view of whoever informed them of the April 12 Central Committee meeting. Therefore, it must also have been Stalin's. The only ones attending the Politburo meeting were Shepilov, A. A. and Yuri Zhdanov, Suslov, other members of the Politburo, and Stalin himself. None present would have gossiped about such a meeting or its consequences. General knowledge of this meeting was made possible only in 1998, after publication of *I primknuvshii k nim Shepilov*, but the doctors (Dr. Etinger at the least) knew almost immediately. Reconstructing the pathway whereby this information leaked out leads back either to Poskrebyshev, Stalin's secretary, or to Vlasik, both of whom had frequent contact with Yegorov. That pathway must ultimately lead back to Stalin himself.

"They decided to send me on vacation," Zhdanov complained to Shepilov after the Central Committee meeting held in early July 1948. He wanted to go to a government dacha in the south, but his wishes were no longer decisive. He was sent to the north, to Valdai instead.

One week before Lysenko's July conference, Zhdanov received the telephone call from Shepilov in Valdai. The conversation made Zhdanov exceedingly agitated, and he suffered his first severe attack. His bodyguard noted that from 11:30 P.M. to midnight Zhdanov "could not breathe."[64] Vinogradov described Zhdanov's agitated condition.

> There was a business call on the telephone. [Zhdanov] became very upset and was not able to finish the conversation because of a severe attack of cardiac asthma. He remained motionless, breath was gurgling, phlegm from the lungs containing some blood, and so forth.[65]

Sophia Karpai took an EKG that showed a "left bundle branch blockage"— not a heart attack. As a result, Zhdanov was put on a protocol of "absolute bed rest" that lasted for approximately two weeks. He died five weeks later.

The conversation between Shepilov and Zhdanov had probably to do with the fact that Malenkov had just then been appointed secretary of the Central Committee in charge of maintaining the party cadres. This was an important role. Zhdanov knew the meaning of these transitions. It was precisely this role that Zhdanov had taken over from Malenkov in 1946.

Zhdanov—eulogized by Molotov as "one of the most beloved sons of the Soviet nation, a true scholar and talented colleague of the Great Stalin"[66]— had lived to see himself cast aside. This is why he choked with panic and rage during the telephone call from Shepilov on the evening of July 23. Zhdanov knew he could never recover Stalin's trust.

Lysenko's conference ended on August 7, 1948. On page 5 of the August 7 issue of *Pravda* the following open letter to Stalin from Yuri Zhdanov appeared:

To the Central Committee of the Party
TO COMRADE I. V. STALIN

I unquestionably made a whole series of serious errors in my presentation at a seminar of lecturers on the controversial questions of contemporary Darwinism.

1. Presenting this report was itself a mistake. I clearly underestimated my new position of worker in the apparatus of the Central Committee; I underestimated my responsibility; I did not imagine that my speech would be evaluated as the official point of view of the Central Committee. Here "the university habit" appeared, when in one or another scholarly argument, I would unthinkingly blurt out my point of view. Therefore when they proposed that I make a speech at a seminar of lecturers, I decided to express my opinions, stipulating that it was my "personal point of view" in order that my speech would not in any way obligate anyone to anything. Undoubtedly, this was "professorial" in a bad sense and not a party position.

2. The essential error in this report was its intention to reconcile opposing sides in biology. . . .

My mistake consisted in the fact that . . . I did not subject to ruthless criticism the essential methodological errors of Mendelev-Morganist genetics. I acknowledge that this narrow-mindedness was like hunting for kopeks.

The struggle between directions in biology often takes monstrous forms, squabbles and scandals. However, it seems to me that in this instance there are in general nothing but petty squabbles and scandals. Consequently, I underestimated the principles involved in the question; I approached the argument ahistorically; I did not analyze its deep causes and essence.

All of this taken together gave rise to the desire "to reconcile" the feuding sides, erase the disagreement, to underline that which united, as opposed to disunited the opponents. In science, as in politics, principles cannot be compromised, but must be victorious; the struggle does not proceed by the path of hushing things up, but by means of disclosing the controversy. The attempt to reconcile principles is at bottom narrow mindedness and shortsighted pragmatism, an underestimation of the theoretical sides of the argument leading to eclecticism to which I confess.

3. My sharp and public criticism of academic Lysenko was my mistake. Academic Lysenko at the present time is the acknowledged leader of the Michurinist tendency in biology. He has defended Michurin and his doctrine from the attack of bourgeois geneticists, and he himself has done much for science and the practice of our economy. Taking this into account, the criticism of Lysenko, his specific inadequacies, ought to lead not to a weakening but to a strengthening of the position of the Michurinists.

I do not agree with several theoretical positions of Academician Lysenko (the denial of intraspecific struggle and mutual cooperation, the underestimation of the internal specificity of the organism); I think that he still makes use of the treasure house of Michurinist doctrine weakly (precisely therefore Lysenko has not produced that many significant sorts of agricultural plants); I think that he weakly leads our agricultural sciences. Led by him, VASKhNIL does not work nearly at its full strength. There is no work on livestock, on the economy and the organization of agriculture. There is only weak work in the area of agro-chemistry. But I should not have criticized all these inadequacies as I did in my report. As the result of my criticism of Lysenko, the formal geneticists were in "seventh heaven."

Being devoted with all my heart to the Michurinist doctrine, I criticized Lysenko not because he was a Michurinist, but because he had insufficiently developed the Michurinist doctrine. However, the form of the criticism was incorrectly chosen. Therefore the Michurinists objectively lost from such criticism and the Mendelev-Morganists won.

4. Lenin often pointed out that the recognition of necessity of this or another phenomenon concealed in itself the danger of falling into objectivism. Quite visibly, I did not escape this danger.

I characterized the place of Weismanism and Mendelev-Morganism (I don't distinguish between them) largely by means of "pimenovsky": taking good and evil indifferently. In place of this, in order to come down sharply against these anti-scientific views (expressed to us by Shmalgauzen and his school), that in theory are a veiled form of popism, theological representations of the appearance of forms as the result of specific acts of creation, and in practice lead to "maximization," to the denial of man's ability to alter nature, livestock, and plants, I mistakenly put before myself the task "to acknowledge" the place [of Mendelev and Morgan] in the development of biological theory, to find in them "a grain of rationality." As result, the criticism of Weismanism came out weak, objectivist, and, in essence—superficial.

In sum, again, the main blow was forced upon academician Lysenko; that is, against the Michurinist tendency.

Such are my errors, as I understand them.

I consider it my duty to assure You, comrade Stalin, and in Your person the Central Committee of the party, that I was and remain an ardent Michurinist. My errors flowed from the fact that I insufficiently analyzed the history of the issue, incorrectly established the front of the struggle for Michurinist doctrine. All of this is the outcome of inexperience and immaturity. I will correct my mistakes with my actions.

Yuri Zhdanov[67]
[July 10, 1948]

Fifty years after Stalin's death, this letter published in *Pravda* will inevitably strike Western readers as strange, if not entirely inexplicable. Dated July 10, 1948, but not published until August 7, the last day of Lysenko's conference, it must have seemed to general readers of *Pravda* both opaque and ominous. Such self-abasing public letters, though a frequent occurrence in the 1930s, had not often appeared since the war. Furthermore, the April Politburo meeting and A. A. Zhdanov's banishment to Valdai would have been unknown to them. To the scientific community, however, the meaning of this letter would have been crystal clear: Lysenko had won total victory.

To the circle of doctors treating Zhdanov in Valdai, it would have had another special meaning concerning the relationship between the elder Zhdanov and Stalin. In his letter Yuri related the details of a previously unknown event in a manner that shed little or no light upon it. We learn almost nothing concerning the actual content of Yuri's lecture, the nature of his actual criticism of Lysenko, or under whose auspices the lecture was given. Yuri wrote, "when *they* proposed that I make a speech," leaving the "they" unidentified. The doctors would readily have surmised that the letter's purpose was not to provide information, clarify an issue, point of view, or official position, but rather to show solidarity with the party position. Written three days before A. A. Zhdanov left for Valdai, the letter was a veiled plea for mercy—not only for himself, but for his father as well. Such mercy was necessary only in the event of a rupture. That was the key, if indirect, piece of intelligence Yuri's letter would have provided those in charge of Zhdanov's health in Valdai: A serious rupture between Zhdanov and Stalin had occurred; Zhdanov had been sent to Valdai; on July 23 Shepilov telephoned with information that greatly upset him. At that point, the doctors might not have known exactly what happened in the Kremlin, but they would have known that it was very serious. If, according to Timashuk, they had underestimated the gravity of Zhdanov's physical condition, they did not underestimate the gravity of his political situation.

August 7 also marks a turning point in Zhdanov's treatment. On this day Sophia Karpai, who had been taking the electrocardiograms of his heart, was sent on vacation, and from this date until August 28 no further EKGs were taken. During these twenty-two days Zhdanov was gradually put on an active regimen; his nursing care became sloppy; his personal physician, Maiorov, went fishing; and he had recurrent attacks of cyanosis and shortness of breath accompanied by considerable discomfort that were not noted in the medical record. It seems that the publication of Yuri Zhdanov's letter, timed with the triumphant conclusion of Lysenko's conference, played a part in whatever unspoken medical deliberations went forward in the troubled atmosphere of Valdai.

What did Stalin want them to do? What was the hidden message they were to take from Yuri's public humiliation? Yegorov might well have checked through Vlasik, or perhaps Poskrebyshev, for guidance. The results would have been indirect, ambiguous, like the Greek oracles. At the very least, though, they well might have told Yegorov that Zhdanov was in disfavor and was being marginalized. Be vigilant. Do what you know is right under the circumstances. All concerned knew what that meant.

From this point on, the doctors took no positive steps to murder Zhdanov, by poisoning him, for instance, but they took a series of negative steps that demonstrated that they had understood their assignment—the EKG technician was not replaced for three weeks, the nursing care was negligent, they did not take standard precautions in treating someone who had recently developed a "left bundle branch block," whose general physical state was clearly deteriorating. Such steps became evident to members of the nursing staff, among whom there were undoubtedly informants like Timashuk. The MGB would have been alerted to this lax care, and this might be the explanation for Timashuk's sudden appearance on August 28. Wheels within wheels of conspiracy were grinding against each other in Valdai.

Stalin prided himself on being an amateur psychologist, and he surely would have calculated the effect of Yuri's letter on the elder Zhdanov, coming at a time when Zhdanov was already aware that he had been displaced on the Central Committee by Malenkov. Given the date Yuri's letter was written, the timing of its publication was not accidental. Coming on the final day of Lysenko's conference, it became the subject of some discussion in the closing speeches of the conference and provided further confirmation to all present that Lysenko enjoyed full support from Stalin and the Central Committee.[68] The obverse was also clear enough: Those who did not support Lysenko were now pariahs.

Though professing inexperience and youth, Yuri demonstrated a sophisticated understanding of political realities. Foremost among these was his repudiation of a "personal point of view." Yuri assumed all blame, paradoxically, attributing his errors to a misguided commitment to Lysenko. His unwavering loyalty to Stalin underlay the whole. It was precisely his "personal point of view" that got the elder Zhdanov into trouble during the siege of Leningrad.

The letter was not written to improve Yuri's position with Stalin. It had another purpose. Stalin's anger was aimed not at Yuri but at the father. This was the exemplary fashion in which the guilty one would be punished. Not only would he lose his position; he would be publicly humiliated and discredited as well. Among those who knew the intricacies of Stalin's Kremlin, it was clear that Andrei Zhdanov could never again be a leader. This ardent fighter for Stalin, whose "annihilating scorn" exposed the mercenary bourgeois literature of the West that poisoned the minds of noble Soviet youth, himself had a son whose mind had somehow become polluted. The shame of this would have been unbearable to Zhdanov, who would have understood completely the extreme psychological sadism now being practiced upon him by his old comrade. This alone might have been enough to kill Stalin's ailing colleague and may have accounted for Zhdanov's apparent fatalism when he refused to recall Maiorov from his fishing trip despite suffering acute difficulty breathing, accompanied by cyanosis and extreme exhaustion. It may also have accounted for his kindness toward the nurse who fell asleep during her watch.

Having read Yuri's letter, the doctors in Valdai were able to gauge Zhdanov's reaction firsthand. Something had happened. This "true son of the Motherland and party" had inexplicably gone astray. In three weeks he would be dead. The last line of Yuri's letter: *Delom ispravlyu oshibki* (I will correct my mistakes with my action) may have had more meaning than he intended. To those who knew the story, it was his father who had become the mistake.

Dr. Yegorov was known "as an unprincipled individual, a petty tyrant, working in the Lechsanupra by shouts and threats." He was a consummate political flunky who valued political position above medical responsibility and was one of those who would have rendered fealty to Lysenko to save his skin. Apparently he read the covert message of Yuri's letter, knowing the importance of suppressing any "personal point of view," and portentously had told the doctors in Valdai that Zhdanov would never recover. Vinogradov remembered Yegorov saying, "A heavy job has fallen to us . . . the fate of Zhdanov [is] predetermined."[69] This pronouncement probably had both a political and a medical meaning. Vinogradov said that he and the other doctors took this to mean that their job was to ensure that Zhdanov did not recover from his illness. "When doctor Timashuk, L. F., attempted to unmask us in the criminal treatment of A. A. Zhdanov, I, Yegorov, Vasilenko and Maiorov took all possible measures to hide the traces of our villainous deeds. We collectively accused Timashuk of ignorance and took vengeance on her. . . . Soon after this, Yegorov removed Timashuk from the Central Polyclinic, transferring her to the 2nd Polyclinic and in this way removed her from taking part in the treatment of leading workers of the Soviet state."[70]

Was this true? The situation is complex and murky. The facts of the case support Vinogradov's version of the events, except that the doctors did nothing to kill Zhdanov. They performed no "villainous deeds." But they were "vigilant," and appear to have acted out of loyalty to what they took to be Stalin's will, much as Yuri Zhdanov confessed he would—to correct their mistakes with their deeds, and to sacrifice their "personal point of view" for the will of the Central Committee. They saw the will of the Central Committee revealed in *Pravda* on August 7. A. A. Zhdanov was humiliated; his rest at Valdai was banishment. Their source in the Kremlin had informed them that Zhdanov, the father, had been harshly treated by Stalin and was to be punished in "exemplary fashion." As a trained political opportunist, a loyal member of the "middle cadres," Yegorov did not need to know more to know what Stalin wanted him and his colleagues in Valdai to do.

Khrushchev once joked that during the time of the Doctors' Plot, it was rumored that the interrogators had gotten poor Dr. Vinogradov to go so far as to confess that it was he who had written *Eugene Onegin*.[71] This was Khrushchev's way of showing the absurdity of all the charges against the doctors. Vinogradov's abject confession that he worked under the influence of the English spy Dr. M. G. Kogan was precisely this absurd:

> . . . and now I confess that I evilly sabotaged the treatment of the
> patient A. A. Zhdanov because of my anti-Soviet disposition and, in
> addition, I was under the influence and pressure of the English spy
> Kogan, M. G., with whom I was connected in criminal actions since
> 1937.[72]

Espionage, premeditated murder, and treason linked to the Jewish doctor Kogan were pure invention. But this does not mean that everything was
fabricated. Khrushchev's desire to discredit the charges against the doctors
in this fashion shows how eager he was to sweep away all scrutiny of the
Doctors' Plot. If the case were all lies and falsehoods, there would be no
need to examine its deeper causes or the system out of which it grew and
of which Khrushchev, for all his denials, remained a part. What seems to
have been true is that Yegorov either told the doctors directly or let them
understand in other ways that, as Vinogradov put it, "the fate of Zhdanov
was predetermined"—predetermined not simply by his physical state but
by the will of the leader of the Soviet government.

When Sofia Karpai first took the EKG on July 25, she found evidence
of an "intraventricular blockage," and she informed the doctors that
although she did not find direct evidence of a heart attack on the EKG, it
was not possible to exclude the possibility one had taken place. The clinical data "was not absolutely typical of a recent infarct," she told them. On
the basis of this ambiguous July 25 evaluation, the doctors "decided to treat
the patient as an infarct patient."[73] Great care was taken despite the absence
of a positive diagnosis.

Karpai took an EKG again on July 31. To her surprise, she found that
the blockage she had previously detected was gone.[74] A week later she
delivered this opinion orally to the doctors, stating that in her opinion
Zhdanov suffered from cardiosclerosis, chronic coronary failure, and progressive deterioration of the coronary vessels. She also told them that on
the basis of her EKGs she concluded that there were small sites of necrosis
in the patient's heart. He had suffered micro-infarcts, she believed, but no
major heart attack.

Karpai's August 7 departure was never explained. There is no record
that she had requested a vacation at that time. It is possible she was removed
because she was Jewish and the anti-Jewish purges were now progressing

violently to their height, but this does not explain why Karpai was not replaced in Valdai for another three weeks. It is more likely that Yuri's letter in *Pravda* pointed the doctors on the path they now knew they would have to take, and they feared that Karpai, who was not part of their inchoate conspiracy, knew too much about Zhdanov's condition to keep in Valdai. She had delivered an opinion that would allow them great latitude in further treatment. It was opportune to let her go. Despite Zhdanov's progressively deteriorating condition over the next three weeks they did not conduct another EKG.

On August 25 while strolling around the park, Zhdanov developed cyanosis of the lips and gasped for breath.[75] His doctor, having gone fishing, was nowhere to be found. No EKG was performed. Three days later Zhdanov had another attack of some sort, and only at this point did the doctors call Timashuk to Valdai. She took the EKG on August 28 and found the same blockage Karpai had detected in July, but which had "vanished" by July 31.[76] In July the doctors reacted appropriately, treating Zhdanov as an "infarct patient" and ordering strict bed rest. Now they did not.

"Karpai . . . is a typical person of the street with the morals of the petty bourgeoisie," Vinogradov told the authorities. On July 25, 1948, having taken Zhdanov's electrocardiogram, Karpai "shyly expressed," Vinogradov said, "the suspicion that A. A. Zhdanov had suffered a myocardial infarct; however, she easily repudiated this opinion and agreed with us in the incorrect diagnosis."[77] Karpai disputed this, stating that the diagnosis had been inconclusive, as in fact it had been.

> The electrocardiograms taken by me of the patient Zhdanov on July 25, 1948 showed an intraventricular blockage. To the question of whether this was an infarct, I answered that although there were no typical signs of a recent myocardial infarct, it could not be ruled out.
>
> I also think that the clinical data was not absolutely typical of a recent infarct. However, I believe that the consultation decided to treat the patient as an infarct patient.[78]

Vinogradov was anxious to prove that Karpai, like the others in Valdai, had conspired to hide the truth of Zhdanov's heart attack. But the truth *was* unclear. Karpai let the doctors understand this. She was dangerous precisely because she knew that the diagnosis of Zhdanov's medical condition was ambiguous; she also knew what Timashuk did not, which is that faced with the identical ambiguity in July, the doctors had reacted with requisite extreme caution. In August they took advantage of the ambiguity not to show such caution. Karpai alone could have provided a measurement of the *alteration* in the doctors' approach to what amounted to the same medical situation. The doctors may have hoped to conceal this foreseeable alteration in the quality of care they provided. On August 7 they did not know that the blockage Karpai had detected on July 25, but which had disappeared by July 31, would return, but they knew well enough what measures not to take if it did.

The seventh of August was therefore a watershed date in the first phase of the conspiracy. It marked the point at which the doctors became trapped in their own eventual undoing. On this date the doctors were not the target of the conspiracy—Andrei Zhdanov was. They understood from Yuri's letter in *Pravda* and the overwhelming victory of Lysenko over the scientific community that it would be convenient for Stalin to remove his former favorite. They would be the means by which this could be most expeditiously achieved. There was no evidence that the doctors themselves—Jewish or otherwise—were or would be conceived as the ultimate victims of Stalin's maneuvers against Zhdanov. All the signals to the doctors up to this point had been positive. Yuri Zhdanov's letter gave them objective confirmation of what they suspected and of the rumors swirling around the Kremlin at the time, which certainly would have reached Yegorov if not the others. They knew of the explosive rupture between Stalin and Zhdanov. Stalin demanded "vigilance." He demanded it from the scientists who pledged unwavering support for Lysenko, and now, without a direct command, he demanded it from the doctors. They saw clearly what happened to those scientists who refused to go along—they were reviled, demoted, despised, arrested. Yegorov, a Stalinist toady and drinking buddy of Vlasik, would not be among them.

In the December 1948 incident at Blizhnyaya dacha we saw Stalin work through Poskrebyshev to "test" Molotov and Mikoyan. In August 1948, by writing "into the archive" at the bottom of Timashuk's letter, Stalin also was working through subordinates, this time through Abakumov to communicate his attitude about Zhdanov and Timashuk to Yegorov. Stalin did not need to speak directly with Yegorov. Abakumov's silence was understood by Vlasik and eventually by Yegorov as a sign of Stalin's tacit approval.

Timashuk, Karpai, Yegorov, Vasilenko, Vinogradov, and Maiorov all played roles in a drama without explicit stage directions or directives. Each attempted to interpret Stalin just as Vladimir and Estragon search out the hidden motives, intentions and plans of the invisible Godot for whom they hopelessly wait. Stalin's world was defined by conspiracies and plots. The dimensions of this conspiratorial world can be judged from Ignatiev's observation that when he became minister of state security in 1951 there were approximately ten million informants in the Soviet Union—some working for money and some of their own free will.[79]

The system compelled people to spy on and denounce one another, often without any idea where their denunciations might lead. Had Timashuk not denounced the doctors, she was in danger of being denounced herself by the others at Valdai. Self-interest was the operative principle in Soviet life. Denunciation was also a means of career advancement, and Timashuk would have known that denouncing her superiors might lead to significant recognition and therefore benefit for herself. She turned out to have been both right and wrong.

Yegorov understood this principle and knew that the entire Kremlin Hospital system was filled with informants who often worked blindly to satisfy what they took to be the interests of the Kremlin. Yegorov thought he knew what those interests were. He, too, was both right and wrong. Abakumov's call on August 30 or 31 gave Yegorov what he thought was a signal of Stalin's intention and illustrated how another side of the system worked: Abakumov called to say that Stalin had received a copy of Timashuk's letter. He said nothing more. He did not need to. Poskrebyshev's approval for the autopsy in Valdai was another illustration. These two signals in combination with whatever information Vlasik had conveyed made him relatively sure of his ground. Yegorov's fury at Timashuk came from his sense that she was stupidly trying to use the system against *him*. The extraordinary September 6 session in his Kremlin Hospital office was a highly coded performance. Both Timashuk and Yegorov were struggling against each other in the dark; each found his certainties in how the system worked upset: Timashuk because she saw her letters go unanswered, Yegorov because Timashuk's insistence *might* have meant that he had in fact misunderstood the nature of Stalin's unspoken directive. Knowing he followed a higher order, Yegorov regarded himself as innocent, though he knew he was guilty for the reasons Timashuk stated. On her side, Timashuk knew that the EKG (though in fact she had misinterpreted its results) justified her accusation against the doctors, but the Kremlin took no notice, as though she were in the wrong.

Timashuk became a formidable menace to the doctors not because she was right and they were wrong, but because at a certain point in time, circumstances changed and her allegations of criminal mistreatment came to suit "the interests of the state"—the *new* interests of the state. Yegorov might have worried about such an eventuality, but there was little he could do about it. At the time, neither Yegorov nor Timashuk understood that they both might be serving the interests of a security service working at cross purposes with itself.

THREE

THE PYGMY AND THE TERRORIST

LUBYANKA-LEFORTOVO, NOVEMBER 1950 — MARCH 1951

In Moscow there live more than a million and a half Jews. They have seized the medical posts, the legal profession, the union of composers and the union of writers. I'm not even speaking of the trade networks. Meanwhile of these Jews only a handful are useful to the State, all the rest—are potential enemies of the State . . .

—RYUMIN TO MAKLYARSKY, FEBRUARY 1952

It seemed to me that only Stalin could correct the historical injustice committed by the Roman kings.

—ITZIK FEFER

Etinger was such a typical Jew and spoke with an accent.

—V. S. ABAKUMOV, AUGUST 8, 1951

The second major stream of events leading to the Doctors' Plot was set in motion with the arrest on November 18, 1950, of Dr. Yakov Etinger who had been the subject of Aizik Krongauz's testimony and a member of the consulting team with whom Vinogradov met on August 31, 1948. Etinger was not arrested, however, in connection with the death of Zhdanov two years earlier, or with the subject of medical sabotage whatsoever. Although Etinger had consulted with Vinogradov about Zhdanov's illness, he had not been directly involved in the case, possibly a sign of the growing anti-Semitism of the government leadership and the fact that he may already have come under suspicion by the authorities. At the time of his arrest, the theme of medical sabotage, despite Timashuk's 1948 letter, or any conspiracy by Jewish doctors to murder Kremlin leaders was absent from the MGB inquiry. The two Jewish doctors arrested in connection with the Jewish Antifascist Committee—Boris Shimeliovich and Lina Shtern—were not interrogated along these lines, despite the fact that Shimeliovich spoke freely to his captors of the problems of the Kremlin Hospital in 1948 and 1949. A month before his execution, Shimeliovich wrote a letter to Malenkov in which he pointedly complained that no one had paid any attention to his analysis of the poor treatment given important patients in the Kremlin Hospital system.[1] No one listened to Shimeliovich, just as no one had listened to Timashuk. Etinger had been arrested not as a doctor, but as a Jew who ardently supported the state of Israel.

The so-called anticosmopolitan campaign, which had been directed almost exclusively against Jews, formally began in 1947 with editorials in *Pravda* and other official government organs, attacking alleged Jewish ideological sabotage of the arts. Initially the charge was that Jews felt themselves superior to Russians and devalued Russian culture.[2] With the founding of the state of Israel and outbreaks throughout the Soviet Union of what Stalin perceived to be dangerous pro-Israel, Zionist fervor in the Soviet Jewish population, it was alleged that Jews could not be trusted to be loyal to the USSR and constituted a potential fifth column in case of a war with the United States. From 1948 to1953 thousands of Jewish intellectuals, scientists, political leaders, state security personnel, and private individuals alike were mercilessly interrogated, thrown out of their positions, publicly ridiculed, taunted, threatened, and imprisoned. Many were shot.[3] In January 1948, Solomon Mikhoels, the internationally respected Soviet Jewish actor and director, founder of the Moscow Yiddish Theater, and head of the Jewish Antifascist Committee formed during the war, was assassinated in Minsk on direct orders from Stalin.[4] Soon thereafter all Jewish theaters in Russia were closed. In 1950–1951 Jews were dismissed en masse from health care facilities and other professions throughout the Soviet Union. Those not dismissed were often demoted or exiled to peripheral positions. Wave after wave of anti-Semitic propaganda swept over Soviet society during these years. Yakov Rapoport, a doctor arrested in February 1953, has depicted the mood of those times.

> Having no mother country of their own, kinless cosmopolitans, it was averred, were unable to appreciate the work of the Russian people, the nature of things Russian and Soviet, and therefore had no right to pronounce judgments on it. . . . In the course of the campaign aspersions were cast on the work of Bagritsky, Svetlov, Vasily Grossman, and many others. The portrait of the composer Felix Mendelssohn was removed from the Grand Hall of the Moscow Conservatory. . . . Even during Hitler's rule in Germany, it did not occur to anyone to take down the portrait of Richard Wagner at the Moscow Conservatory.[5]

Etinger's arrest, and therefore the eventual shape of the Doctors' Plot, is inextricably connected with Stalin's action against the Jewish population of the Soviet Union at this time and in particular against the Jewish Antifascist Committee, the most visible and influential Jewish organization in the Soviet Union at the end of the war. On August 12, 1952, four years after Yuri Zhdanov's letter in *Pravda*, fourteen members of the Jewish Antifascist Committee were convicted of anti-Soviet activity in the service of American intelligence.[6] Thirteen of the defendants were shot in Lubyanka prison, in downtown Moscow.[7] As Yakov Rapoport put it, "Jewish culture in the Soviet Union was put before the firing squad, its finest representatives physically destroyed."[8] The elderly, distinguished Dr. Lina Shtern was sentenced to five years of exile in the gulag. She alone survived. A fifteenth member collapsed before the trial began and died in prison, unable to bear the torturous interrogations.

The cruel persecution of these fifteen innocent Soviet Jewish leaders (among them prominent doctors, journalists, government figures, lawyers, writers, and actors), culminating in what has come to be called their "Unjust Trial,"[9] was the outcome of many tendencies in postwar Soviet society and internal politics that flowed together with Stalin's own murderous intentions. It also reflected the growing tensions of the Cold War.

The Cold War was now some four or five years old, and the arms race filled newspapers on both sides of the Iron Curtain with daily portents.

The Berlin blockade had led to an international standoff between the United States and the USSR in 1948–1949; the Soviets tested their first atom bomb in the summer of 1949; and the Korean conflict had turned into a hot war in June 1950. Stalin had reason to be concerned about U.S. political intentions and military power, and he perceived correctly that many Soviet Jews had a special relationship with the United States either through family ties or because of U.S. support for Israel. When Golda Meir visited Moscow in the fall of 1948, Stalin saw thousands of fervid supporters fill the streets and cram the Moscow synagogue shouting *Am Yisroel chai*—"The People of Israel lives!" To Jews throughout the world this was a traditional, deeply resonant, affirmation of national renewal; to Stalin it looked like dangerous bourgeois Jewish nationalism that undermined the authority of the Soviet state. The anti–Semitic campaign already under way gained new intensity as a result.

By 1950, as relations between Israel and the USSR had cooled, and it became clear to Stalin that Israel would be just another weapon in the U.S. global strategy, official anti–Semitism became ever more vicious. The authorities watched for any and every sign of possible disloyalty among prominent Jewish figures. It was in this context that Dr. Etinger came into their sights. Etinger was a highly educated, influential physician, widely regarded as the greatest Soviet diagnostician of his time. He was strong-willed, intelligent, and well read. He had visited medical clinics in Rochester, Chicago, Boston, and New York in the 1920s, and had received medical training in Berlin before the First World War.[10] He was a man of considerable learning with a generous intellectual horizon and an aggressive bearing who engaged in vigorous political conversations with all his acquaintances. Literate in English and German, Etinger read foreign literature voraciously and knew about the world outside the boundaries of the Soviet Union. He listened to contraband foreign radio broadcasts, in particular Voice of America, on his shortwave radio and was not shy about expressing his opinions. His brother, with whom he had little correspondence, lived in Israel.[11] He had precisely the kind of intellectual and social profile Stalin wished to eliminate from the Soviet body politic. He was a perfect enemy.

Etinger was arrested November 18, 1950, by order of the Ministry of Security. Etinger's activities and opinions had been known to the authorities for some time, and the case against him had been building for several years.

One of the members of the Presidium of the Jewish Antifascist Committee, the Yiddish poet Itzik Fefer, was among those ten million informants Ignatiev spoke of, and on April 22, 1949, already himself under arrest, he denounced Etinger during an interrogation session.[12] The committee itself had recommended Fefer to the MGB as an informant, and in that capacity he had accompanied Mikhoels to America on their famous goodwill trip in 1943.[13] At his trial in May 1952 Fefer gave the following account of himself:

> My nationalistic tendencies came out in the following ways: I said that I love my people. Is there anyone who does not love his people? I wanted my people to have what all others had. . . . It seemed to me that only Stalin could correct the historical injustice committed by the Roman kings. . . . I had nothing against the Soviet system. I am the son of a poor schoolteacher. Soviet power made a human being out of me and a fairly well known poet as well.[14]

In August 1952 he would be shot along with his colleagues.

Fefer had been introduced to Etinger in the summer of 1944 in the Moscow offices of the Jewish Antifascist Committee through their mutual friend, the journalist Shakhno Epshteyn, who was executive secretary of the committee. Epshteyn told Fefer that Etinger, a professor at the second Moscow Medical Institute, "was a real Jew," a "Jewish nationalist," and a close acquaintance of Mikhoels. Epshteyn described Etinger as having a "lively interest in the work of the Jewish Antifascist Committee," and assured Fefer that he would be "ready to help . . . in the creation of a Jewish republic in the Crimea"—the long-cherished ambition of the Jewish Antifascist Committee that played a key role in its post-war liquidation.[15] Fefer "thanked Etinger for his concern about the fate of the Jewish people," and they agreed to meet again. Before he left, Etinger received copies of the "bourgeois, Anglo-Jewish magazine" *Jewish Chronicle*. They subsequently met on many occasions at the JAC offices and Etinger, according to Fefer, would "as a rule . . . leave with packets of foreign literature."

Fefer said that Etinger often described "the mood in the highest strata of the Jewish intelligentsia," and revealed his own bitter resentment over "the alleged expulsion of Jews from ministries and institutions of higher education." According to Fefer, Etinger "maliciously" stated that these were not isolated instances but part of a system dictated from above. If the government wished it, Etinger said, then all these persecutions of the Jews would immediately cease.

Saying that anti-Semitism was a state phenomenon, Etinger invariably repeated, "If the representatives of local power knew that the government would punish them for their anti-Semitism, they wouldn't do it."

Etinger articulated clearly what many in "the highest strata" must have felt—that those at the top with real power controlled general social conditions. If Stalin had objected, the dismissals of Jews undoubtedly would have ceased. By 1948 Etinger was dismayed at the Soviet Union's lack of support for Israel and by the fact that the Soviet government prevented its Jewish citizens from helping Israel defend itself: "If the Soviet government does not wish to help the Israeli Jews," he told Fefer, "then [they should] allow us to." Etinger paid close attention to all news connected with Israel, and in particular, he knew about the help the Americans gave Israel in the War of Independence. He defended Israel's stance toward the USSR, saying, "They see that the USSR won't help them and therefore they lean toward the Americans." He told Fefer that he got his information by listening regularly to foreign radio broadcasts.

Fefer also revealed that, like thousands of other Soviet scientists, Etinger was appalled by the 1948 Lysenko conference. "As early as 1948," Fefer told his interrogator, "Etinger maliciously slandered the Central Committee of the Party in connection with a discussion of the results of the [1948] session of the Academy of Agricultural Science."

> [Etinger], for instance, stated that it was impossible to build science in this way and he labeled the past session a spectacle that would produce an allegedly horrible impression on people of science. Etinger slanderously insisted that scientists were deeply confused by this pressure of the Central Committee of the party on science but were afraid to say anything about it.[16]

The result of Fefer's denunciation was that the security service bugged Etinger's house and taped his conversations with friends and relatives. This bugging was productive, and the government amassed considerable evidence of Etinger's dissatisfaction with the Soviet state and Stalin himself.

By early 1950 Etinger had several decisive marks against him, but the most important was his hatred of Stalin and his continued, relentless criticism of the Soviet regime. However, when the question of arresting Etinger came up at a spring 1950 meeting of the Central Committee, Stalin withheld the order to do so. "Investigate further," he instructed.[17] According to Yevgeny Pitovranov, a deputy minister of state security and head of the Second Chief Directorate (counterintelligence),[18] who was present at that meeting, one of the bugged conversations between Etinger and his son yielded Etinger's seditious opinion, "Whoever could liberate the country from such a monster as Stalin would be a hero."[19] More than enough for his immediate arrest.

Stalin waited for more material to be gathered. More names. More connections. More crimes. As it happened, Etinger spoke freely to many people and proved to be a great resource to the investigation. The question of arresting him was brought before the Central Committee for a second time, and Stalin gave his approval. On November 18 the sixty-three-year-old cardiologist found himself in a cell in Lubyanka. Stalin's name does not appear on the arrest warrant issued by the MGB, and without Pitovranov's account his role in delaying the arrest would be nothing more than speculation. Stalin now stood behind Abakumov, the minister of state security, just as he stood behind Abakumov earlier in dealing with Timashuk's letter. Abakumov did not act independently at any stage in either of the two cases.

Etinger's adopted son Yakov Yakovlevich Etinger had been arrested one month earlier, on October 17, for anti-Soviet agitation. Born in Dr. Etinger's native city of Minsk in 1929, Yakov Yakovlevich was the son of Etinger's intimate friend, Ya. Ya. Siterman, who was murdered during the German occupation.[20] The boy was saved by a Russian woman, Maria Kharyetskaya, who managed to take him out of the ghetto.[21] Etinger returned to Minsk in 1945. He witnessed the total destruction of the city and the horrifying results of German anti-Semitism with his own eyes. He found the boy and adopted him. What he saw deeply disturbed him. Because he was not as naïve as Fefer, it is doubtful that he ever thought Stalin would right the "historical injustice committed by the Roman kings," or that he "drank [his] ideals from the goblet of Stalinism," as Fefer put it,[22] but Etinger understood that the Soviet Union and many individual Russians, such as Kharyetskaya, had prevented the total annihilation of the European Jews. When the anti-Semitic campaign began, he, like many other Soviet Jews, felt betrayed.

At the time of his arrest Yakov Yakovlevich was a student in the history faculty of Moscow State University. "We criticized everything Soviet," he told interrogators, "the culture, science, literature, the internal and external politics of the government, slanderously calling the Soviet Union a 'police' state, a 'fascistic' state. And we accused the Soviet government of conducting an aggressive foreign policy that would unleash a new world war."[23]

Although not formally a member of the Jewish Antifascist Committee, Dr. Etinger was arrested for the same general reasons for which the members of the committee had been arrested, and at first it appeared that he would suffer a similar fate. Another of the defendants at the trial, Emilia Teumin, an editor working in the Sovinformburo,[24] had, like Etinger, also not formally been a member. Nevertheless, she was tortured, tried, and shot with the others. Her comment to the court about her alleged anti-Soviet, bourgeois nationalism closely reflected Etinger's own views.

> In conversations with me Fefer spoke of the supposedly unfair treatment Jews were receiving in the Soviet Union and accused the Soviet government of supposedly encouraging anti-Semitism. He spoke of the need for Jews to unite to struggle for their independence. . . .
>
> I feel that I supported Fefer because I was silent instead of rebuking him. I was poisoned with bourgeois nationalism, and since I responded to these conversations with silence, that means that I supported him.[25]

Fefer's testimony against Etinger initiated the bugging of his telephone and apartment and was the basis for his questioning by MGB interrogators in November and December 1950 who attempted to link Dr. Etinger to the same Jewish nationalism that had condemned the Jewish Antifascist Committee. It appears that Etinger was initially being prepared to play the same role as Teumin in the JAC trial. This did not happen. Despite the fact that there was more than enough material to implicate Etinger in the committee's affairs, bring him to trial, and execute him along with the other defendants, Etinger was never questioned about the committee or his association with Fefer. In January the MGB ceased officially interrogating Etinger altogether.

Just as Stalin had waited to arrest Etinger in the spring of 1950, in January 1951 something caused him to wait again. Something had happened, and it seems that the point at which Etinger was temporarily set aside was the point at which Stalin took his first step toward conceptualizing the Doctors' Plot.

In January 1950 Etinger was moved from Lubyanka to the inmost part of the Soviet inferno—Lefortovo—where the marble stairs have been so worn down by the myriads of doomed prisoners of the 1930s and 1940s that today these stairs can be climbed only by placing one's feet on the outermost margins of each step. At the end of January 1951 word was passed through the chain of command in the Ministry of Security, in a manner that remains somewhat unclear,[26] to A. G. Leonov, the senior investigative officer in charge of Etinger's interrogation, that the Etinger case, for the moment at least, was to be shelved.[27] "Let him sit," Stalin allegedly had told Abakumov. No further interrogations were to take place.

From November 18, 1950, to January 4, 1951, Etinger was brought up for interrogation thirty-seven times. These interrogations produced only five written protocols signed by M. D. Ryumin—dated November 20, and December 1, 6, 11, and 27, as well as a Protocol of Review dated July 30, 1951.[28] Despite Abakumov's direct order to desist questioning Etinger and a warning from the medical staff that Etinger might die from the stress of interrogation, he was brought up "unofficially" for questioning another thirty-nine times between January and March 2, 1951, when he died. No protocols were written down after Etinger was moved to Lefortovo. There is no written record of the questions he was asked or the direction the interrogation took.

What happened during the period from January to March 1951 became the subject of intense debate and intrigue at the highest level of the Soviet government. The documents relating to Etinger's illegal interrogations and death are filled with vituperation and denunciation on all sides, reflecting the larger upheavals beginning to shake Kremlin leadership and Soviet society at large. In September 1950, two months before Etinger's arrest, the leaders of the Leningrad party were tried and executed. This was followed nine months later, in July 1951, with the purging of Abakumov himself. A massive inquiry into corruption and mismanagement rocked the MGB, leading to the expulsion of many other leading MGB personnel, most of whom were Jewish. Pavel Sudoplatov recalled that "Stalin ordered the arrest of all Jewish colonels and generals in the Ministry of Security. A total of some fifty senior officers and generals were arrested."[29] "In those days," Khrushchev wrote, "anything could have happened to any one of us. Everything depended on what Stalin happened to be thinking when he glanced in your direction."[30] It was clear to many in positions of power that Stalin was preparing "a replay of the mass repressions of the 1930s."[31]

Stalin's decision to let Etinger "sit" may well have had something to do with the larger political upheaval, suggesting that the anti-Semitic campaign was more of a political weapon than an end in itself.[32] Nearly two hundred members of the Leningrad party (most of whom were non-Jews), including Voznesensky, Kuznetsov, and Popkov, who had been present at Zhdanov's deathbed in August 1948, became implicated in the "Leningrad affair."[33]

The immediate cause of Stalin's wrath was an "All-Russia fair," allowed by the Leningrad party bosses without permission from the Central Committee.[34] The fair, a routine wholesale trade market, was only a pretext. The purge of the Leningrad party had deeper roots in Stalin's intolerance of any alleged "independence" by members of his government, such as Zhdanov exhibited during the siege of Leningrad and, putatively, in connection with Lysenko.[35] Acting with Stalin's full authorization, Malenkov denounced the Leningrad fair as "anti-party." "There have been too many danger signals about the Leningrad leadership for us not to react," Stalin had told Malenkov, whom he dispatched to Leningrad in January 1949 to investigate supposed irregularities within the party.[36]

Throughout his life Stalin showed his willingness to sacrifice the best and the brightest of his regime in the interests of absolute control. Nothing had changed in this regard after the war. Voznesenksy was a talented economist who had written a book on the Soviet economy during World War

II. Not long before the Leningrad purge Stalin had made him first deputy minister of the Council of Ministers.[37] Kuznetsov had shown exemplary heroism during the Siege of Leningrad and had earned Stalin's respect and trust, however short-lived. According to the testimony of his family and others, in 1941 Stalin wrote to Kuznetsov, "Alexei, I lay my hopes in you. Be strong and resolute. I will always support you." After Kuznetsov was arrested in 1949, Stalin tried desperately to retrieve this letter, going so far as to have the MGB check the schoolbags of Kuznetsov's son on his way to school each morning. It was never recovered and is presumed destroyed. There could be no deviation from Stalin's will. All the victims of the Leningrad affair were rehabilitated after Stalin's death.

The trial of the Jewish Antifascist Committee, the arrest of Etinger, the Leningrad affair, and the subsequent purge of the MGB were part of a larger pattern of events unfolding in the Soviet Union in the years leading up to Stalin's death that also included the mass arrests of those who had been held as prisoners of war in Germany and subsequently returned to the Soviet Union, the so-called *povtorniki*. Stalin may well have believed that war with the United States might not come until the early or mid 1960s.[38] But he had no doubt it would come. He was cautious but kept this end in view. His public pronouncements concerning relations with the United States in the early stages of the Cold War projected a desire for peace and accommodation. He acted with extreme caution in this respect, going so far as to remove all Soviet advisers to North Korea at the beginning of the conflict. When Khrushchev suggested that the Soviet advisers should be reinstated, Stalin characteristically snapped: "'It's too dangerous to keep our advisers there. They might be taken prisoner. We don't want there to be evidence for accusing us of taking part in this business. It's Kim Il-sung's affair.'"[39]

On the other hand, Stalin lost no opportunity to attack the United States and to build an anti-American consensus within the government and the nation as a whole. The initial accusation against the Leningrad party leaders was that they were counter-revolutionaries, saboteurs, and members of an anti-Soviet group, working to undermine Kremlin leadership.[40] Eventually Kuznetsov would be posthumously accused of working for American intelligence and plotting the murder of his old mentor and colleague, A. A. Zhdanov. Stalin waged the war against America through those he alleged to be America's proxies. If he could not win in Korea or Berlin, he could crush the alleged representatives of his enemies at home.

As the Cold War intensified, the simple charge of bourgeois national-ism brought against the Jewish Antifascist Committee was not sufficient for Stalin's purposes.[41] Being anti-Soviet, whether against the Jewish Anti-fascist Committee or the Leningrad party, was not enough. The charge against the committee was that, as Jewish nationalists, they were engaged in undermining the Soviet state at the direction of United States intelligence largely by spreading American and Zionist propaganda. Behind the com-mittee stood the United States, but the intent of their anti-Soviet activity was mainly confined to spreading "nationalist sentiment." This might pose a threat to the Soviet leadership, but it was at best distant and general. Though criminal, the charge lacked immediacy and sufficient force to turn the country against America. After all, the USSR and America had long been engaged in a war for public opinion.

Furthermore, the government's case against the JAC was not strong enough to go to an open, or "show," trial. Public opinion would remain unmoved by the execution of the fifteen defendants.[42] In March 1950, just before Etinger's arrest, the defendants in the Jewish Antifascist Committee trial were told that the investigation was over and that the trial would soon begin—it did not. The MGB possessed no documentary proof of the defendants' guilt, their confessions were often self-contradictory or not credible, and the MGB feared that the defendants might retract them in court.[43] By November 1950 Stalin had become concerned that this trial would not produce the necessary effect. In the end, a closed trial was held, and its results remained unknown beyond the Kremlin walls until November 1955, when only the families of the defendants were told that their relatives had been executed.[44] The actual transcript of the trial was not made public until 1994.

By mid-1950 Stalin realized that the connection between the anti-Soviet, bourgeois, Jewish nationalists and the United States had to become more pointed than in the charges against the JAC members. In the 1930s the anti-Soviet plotters and saboteurs were alleged to have been working for German, French, or Japanese intelligence services. Now the enemy was the United States. M. D. Ryumin, the MGB interrogator of Etinger who will come to play a decisive role in the rest of this narrative, concisely summarized the new situation when he told one of his victims, "in our time hostile actions and plans could not exist without ties with America."[45] Stalin's decision temporarily to "shelve" the Etinger case in January 1951 coincided with his attempt at redefining the nature of these "hostile actions and plans."

Underlying Stalin's strategy was the deeply rooted principle, inherited from Lenin, that enemies were more useful to Soviet power than friends. From his prison cell in Lubyanka on December 10, 1937, Nikolai Bukharin gave astonishing clarity to this premise of the Soviet system while pleading for his life in a letter addressed to Koba,[46] his former comrade and friend:

> There is something *great and bold about the political idea* of a general purge. It is a) connected with the prewar situation and b) connected with the transition to democracy. This purge encompasses 1) the guilty; 2) persons under suspicion; and 3) persons potentially under suspicion. This business could not have been managed without me. Some are neutralized one way, others in another way, and a third group in yet another way. What serves as a guarantee for all this is the fact that people inescapably talk about each other and in doing so arouse an *everlasting* distrust in each other. . . . In this way, the leadership is bringing about a *full guarantee* for itself.[47]

Stalin kept this letter, written not long before Bukharin's show trial and execution, in a drawer of his desk, where it was found after the dictator's death. The *utility* of the purge was that it was a vehicle not for destroying enemies but for creating them. It did not produce stability. Rather, its function was permanently to destabilize the country, because supreme power could be achieved and held only in crisis conditions. Only in circumstances of *"everlasting* distrust" in the population could the Bolshevik regime secure a *"full guarantee* for itself." Enemies were essential to this system of government that lacked true political legitimacy or other institutions that balanced the power of the dictator, the Vozhd, the Great Leader. No wonder, then, that by 1951, according to Ignatiev, there were ten million informants operating in the Soviet Union.

Stalin may have calculated that he had much to gain from causing the

Soviet population to think that the threat of Jewish nationalism cloaked the much greater danger of American aggression. Kirov's assassination in December 1934 was presented as the work of Kamenev, Zinoviev, and others working for Trotsky against the Soviet state. It culminated in the show trials of 1936–1938. The Leningrad affair and the Doctors' Plot followed from the death of Zhdanov that was to have been presented to the Soviet population as a political assassination directed not by Trotsky but by Kuznetsov and American intelligence.[48]

The protocols of Etinger's interrogation in November and December 1950 tell us little about what the "hostile actions and plans" connecting Etinger to U.S. aggression might have been. The skimpiness of the written record when compared with the duration of the individual interrogations suggests a lack of clarity in the interrogation process. We know that Stalin read the interrogations of the JAC members and provided the interrogators with questions, and we know that he took personal interest in Etinger. Stalin himself may not have known at this juncture what direction the investigation should take. The first interrogation lasted one hour and produced barely two pages of testimony; the second interrogation lasted five hours and produced little over three pages; the third lasted four hours and forty minutes and produced not quite three pages; the fourth lasted five and a half hours and produced three pages; the last interrogation on December 27 lasted from 11:10 P.M. to 5:30 A.M., six hours and twenty minutes, and produced a protocol barely two pages in length!

The preserved informational memoranda listing the dates and times of Etinger's interrogations reveal many lengthy sessions that produced no written record at all. For instance, the record shows that on December 1, in addition to the recorded interrogation lasting from 1 P.M. to 6 P.M., Etinger was subjected to a second interrogation that lasted from 11:45 P.M. to 3:55 the next morning.[49] Both interrogations were performed by M. D. Ryumin, but only one was recorded. Not a single protocol was prepared of the unofficial interrogations conducted by Ryumin after Etinger had been transferred to Lefortovo and the investigation temporarily "shelved."

The portrait of Etinger contained in his KGB dossier is of a powerful, decisive man—about six feet tall with broad shoulders. He had a round face, a high forehead, a prominent nose, and a small mouth and thin lips, set above a strong, direct chin. A fine sense of irony flickered in his gray eyes. Nevertheless, Etinger was old and far from well. At his medical examination upon admission to Lubyanka in November 1950, he was diagnosed with what was termed "paroxysmal taxicardia."[50] He asked for medical help twenty-nine times after his transfer to Lefortovo.

The official record noted that "during his time in Lefortovo prison, the prisoner Etinger experienced sudden pain in the region of the heart: January 9, 11, 12, 13, 15, 16, 17, 18. The last two attacks accompanied heart weakness. After urgent medical attention, the attacks subsided." But, the attending physician warned, "in the future, each successive attack of angina pectoris may lead to an unfortunate outcome." On January 19 he had another powerful attack of angina pectoris, and again the security service was warned that Etinger might "die in the interrogation."[51] On March 2 he died, it was said, returning from one such interrogation, a record of which does not exist. The official inquiry reported, "Thus there are grounds to believe that an incorrect regime created for Etinger at Lefortovo prison facilitated the severity of his heart-vascular ailment, terminating in his death."[52]

Etinger was hardly the only prisoner accused of Jewish nationalism who was mistreated and tortured during interrogation. Solomon Bregman could not be properly executed because he had collapsed in a coma during the trial of the Jewish Antifascist Committee and died in prison in January 1953; Boris Shimeliovich was so badly beaten that he had to be carried on a stretcher to his interrogations.[53]

The difference between the interrogations of the JAC defendants and the present case is that Etinger died after an unauthorized interrogation conducted despite a clear warning that further questioning might kill him. It was rare for a potentially important witness to be treated in this manner. Furthermore, the nature of the unauthorized interrogations can only be surmised from subsequent testimony provided by M. D. Ryumin, who conducted them.

Born in 1913, Mikhail Dimitrievich Ryumin was a pure product of Stalin's system. Described by one colleague as "half-educated and dim-witted, an egoist by nature, a deceiver, capable of showing any fact in a way that served his own plan and not as it was in itself," Ryumin was both instigator and victim in this story.[54] According to Sudoplatov, Ryumin was known to be a "primitive anti-Semite."[55] Stalin called him a *shibsdik*— (pygmy)—but for a while Ryumin was a pygmy with lethal power.

Ryumin came from peasant stock and had to be sent to a neighboring district for schooling. He completed the eighth grade in 1930, at the age of seventeen. Afterward he returned home to his parents in Kuban and worked on the kolkhoz for a year. Later he became a bookkeeper at the Udarnik cooperative and was sent by local authorities to take formal classes in bookkeeping. He employed his skills in the Urals from 1931 to 1937. In 1937 he was denounced (according to him, falsely) for having misappropriated administrative funds and was dismissed from his duties. He relocated to Moscow. Through connections he came to work in the financial planning department of the NKVD (the MGB of the time) until the mobilization in 1941, after which he eventually received an appointment in Arkhangelsk where he worked as a senior investigator in Abakumov's famous SMERSH unit (Death to Spies) until the end of 1944.[56] Then he was dispatched back to Moscow, where in 1948 he found work in the Investigative Unit for Especially Important Cases—a special branch within the MGB. In July 1951 he was appointed acting head of this investigative section, and in November of that year he became deputy minister of state security of the USSR as well as head of the Investigative Unit for Especially Important Cases. This was a rapid rise for an egoistic, "half-educated, dim-witted" pygmy.[57]

Short, balding, with a slight paunch but otherwise nondescript, Ryumin was cruel, small-minded, vulgar, and vindictive. His MGB colleagues feared and hated him for his arrogance.[58] Semyon D. Ignatiev, the minister of state security, who succeeded Abakumov, loathed Ryumin as well but was compelled by Stalin to accept the *shibsdik* as his deputy.[59] Ryumin took an active part in torturing and interrogating prisoners in his charge. In due course this included Abakumov, his former boss, whom he was said to have punched in the mouth at their first interrogation. He was adept at falsifying confessions when necessary to suit political realities. His greatest talent as an interrogator seems to have been that he could assume a courteous, almost Jesuitical demeanor as a form of intimidation. He put the case succinctly to one of his victims, Lev Romanovich Sheinin, a Jewish former MGB officer who had been arrested in 1951:

> **Really, you, Lev Romanovich, a man with great experience in investigative work—who, if not you, should understand that confessions**

in the MGB must be politically pointed and sharp, that you must consider the political circumstances in the country and the international situation. . . . That means that you must give a political response and, mainly, *you must provide us with the connection to America, in as much as that is now our main enemy.* The time when there were nationalistic conversations signifies a time when there were nationalistic tendencies, when there were nationalistic tendencies there could not but also have been nationalistic, hostile actions and plans, *and in our time hostile actions and plans could not exist without ties with America.* [Emphasis added.][60]

Ryumin understood the logic of the interrogation process and was able to reduce it to the syllogism: nationalistic conversations = nationalistic tendencies = hostile plans and actions = connection to America. He also understood the larger bureaucratic system and was entirely a creature of it: his actions mirrored the actions of those more powerful than himself. When asked at his trial why he had falsified material sent directly to Stalin, Ryumin responded, "I feared that Ignatiev [the minister of state security] would do to me what I had in my time done to Abakumov. . . . Working closely with Ignatiev, I knew him to be a clever and hypocritical man, capable of [anything] . . . "[61] Ryumin lived in the world of distrust and denunciation depicted in Bukharin's letter. His success always came at the expense of others. He feared that what he did to others would be done to him. Eventually it was.

Ryumin's fall was as precipitous as his ascent. On November 13, 1952, one year after becoming deputy minister, he was removed from his job; he was arrested on March 17, 1953, twelve days after Stalin's death, by Stalin's successors. Ryumin was accused of falsifying documents sent to the Central Committee, illegal use of torture, instigating the Doctors' Plot for his own ends, deceiving the government, and other heinous crimes. He was executed in 1954. Ryumin was guilty of much, but he was a conduit and not a cause, a creature not the creator. The new Soviet government made use of him much as Ryumin had made use of his prisoners, thereby replicating essential parts of the system they sought to reform by punishing Ryumin. The serpent devoured its own tail.

Ryumin's astounding rise in the world began on November 20, 1950, when he became Etinger's interrogator. Eventually he would take over the direction of the entire case against the Jewish Antifascist Committee, boasting that he was the "plenipotentiary of the Central Committee to uncover the Jewish nationalistic center."[62]

"You are arrested for conduct hostile to Soviet power," Ryumin charged at his first encounter with Etinger. "Do you acknowledge yourself guilty of this?"

"No," Etinger replied. "I do not acknowledge this in so far as I have not conducted any action of the sort."[63]

Until the first element of the syllogism could be established, Ryumin could not establish the tie to the United States. Therefore, at the first interrogation Ryumin sought to establish all those elements in Etinger's background that would have induced him to hold "nationalistic conversations" leading to "hostile plans and actions." The bugged conversations were not enough. They needed a confession. Etinger's bourgeois background was a start. His father had owned a large house in Minsk and was a leading timber merchant before the revolution. His wife's family was also connected to the timber business. What was worse, Etinger's brother, Simkha Gilyariyevich, had resided in Israel since 1933, having lived ten years in Poland and Germany.

Despite these marks against him, Etinger insisted that he had never acted against Soviet interests and that he had had no contact with his brother: "I am able to explain this by saying that my brother lived abroad and I thought that a tie to him, even by letter, might compromise me as a Soviet citizen."[64]

Etinger told Ryumin that he had always worked honorably for the interests of the motherland. "Being a scholar," he said, "I have invested a good deal in Soviet medical science."

"Don't wrap yourself up in the toga of Soviet scholarship," Ryumin shot back, "you can do that later. Just tell us more about your enemy activities conducted against the Soviet State."

"Once more I repeat that I have never committed any crimes against Soviet power."

"You—liar!"[65]

Etinger revealed that he had studied medicine in Berlin before the revolution and in 1926 went on a tour of Germany and America where he visited Chicago, New York, Boston, and Rochester. He was very favorably impressed with American medical practice. It was not until his last recorded interrogation on December 27 that Etinger admitted that he was guilty of slandering "the national politics of the Soviet government" and expressing in the presence of his wife and adopted son his "hatred toward the leadership of the party and . . . the leader of the Soviet people." Ryumin wanted more: "Your criminal work was not circumscribed by these anti-Soviet fabrications," he told Etinger. "Tell us about your other enemy plans and give us your enemy ties."

"I had no enemy plans," Etinger insisted.

"In that case, the investigation will be forced to unmask you with concrete facts in subsequent interrogations."[66]

There were no officially sanctioned "subsequent interrogations." A week later Etinger was moved to Lefortovo and placed in cell number 15. The accompanying order stated that Etinger was not "to be given books, walks, or allowed access to the prison store. It [was] necessary," the order read, "to establish strict observation of him."[67] At the end of January Abakumov told Ryumin to shelve the investigation and assigned him to another case.

It was here that Ryumin took the initiative that would earn him his unprecedented rise. Instead of dropping the Etinger case, Ryumin secretly made twenty-four trips to Lefortovo, interrogating Etinger thirty-nine times, between January 2 and the evening of March 2, to extract (or invent) further confessions. Contravening Abakumov's order, Ryumin subjected his elderly terrorist to the "conveyor system," forcing him to stand for hours on end, with little sleep. Typically Ryumin would conduct two interrogations on a single day. The first would begin usually at 12:30 or 1:40 P.M., ending a little after 5 or 6 P.M. Etinger would then be led back to his cell and allowed to sleep until approximately 11:30 P.M. when he would be returned to Ryumin for another session that could last until 2:30 A.M., though sometimes the sessions lasted until 4 A.M.[68]

On March 2 Etinger returned from one such unauthorized interrogation to his cell at 5:15 P.M., according to the official summary. He tasted a piece of bread, made several steps in the direction of the door, and fell unconscious to the floor.[69] The official notice reads that Etinger's "death came suddenly from paralysis of the heart resulting from a thrombosis of the coronary artery, arteriosclerosis and angina pectoris."[70] The head of the medical unit of Lefortovo prison, a Lieutenant Colonel Mikhail Yanshin, who wrote out Etinger's death certificate, testified that the investigation had speeded up and intensified Etinger's systematic interrogations day and night, which, according to Yanshin, "severely traumatized and weakened [Etinger's] cardiac functioning." Yanshin stated that he had officially notified the Investigative Unit for Especially Important Cases of the seriousness of Etinger's physical condition, cautioning that he might die under further interrogations.[71]

There is a bitter irony here. Etinger was an old man with a weak heart, deprived of sleep, the victim of an "incorrect" regimen, the charge brought against the doctors in Valdai. Yanshin noted that on January 16, 1951, considerable swelling was observable in Etinger's legs and that in his opinion, "this was not the result of the heart ailment, but simply a result of the interrogation during which Etinger was forced to stand for long periods of time."[72]

In the wake of Etinger's death in March 1951 Ryumin received a formal reprimand by the party—not because Etinger died, but because Ryumin had not expeditiously fulfilled Abakumov's order to compose the general protocols of Etinger's official interrogations. Mikhail T. Likhachev, deputy head of the Investigative Unit for Especially Important Cases and Ryumin's immediate superior, gave this account of Ryumin's difficulties at the time.

> ...RYUMIN did not keep his word, and I once again had to remind him that he had to complete the ETINGER protocol, and several days after this last conversation, at a session of the party bureau, I learned that ETINGER had died. Moreover, prior to the party bureau meeting neither RYUMIN nor LEONOV [Head of the Investigative Unit for Especially Important Cases] had said anything to me about ETINGER'S death.
>
> At the party bureau session, LEONOV gave a short statement about what had happened and proposed consideration of RYU-MIN'S conduct.
>
> The communists who attended the party bureau meeting, including myself, scolded RYUMIN for not fulfilling LEONOV'S

order [to complete the protocols] and demanded that he be punished by the party.

RYUMIN confessed his guilt. I also acknowledged my responsibility as the leader of the investigative group.

It's necessary to say concerning all this that ABAKUMOV's order concerning the ETINGER case was not concealed at the party bureau.

The result of the discussion of RYUMIN'S mistake was that a rebuke was declared against him.

Two or three days after the party bureau meeting, LEONOV ordered that ETINGER'S confessions be set out in a detailed informational memorandum for a report to ABAKUMOV.

RYUMIN composed this memorandum, but ETINGER'S confession about terror was not reflected in this memorandum.[73]

Although no action other than this official reprimand was taken at the time, it must have registered keenly on Ryumin that he had now come under intense scrutiny by the party for lax work and misconduct. At the party meeting Abakumov's order that the case be "shelved" was specifically brought up for discussion. The "confessions" Likhachev mentions were those Etinger supposedly gave to Ryumin while he was illegally interrogated in Lefortovo. Only by producing the protocols of those illegal interrogations could Ryumin vindicate himself for having conducted the interrogations in the first place. He pledged to prepare the protocols for Abakumov expeditiously, but in the end was unable to produce them. The pygmy must have begun to feel surrounded by enemies. His future was once again in grave danger.

Other danger signals for Ryumin had begun to appear when the procurator's office of the MGB initiated an inquiry into the Etinger case. On July 30 this inquiry would issue a report implicating Ryumin directly in Etinger's death. The report noted that Ryumin was responsible for the Etinger case, observing that the medical staff of Lefortovo prison had warned the MGB on several occasions about Etinger's fragile health. It concluded: "there are grounds to consider that an incorrect regimen," created for Etinger at Lefortovo prison, had "facilitated the severity of his heart-vascular ailment, ending in his death." Not having completed the protocols of Etinger's Lefortovo interrogations, while violating Abakumov's direct order to cease the interrogations altogether, had earned Ryumin party censure. Now he would be blamed for Etinger's apparently accidental, but foreseeable death as well. Negligence would be added to inefficiency and insubordination. He might have been a dim-witted egoist, but he was intelligent enough to know that his credibility within the MGB could not have withstood this double blow. He would be denounced again. Expulsion if not imprisonment almost certainly awaited him.

In early July 1951, before the procurator's office could complete its inquiry, Ryumin, who had worked his way up the Soviet bureaucratic ladder from village bookkeeper to senior MGB investigator, took a fateful step. He denounced others before they could denounce him. Timashuk had taken such a step in August 1948, but nothing to date had come of her letter to Vlasik. Ryumin's action brought a nearly instantaneous response.

There are different accounts of what precisely Ryumin did and how he accomplished it. The simple account, supported by Ryumin's own testimony, is that on July 2 he wrote a letter to Stalin denouncing V. S. Abakumov, minister of state security, for concealing the terrorist aims of Dr. Etinger's medical treatment of Kremlin leaders. That he wrote a letter is not in question. However, a new element has been added to the story by the publication of Pavel Sudoplatov's recent memoirs. Although Sudoplatov's accounts of many things have been shown to be unreliable, partial, or simply false, what he has said about Ryumin's letter finds some support in another account as well, published in Moscow in 1992.[74]

According to Sudoplatov, Ryumin did not write the letter himself, but was "inspired by Malenkov." Sudoplatov claims that Ryumin met with Malenkov's deputy Dmitri Sukhanov to work out the details and that he "readily accommodated Sukhanov's demand that he write a letter to Stalin denouncing Abakumov."[75] In addition to Sudoplatov's account, Malenkov's son revealed in 1992 that his father had once told him that Ryumin had come to Sukhanov in June 1951 with a note alleging that the original investigator of the Etinger case had been guilty of forgery. According to Malenkov's son, this note was passed from Sukhanov to Malenkov to Poskrebyshev to Stalin. Ryumin was then summoned to the Kremlin for a meeting that ultimately produced the letter denouncing Abakumov.

Sudoplatov related that his sister-in-law, a typist working in Malenkov's office at the time, said that Ryumin had to rewrite his denunciation eleven times and that Sukhanov "kept him waiting in the reception room for almost ten hours while he conferred with Malenkov on the contents of the letter."[76] We will never know precisely what transpired, and other parts of Sudoplatov's story about Ryumin may be unreliable.[77] Nevertheless, some connivance between Ryumin and Malenkov appears to have taken place, implying Stalin's direct foreknowledge. Timashuk's 1948 letter disclosed what she took to be a conspiracy, but Ryumin's letter was part of a conspiracy that had already begun to mature. What, in fact, may have happened is that when Ryumin was summoned to the Central Committee to receive his official reprimand, he used the occasion to present his "note" that was then embellished through eleven drafts into the letter to Stalin.

Dated July 2, 1951, this letter deserves to be reproduced in full.

To comrade Stalin from the Senior Investigator of the MGB of the USSR, Lieutenant Colonel, RYUMIN, M. D.
July 2, 1951

In November 1950, I was commissioned to conduct the investigation in the case of the arrested doctor of medicine, Professor ETINGER.

At the interrogation, ETINGER acknowledged that he was a confirmed Jewish nationalist and in consequence of this bore hatred toward the Party and the Soviet government.

Further, speaking in detail about his hostile activity, ETINGER acknowledged that when he was commissioned to cure comrade SHCHERBAKOV in 1945 he acted so as to shorten his life.

I set the confessions of ETINGER on this question before the assistant head of the Investigative Unit comrade LIKHACHEV and soon after this I and comrade LIKHACHEV together with the prisoner ETINGER were summoned to comrade ABAKUMOV.

During the "questioning," rather, the conversation with ETINGER, comrade ABAKUMOV mentioned several times that he had retracted his confession about the villainous murder of comrade SHCHERBAKOV. Then, when we were leading ETINGER from the office, comrade ABAKUMOV forbade me from questioning ETINGER along the lines of unmasking his practical activities and terrorist plans, saying that he—ETINGER—was leading us astray. ETINGER understood the wish of comrade ABAKUMOV and, after leaving him, in subsequent interrogations denied all of his acknowledged confessions, although his hostile relationship to the party was incontrovertibly confirmed by materials secretly overheard and by the confessions of his adherent, the arrested YEROZALIMSKY, who, by the way, in the investigation spoke about the fact that ETINGER expressed hostility concerning comrade SHCHERBAKOV.

Using these and other evidentiary materials, I continued to interrogate ETINGER and he gradually came to reinstate his earlier confessions about which everyday I wrote memoranda for a report to the leadership.

Approximately, January 28–29, 1951, the head of the Investigative Unit for Especially Important Cases, comrade LEONOV, summoned me and, referring to instructions of comrade ABAKUMOV, told me to interrupt the work with the arrested ETINGER, and to shelve the case, in the words of comrade LEONOV.

In addition I must remark that after the summons by comrade ABAKUMOV of the arrested ETINGER a more severe regime was established for him and he was transferred to Lefortovo prison to the coldest and dampest cell. ETINGER was elderly—64 years old—and he suffered from angina pectoris, about which on January 20 an official medical document was sent to the Investigative Unit. In this document it was indicated that, "in the future, each subsequent attack of angina pectoris might lead to an unfortunate outcome."

Taking this circumstance into account, I put the question several times before the leadership of the Investigative Unit whether they would allow me to be included in future interrogations of the arrested ETINGER, but they refused. It was all ended by the fact that in the first part of March ETINGER suddenly died and his terrorist activity remained uninvestigated.

Meanwhile, ETINGER had wide contacts, in which number there were leading physicians, including several who had a relationship to the terrorist activity of ETINGER.

I consider it my duty to inform You that comrade ABAKUMOV, as far as my observation goes, has an inclination to deceive the government organs by keeping quiet about serious defects in the work of the organs of the MGB.

Thus, at the present time, I am responsible for the investigative case concerning the former deputy of the General Director of the Joint Stock Company, "Vismut"[78] in Germany, SALIMANOV, who in May 1950 fled to the Americans, and then after 3 months returned to the Soviet Zone of occupation where he was detained and arrested.

SALIMANOV confessed that in May 1950 he was fired and he had to return to the USSR; however, he did not to do this and utilizing the absence of surveillance by the organs of the MGB, he fled to the Americans.

Further, SALIMANOV said that betraying his Motherland, he fell into the hands of American intelligence agents and, associating with them, confirmed that American intelligence had at its disposal detailed information about the activity of the Joint Stock Company, "Vismut," which was engaged in the extraction of uranium ore.

This confession of SALIMANOV demonstrates that the MGB organs are poorly organized for counterintelligence service in Germany.

Instead of informing the government authorities about this and using the confession of the arrested SALIMANOV for the elimination of serious inadequacies in the work of the organs of the MGB in Germany, comrade ABAKUMOV forbade attaching the confession of SALIMANOV to the protocols of the interrogations.

Agents of American and English intelligence have been arrested at various times by the Ministry of State Security; moreover, many of them before their arrest were secret workers in the MGB and were double-dealing.

In the information concerning these cases, comrade ABAKU-

MOV would write: "We caught them, we unmasked them." But in reality—they caught us, they unmasked us, and for a long time they led us around by the nose.

In passing, some words about the methods of investigation:

In the Investigative Unit for Especially Important Cases, the decree of the Central Committee and the Soviet government concerning the work of the organs of the MGB is systematically and crudely violated in relation to the recording of the summons to questioning of the prisoners by the protocols of interrogations, which, by the way, in almost every case were accomplished in an irregular manner and in a series of unobjective circumstances.

In connection with this ABAKUMOV violated other Soviet laws and in addition followed a line the result of which, particularly concerning cases of interest for the government, were recorded by necessity with impermissible generalizations, which often misrepresented actuality.

I won't relate concrete facts, although there are many, in so far as the fullest picture in this connection might be provided by the special verification of the cases with a reinterrogation of the prisoners.

In conclusion I will allow myself to say that in my opinion, comrade ABAKUMOV did not always strengthen his position in the government apparatus by honorable means and that he has turned out to be a dangerous person to the government, all the more so in so important a position as Minister of State Security.

He is particularly dangerous because the Internal Ministry has placed many "reliable"—from their point of view—individuals in the most key positions and in particular in the Investigative Unit for Especially Important Cases, people who receiving their careers from his hand, gradually lose their party spirit, and turn to toadying and obsequiously fulfill everything that comrade ABAKUMOV wants.[79]

For the most part, the charges Ryumin brought against Abakumov were pure provocation. Even Khrushchev, who had Abakumov executed in 1954, admitted this. A year after Stalin's death, Khrushchev said that Abakumov was a "hardened criminal, a conspirator—but in other matters."[80] He admitted that although Abakumov deserved to be burned at the stake for all the evils he had accomplished in connivance with Beria, particularly the Leningrad affair, he was innocent of covering up the terrorist plots of Dr. Etinger.

Two questions have haunted all accounts of the Doctors' Plot: Who conceived it and when precisely did it originate? If the story told by Malenkov's son is true—that Ryumin's initial accusation was brought first to Sukhanov, who relayed it through Malenkov and Poskrebyshev to Stalin[81]—then it may be possible to date the inception of the plot to destroy the Jewish doctors simultaneously with the plot to destroy Abakumov and purge the MGB. Many other facts of the case point to this conclusion. If true, then the conspiracy against the Jewish doctors must be viewed in a much wider context than that of Stalin's personal anti-Semitism. It became a tool of his foreign and domestic policy. Far from being an aberrant expression of Stalin's paranoid vengeance, it reflected the essential character of his system. Before turning to these larger considerations, however, we should look at Ryumin's specific charges against Abakumov in connection with Etinger. Only someone ready to believe them could have been persuaded.

The most important point is that there was no hint of the linkage essential to the Doctors' Plot between Jewish nationalism and the medical sabotage of Soviet leaders in Dr. Etinger's own recorded admissions, in the MGB procurator's report of July 1951, or in anything that Etinger's son said to the authorities. In January 1951 there was no Doctors' Plot. In July 1951 it had taken root in Ryumin's denunciation of Abakumov but was absent in the Justice Ministry's report of July 30, which said nothing of terrorism or sabotage and noted only that the sick, elderly Dr. Etinger had "anti-Soviet feelings" and had been called up for interrogation thirty-seven times between November 18, 1950, and January 4, 1951; and thirty-nine times from January 5, 1951, to his death on March 2. In the written protocols of these interrogations, Etinger admitted his hostility toward the Soviet Union, but they contained no confessions of medical sabotage; nor did they demonstrate that he was even asked about his or other doctors' treatment of leading Politburo figures. The interrogations conducted by Ryumin himself only attempted to establish Etinger's "anti-Soviet disposition." The question of medical sabotage simply did not come up.

Nevertheless, matters were not as simple as they appeared. We know that Etinger had said *something* about A. S. Shcherbakov, and Abakumov reacted to it. What precisely Etinger said remains in doubt because it was never written down. Why Abakumov reacted as he did became the basis of Ryumin's denunciation. Khrushchev revealed that the principal charge against Abakumov leading to his arrest was this accusation that he had covered up the Etinger case, not the secondary accusation of general mismanagement.[82]

Ryumin wrote that "ETINGER acknowledged that he was a confirmed Jewish nationalist and in consequence of this bore hatred toward the party and the Soviet government." This was true. On December 27, 1950, Etinger told Ryumin that his guilt "consists in the fact that in my domestic circumstances, in the presence of my wife and my adopted son, I slandered the national politics of the Soviet government; I expressed hatred toward the leadership of the VKP(b) and spread evil conceits about the leader of the Soviet people."[83] When Ryumin pressed him to confess that his "criminal work was not circumscribed by these anti-Soviet fabrications," Etinger firmly denied this. "I had no enemy plans" was Etinger's last recorded statement.[84]

What followed in Ryumin's letter was a characteristic amalgam of falsehoods, facts, and half-truths. Ryumin's statement that "ETINGER acknowledged that when he was commissioned to cure comrade SHCHERBAKOV in 1945 he acted so as to shorten his life" is not true. Etinger did not say this, or if he eventually confessed something of the sort, it was because of extreme psychological pressure and physical exhaustion. Etinger had taken part in the treatment of Shcherbakov but only as a consultant; he never directly dealt with the patient. There are different accounts of the matter.[85]

Likhachev, deputy head of the Investigative Unit for Especially Important Cases, remembered that one or two days after Ryumin took over the case, he had heard that Etinger had given a confession concerning the murder of Shcherbakov. He immediately reported this to Abakumov, who summoned Etinger. Apparently Etinger told them that Shcherbakov was a doomed man whose life might have been marginally prolonged if different doses of certain medicines had been administered and if he had been forced to stay in bed. But, as Etinger said, "judge for yourself how it was possible to force such a patient [Shcherbakov] to observe a rigorous bed rest regimen."[86] A. S. Shcherbakov (1901–1945) was first secretary of the Moscow region and Moscow party, a candidate member to the Politburo, and a member of the Central Committee. He died of a heart attack on May 10, 1945, one day after the celebration of the victory over Nazi Germany took place in Moscow. Though a known anti-Semite, he had welcomed Etinger in his house as a consulting physician to him for years. He had a stubborn and domineering manner.

Likhachev did not think Etinger's "confession" was credible. Nor did Abakumov. At the interview Likhachev arranged, Etinger told Abakumov that he was upset that the Soviet regime had "kept the Jews down," but that he had committed no crimes. Abakumov apparently raised the subject of Shcherbakov's death, asking whether Etinger had been involved in his medical treatment. Etinger said he had but had committed no crime.

> In connection with this or before it, I said to him [Etinger] that he must speak about how he had treated SHCHERBAKOV. He presented himself as though he was guilty of nothing, he became anxious to prove to me in detail that SHCHERBAKOV was very ill, and he began to explain what his sickness was, pointing to the region of the heart, and [he said] that SHCHERBAKOV was doomed. He said that they treated him in the Kremlin polyclinic. In all the years he was there under observation he, ETINGER, was invited only for consultation. . . . ETINGER said that he helped as far as he could. He acted as an honest man in the matter. I said to ETINGER that he had all the same to tell everything, that it would be better for him to think and to speak about all his anti-Soviet ties, nationalistic sentiments, that an investigator would summon him and he would have to tell everything.[87]

When Abakumov was asked why this interrogation had not been formalized into a protocol, he answered, "It wasn't formalized because in actuality there was no such interrogation. Reinterrogation or rechecking of facts usually happened [later]. I repeat, the protocol was not composed because there was nothing to write down. I only began to interrogate him and I suggested that he conduct himself honorably and speak about his crimes; I then said to him that the minister of state security is interrogating him and I thought that maybe he would immediately speak. I predicted that it would be better for him if he conducted himself honorably and confessed everything."[88] Abakumov was convinced that Etinger was guilty of nothing more than harboring anti-Soviet feelings.

Ryumin apparently recognized—or someone (perhaps Malenkov) helped him recognize—the potential political value in Etinger's earliest statements, reported to Likhachev, concerning the medical treatment of Shcherbakov. The "confessions" were tenuous at best, and no one except Ryumin appeared to believe them. By July 1951 Ryumin had much to gain from exploiting this new motif. Reprimanded by the party after Etinger's death, Ryumin was looking for a way of restoring his credibility and deflecting responsibility for the doctor's death.[89] Enemies were all around him. They had been all around him ever since he himself had been denounced in 1937. He understood well how the system worked.

Ryumin struck first. On July 2, 1951, he wrote the secret letter to Stalin. In doing so, he would not be ranked among those Stalin customarily called the MGB's inept nincompoops and hippopotamuses.[90] Unlike Timashuk's letter to Vlasik, Ryumin's clearly wanted blood. And blood Ryumin got. On July 4 Abakumov was removed from office. On July 12 he was arrested on charges of betraying the motherland.

The principal charge against Abakumov in Ryumin's July letter flatly contradicted all the official documents pertaining to the Etinger case. On January 17, 1951, Ryumin himself had signed the *Decree Concerning the Extension of Etinger's Custody to 18 February 1951*, stating only that Etinger had acknowledged himself to be a Jewish nationalist and until the day of his arrest had "conducted anti-Soviet activity" with the "intention to betray the Motherland—to flee abroad."[91] Ryumin did not mention medical "wrecking" in the death of Shcherbakov. Nor was it mentioned in the *Decree Concerning the Termination of the Investigation*, dated April 8, 1951, also signed by Ryumin, which merely restated the previous charges. Therefore, in *all* official documents relating to Etinger's imprisonment and interrogation, up to the spring of 1951, there was no official mention of what would explode less than two years later as the *dyelo vrachey*—the Doctors' Plot.

Ryumin explained the absence of written confirmation by saying that he was explicitly forbidden by Abakumov even to hint that medical "wrecking" was a possibility in the Etinger case, thus confirming Abakumov's treachery. It is more likely that, as Abakumov said, there was nothing to write down. Nevertheless at his trial Ryumin claimed that he regularly informed Leonov, the head of the Investigative Unit for Especially Important Cases, and A. V. Putintsev, senior investigator and assistant head of this same department in the MGB, of Etinger's admissions. Ryumin said that he showed Putintsev his working notes on a daily basis. It is therefore particularly significant that both Leonov and Putintsev also signed the *Decree Concerning the Termination of the Investigation* of April 8, 1951, terminating the Etinger case, in which the subject of medical sabotage is not mentioned.[92]

Putintsev supervised Ryumin's conduct of the Etinger interrogations, which Ryumin claimed contained explosive confessions of medical sabotage.[93] Ryumin could not produce his working notes from these interrogations at his trial, and they have never subsequently come to light. If Ryumin's assertions were true, then Putintsev and Leonov, in addition to Ryumin himself, signed official documents withholding vitally important state information. He, along with Leonov and Putintsev, would have been guilty of conspiring with Abakumov in withholding vital information from the Central Committee, yet only Leonov, who was arrested in December 1951, was ever charged for such. Putintsev was arrested for other reasons after Stalin's death when the Doctors' Plot had been officially repudiated. He was fired from his job, but not otherwise punished. Leonov was shot along with Abakumov in 1954, well after Etinger's innocence had been established. Neither was ever rehabilitated.

At his trial in 1953 Ryumin sharpened his attack on Abakumov, claiming that Abakumov withheld Etinger's confessions of murdering Shcherbakov because he had pro-Jewish sympathies. Nothing but falsified and forced confessions could suggest that Abakumov's sympathies ever inclined toward the Jews. It had been Abakumov who presided over the postwar anti-Semitic campaign, the murder of Mikhoels, and the arrest of the members of the Jewish Antifascist Committee. There is no reason to believe Abakumov would not have pursued Dr. Etinger as mercilessly as he pursued Dr. Shimeliovich or Dr. Shtern, if he thought there was any credibility to the charge or any political gain to be had by it. Under Abakumov the MGB already had falsified and illegally extracted false confessions from the defendants in the Jewish Antifascist Committee affair.[94] Abakumov was not above further falsification if conditions warranted. Furthermore, his handling of the Timashuk letter demonstrated that he worked through proper channels. Although Stalin accused him of it later on and various accounts have perpetuated differing versions of accountability and motive,[95] Abakumov did not suppress Timashuk's letter accusing the doctors of incorrectly diagnosing Zhdanov's illness. It would have been uncharacteristic of him to have suppressed Etinger's confession of medical sabotage, or not to have invented it if he was told to do so. Abakumov did not investigate Timashuk's denunciation because Stalin had written "into the archive" at the bottom of his cover memo, and he did not investigate Etinger after Stalin had given the order in January 1951 to desist.

In November 1950 Abakumov appears to have been genuinely unaware of the new direction Stalin's thinking began to take. It is possible that Abakumov lacked the imagination or initiative to comprehend Stalin's larger political vision or what role the persecution of the Jewish Antifascist Committee and those connected with it might eventually play. He may have been purposely kept in the dark. The Jews, it must have seemed to him, were to be punished *as Jews*, not because they served a larger political purpose. Therefore, Abakumov's inaction may have resulted from the simple fact that Stalin did not order him to take action. Like so many others in this narrative, Abakumov was doomed by the very system he served, the target of a conspiracy he himself was constructing.

The embryonic conspiracy against the doctors coalesced in July 1951 with a conspiracy against Abakumov. Although no documentary proof exists of Stalin's direct involvement with Ryumin's denunciation of Abakumov, his behavior suggests he had been expecting it. As we have seen in the case of Etinger's arrest, Stalin would delay action until he thought a case had sufficiently developed.

In the case of Ryumin's denunciation of Abakumov, Stalin acted at once, even though circumstances might have suggested that he could have waited, as he had with Etinger. But now Stalin needed no further evidence. He did not need evidence at all. He needed to link a conspiracy in the MGB with the conspiracy among the doctors. The political, if not the legal, imperative was clear.

The Russian journalist Arkady Vaksberg is no doubt correct in raising the question of whether Ryumin's letter "was the act of a single madman who decided to risk everything."[96] But the question of whether Ryumin wrote the letter with Sukhanov's assistance and Malenkov's inspiration is less important than why at this particular moment Stalin acted on it.

Stalin may have wanted to rid himself of Abakumov at the point of inaugurating the conspiracy against the doctors because he needed complete secrecy and control. If the case against the Jewish doctors was to become a public trial, it was essential to have complete credibility throughout the Ministry of Security and among all the individuals involved. Abakumov would have known too much; Leonov would have known too much. As Abakumov said again and again during his interrogations, there was nothing to the confessions of Etinger. This being so, the entire case had to be fabricated. Much had to be done in secret; new personnel were needed. Furthermore, in ridding himself of Abakumov, Stalin also destroyed the man who knew in detail how the Leningrad affair, as well as the entire anti-Jewish campaign, came into being. It may be that at precisely the moment when medical sabotage loomed on the horizon of Stalin's thinking, he came to see the usefulness of Timashuk's letter that he had sent into the archive three years before. If medical sabotage and terrorism were to be the core of a new conspiracy through which his enemies could be punished and the country mobilized against the West, then this letter would hold special significance, even though the doctors involved were not Jewish. Only Abakumov and perhaps Poskrebyshev knew for certain that Stalin had seen Timashuk's letter in August 1948. It would be awkward, if not dangerous, then, to produce it in the future if Abakumov were still at the head of the Ministry of Security.

Another factor undoubtedly came into play. This was that the conspiracy against the doctors was *only a part* of a larger conspiracy. The threat of the doctors' medical sabotage could achieve the maximum effect only if the entire Soviet state were threatened by it. For this to be the case, the Ministry of Security had to become as deeply implicated as the doctors.

Almost every high-ranking member of the Ministry of State Security involved with the Doctors' Plot was eventually imprisoned. Most were shot, including Ryumin, Abakumov, Leonov, Likhachev, as well as S. A. Goglidze, deputy minister of state security, and V. I. Komarov, a deputy head of the Investigative Unit for Especially Important Cases. Stalin arrested Abakumov, Leonov, Likhachev and Komarov; the rest were arrested after his death by his successors who also wanted the truth of these interlocking conspiracies consigned to oblivion. The bullying, anti-Semitic Komarov was particularly confused by his arrest. He wrote to Stalin from Lefortovo in February 1953 that "defendants literally trembled before me. They feared me like the plague, feared me more than they did the other investigators . . . I especially hated and was pitiless toward Jewish nationalists, whom I saw as the most dangerous and evil enemies. Because of my hatred I was considered an anti-Semite not only by the defendants but by former employees of the MGB who were of Jewish nationality."[97] Even the interrogators could not see the logic behind their plots.

Ryumin's denunciation of Abakumov fell on well-prepared, but not fully prepared ground. Although Etinger had participated in the Kremlin Hospital special session on August 31, 1948, held by Vinogradov to discuss the diagnosis of Zhdanov's illness, the subject of Zhdanov's death or Timashuk's letter never came up. Therefore, if Stalin had begun to consider linking medical terrorism to Jewish nationalism, he evidently did not want Etinger interrogated along those lines or had not yet made the crucial connection between Zhdanov's death and the case against Etinger. Stalin, in any case, would have understood that interrogating Etinger on this subject before Abakumov was removed might have proved risky.

Leonov and Likhachev, his deputy, were arrested in October 1951, three months after Abakumov, their boss. Ryumin reported to Likhachev, Likhachev reported to Leonov, and Leonov reported to Abakumov. On October 29 Yevgeny Pitovranov, the head of the Second Counterintelligence Directorate, was arrested. The chain of command within the MGB was being dismantled so that a new chain could be established for which an alleged connection between the terrorist Dr. Etinger and Andrei Zhdanov's death in 1948 would not have aroused incredulity.

Both Leonov and Likhachev knew the substance of Ryumin's interrogations, as well as Abakumov's view of the case. They also knew about Ryumin's illegal interrogations, his lack of efficiency, and his responsibility for Etinger's death; they had been present at the party bureau meeting at which Ryumin was reprimanded and knew that his situation in the MGB was precarious. By October 1951, in the wake of Ryumin's July letter to Stalin, no one was left in a position of authority within the Investigative

Unit for Especially Important Cases to challenge Ryumin personally or his account of Etinger's confessions and death.

Individuals, such as Abakumov, Leonov, and Likhachev, were removed, but party procedures were also quietly undermined after Ryumin's letter. The inquiry into Ryumin's conduct that produced the party bureau reprimand was dropped even though Ryumin's allegations about Etinger's terrorism found no support in any official documents and memoranda that he himself had composed or signed, and Ryumin never produced the "working notes" he claimed he had sent daily to Putintsev.

In addition, the procurator's office of the MGB never inquired further into Ryumin's responsibility for Etinger's death and never intervened in subsequent proceedings despite its finding that Etinger had been guilty of nothing more than Jewish nationalist and anti-Soviet feelings. A cursory examination of the official interrogation record showed that Ryumin continued subjecting Etinger to longer and longer sessions after explicit warnings from the medical staff of Lefortovo. The procurator's report of July 30 unequivocally concluded that an "incorrect regimen" imposed on Etinger at Lefortovo led to his death, observing that Etinger was under the supervision of "the senior investigator of the Investigative Unit for Especially Important Cases, lieutenant-colonel Ryumin, who allowed Etinger walks and the use of books and the prison store."[98] One witness reported that when Ryumin was informed that Etinger might well be dying and that his regimen needed to be made less strenuous, Ryumin shouted: "To Hell with him."[99] Nevertheless, it was Abakumov, not Ryumin, who was held accountable for Etinger's death, allegedly to cover up the evidence of his crimes.

The procurator's report produced no unpleasant consequences for Ryumin—no official reprimand or inquiry—just as Timashuk's denunciation produced no immediate consequences for Yegorov, Vinogradov, Vasilenko, and Maiorov. Nothing came until much later—after Stalin's death and the new regime began to search for evidence of its own to silence Ryumin and condemn him to death.

In July 1951, however, the disposition of the government was quite otherwise. Ryumin denounced Abakumov, and through these means was able to incriminate many of the very people who had been most intimately involved in the construction of the Etinger case besides Abakumov,

Leonov, and Likhachev. Scores of other MGB operatives thought to be loyal to Abakumov—many of whom were Jews and had been directly involved in the falsifications of the Jewish Antifascist Committee case and the Leningrad affair—were also imprisoned. In typical fashion, those responsible for carrying out Stalin's crimes against his own people would themselves be declared enemies of the people, usurpers, wreckers, scoundrels, and traitors.

With Abakumov, Leonov, and Likhachev behind bars, the conspiracy could take on a life of its own and develop unimpeded in many different directions beyond the simple linkage of Etinger to terrorism and to Zhdanov. One direction it now took was that A. A. Kuznetsov was posthumously placed at the center of the conspiracy. Kuznetsov had been shot in October 1950 on charges that he and other Leningrad party leaders had been working to overthrow the Moscow center. By the summer of 1952, however, Kuznetsov became implicated in what by then had grown into the Abakumov–Doctors' Plot.

The draft report on the Abakumov case, drawn up by Pavel Grishaev, a senior MGB investigator, in the summer of 1952, concluded, "With the help of Kuznetsov . . . Abakumov appropriated the former manager of the administrative department of the Central Committee who instantly fulfilled Abakumov's demands, advancing all of his proposals, and in conversations with Chekists put Abakumov forward as 'a talented state worker,' saying that he would do whatever Abakumov told him to do."[100] In fact, there would have been nothing unusual about this kind of close cooperation between Kuznetsov and Abakumov. As the secretary of the Central Committee in charge of the Administrative Organs Department, Kuznetsov worked with the Ministry of Security and in the normal course of events routinely cooperated with Abakumov on many matters. While Abakumov was still in favor with Stalin, there would have been no reason not to "put Abakumov forward as a 'talented state worker.'" Once Abakumov fell out of favor, however, all this inevitably changed. By gratuitously blackening Kuznetsov's name in this way, the larger plot widened to encompass Central Committee members as well.

By the time the official report was prepared for Stalin in November 1952, the charges against Kuznetsov had grown even beyond a conspiratorial association with Abakumov. Kuznetsov, Zhdanov's protégé and a hero of the siege of Leningrad, would be charged not only with conspiring with Abakumov but also with having inspired Yegorov to murder Zhdanov at the instigation of American and British intelligence.[101] At this point, the full linkage of Etinger-Abakumov-Shcherbakov-Zhdanov was articulated.

Stalin made use of Kuznetsov in another way as well. Tying him back to Abakumov, who two years earlier had had Kuznetsov executed, as well as to the doctors, Stalin could widen the scope of the conspiracy both internally and externally. Through Kuznetsov, the conspiracy now could link factions within the MGB to the Kremlin doctors, who could be linked to the Jews and the Leningrad party, who now were linked to elements within the Central Committee. This internal conspiratorial group could then be subsumed in the larger political sphere that included the United States and Great Britain. Other sectors of Soviet society and government would eventually also come into the expanding orbit of Stalin's conspiracy—the military, in particular.[102] The handwritten headnote to the 1952 draft memorandum referred to the "Abakumov-Shvartsman" case, as if the original charges against Abakumov had included any reference to Lev Leonidovich Shvartsman, who was not mentioned in Ryumin's July 1951 letter to Stalin. A year later, however, the link was definitively made, thereby associating Abakumov inextricably with the name of one of the leading Jewish members of the MGB.

Lev Shvartsman, whose original patronymic was Aronovich, had been arrested on July 13, 1951,[103] one day after Abakumov. Lev Leonidovich Shvartsman was born in 1906 in the shtetl of Shpola near Kiev. He attended a Jewish *cheder* in Shpola, studied the Talmud, and learned the Hebrew language. In 1922 his mother and stepfather (his father had died) moved to Kiev. In Kiev he worked in the editorial office of the *Kievsky Proletariat* and from there he went to work in Moscow as a literary assistant in the editorial office of the newspaper *Moscow Komsomolets*. He began work in the NKVD before the war and had risen to the rank of deputy head of the Investigative Unit for Especially Important Cases in the MGB. He was denounced by Ryumin and was arrested on the grounds of being an ardent Jewish nationalist who had enabled Abakumov to achieve various subversive ends. Shvartsman had a reputation for being well educated and brilliant, and he is said to have entertained his interrogators with outrageous confessions. According to Sudoplatov, Shvartsman admitted that he had become a Jewish terrorist after drinking his aunt's Zionist soup.[104] Hoping to be put under psychiatric observation, he also confessed to "sleeping with his stepdaughter" and "having homosexual relations with his son," in addition to Abakumov.[105] But the key piece of information the MGB wanted had nothing to do with his sexual proclivities or his aunt's soup. It had to do with his absurd confession that "Abakumov, for some reason, was well disposed to individuals of Jewish nationality . . . "[106] Thus Abakumov was transformed from being Haman, the archetypal destroyer of the Jews, into a protector of the Jews, promoting their traitorous plots and causing numerous Jews to advance throughout the MGB at the expense of their Russian colleagues.

At a certain point, however, Shvartsman alarmed his interrogators with the threatening words: "Abakumov was a conspirator . . . *his threads went higher.*"[107] By this, Shvartsman was thought to mean that if the interrogators kept pressing him, he would reveal what he knew about top Soviet leaders, including Stalin. Ryumin dismissed this at the time as little more than a provocation, but the threads did go higher, to the apex of the Soviet government.

After Stalin's death, Ignatiev offered important information on how tangled these threads were and how high they went. Three weeks after Stalin's death, Beria, who assumed Ignatiev's position as head of Soviet security, ordered Ignatiev to explain the origins of the Doctors' Plot and other matters to him in a written statement. The letter Ignatiev wrote has remained a top secret document to the present day, but Khrushchev's knowledge of it probably lay behind his own account of Stalin's conspiracy against the doctors. Ignatiev wrote that Stalin ordered that all materials pertaining to the progress of the Doctors' Plot had to be sent directly to him, bypassing the Politburo. At his discretion, Stalin would distribute MGB documents to the members of the Politburo himself. He ordered that nothing be altered in any way by Ignatiev in the documents because, as Stalin put it, *"we ourselves will be able to determine what is true and what is not true,* what is important and what is not important"[108] (emphasis added). Ignatiev was not to amend, annotate, or delete any part of any confession or statement, extracted under torture or otherwise, as outrageous or patently false as it might appear. Stalin, not the investigators or the Politburo, would decide whether it was true or not.

In one respect this statement reveals the depths of Stalin's political plotting as he self-consciously manipulated a huge political frame-up. But this statement reveals something more important. Unlike a simple frame-up in which evidence was fraudulently adduced to justify false charges, Stalin was not concerned with false charges or fraudulent evidence. He was concerned with creating a reality in which these charges and evidence would appear normal, logical, and consistent. His concerns were not judicial. They were political. As such, facts did not matter and truth was subject to the "interests of the state," as he defined them. Stalin did not simply wish to eliminate specific enemies. He wanted to construct an internally consistent reality that served his political purposes. This was a reality that knit together Abakumov–Kuznetsov–Shvartsman–Etinger–Kremlin doctors–Jews–Central Committee members–American and British intelligence into a seamless whole. In Ignatiev's statement we see into the depths of Stalin's Nietzschean will to power: He (not even the party any longer) determined the truth.[109] He alone determined reality.

Stalin was not delusional. He *knew* that some or all of the confessions were not necessarily true in the pedestrian meaning of the term in which statements corresponded to observable "facts." For him this distinction made no difference. If certain "facts" were not empirically true, they became functionally true to suit political purposes that, in Stalin's universe, represented a higher reality. Stalin held *both* realities in view at the same time—the quotidian and the political—and this ability made him both monster and master, creator and destroyer, a titanic figure in the history of the twentieth century.

Before that meeting with Stalin it had not occurred to Ignatiev "that the information presented to com. STALIN from the MGB was not always made known to the leaders of the party and government." Ignatiev wrote that he was shaken in February 1953 to discover this was indeed the case, and he illustrated his point with an incident that took place a month before the dictator's death.

Comrades OGOLTSOV and PITOVRANOV informed me that they had the opportunity to attempt to conduct special measures in connection with TITO. I told them not to undertake any measures of this sort and I reported it to com. STALIN, who summoned com.

OGOLTSOV and me to see him. He gave the order that it was not necessary to hurry and he commissioned me to summon to him the principal worker of the 1st Chief Directorate for this case, com. KOROTKOV, who was then in Austria. After 3–4 days, comrades OGOLTSOV, KOROTKOV and I again were summoned to com. STALIN. Listening to com. KOROTKOV, com. STALIN told him to think this question through thoroughly and to all of us he ordered, "don't enlist anybody else in this affair, even members of the Politburo." Com. STALIN added: "If this matter is successful, I myself will tell them."

What I have said above caused me to think that a great deal of information of the MGB sent to com. STALIN, and by the same token several of his important orders to us, might be unknown to other leaders of the party and government.[110]

Stalin did not feel compelled to work through the Politburo; he felt equally free to circumvent the minister of state security or anyone else as the need arose, maintaining a government within a government, based on terror, patronage, and conspiracy. Stalin's government answered to no one. Those most loyal to the Vozhd were part of it, but they were also the ones most in danger.

"If this matter is successful, I myself will tell them." This is what Stalin did in January 1953 when the Doctors' Plot was finally revealed to the world. From 1951 to 1953, through secret channels, links were being forged to connect the death of Zhdanov to the alleged terrorist plot of Etinger; to the treason of Kuznetsov; to the sabotage of Abakumov; to the betrayals of Mikoyan, Molotov, Kaganovich and Voroshilov; to the imperialist aggression of America and the international conspiracy of the Jews. Stalin's little pygmy in Lubyanka was turning out these links one by one.

The absence of these connections in any official documents up to July 1951 is not surprising. The "plot" was not yet a plan. Although Stalin knew what his long-range goals must be, the plan to achieve them evolved out of changing historical circumstances to which Stalin continually adapted his thinking. Nor would specific details be committed to paper in any form until many little pieces had fallen into place and Stalin was more certain of success. The continual widening of the conspiracy demonstrates the comprehensive, totalizing habit of Stalin's mind. But it also demonstrates his immense strategic capabilities and grasp of underlying processes. Stalin thought strategically about all things, pressing advantages, cultivating opportunities, searching continually for ways of causing circumstances and his underlying political intentions to coalesce.

Stalin's approach to the "special measures" against Tito, who by 1953 had become an adversary, demonstrates his characteristic circumspection. Kuznetsov's evolving role in the conspiracy against the doctors reveals Stalin's strategic thinking, as he continually revised the basic elements of the "plot" until he found the right combination of elements to suit his political needs. In 1950 Kuznetsov was shot with others for allegedly seeking to usurp state power; in the summer of 1952 Kuznetsov was made a coconspirator with Abakumov, the man who had him shot, in planning a coup d'etat; by November 1952, however, Kuznetsov became a tool of American and British intelligence conspiring with Abakumov and Jewish doctors to murder Zhdanov and other Soviet leaders. From a relatively isolated political intriguer with the general aim to usurp Kremlin power, Kuznetsov became the tool of an international conspiracy whose purpose was to murder specific Kremlin leaders and destroy the entire Soviet Union. Stalin may have been paranoid and vengeful, but he was not *merely* paranoid and vengeful. This careful development represented a confluence of foreign policy objectives with perceived internal political imperatives, in combination with a mind for which those objectives and imperatives constituted the highest reality.

When Stalin instructed Abakumov to let Etinger "sit" in prison, he probably did not have a fully formed plan, though he may already have had a fully formed intention to purge his government, reassert political mastery over his country, and turn the Soviet Union against the West. Ryumin became one of his most important instruments in transforming this intention into a plan, but neither Ryumin nor anyone else in the Politburo or Ministry of State Security could have developed the plan himself. It had too many interlocking parts over which no one except Stalin could have exercised control. The seemingly insignificant role of Lysenko was only a catalyzing element; the death of Zhdanov that loomed much larger was also only an element. The same was true for the threat of a Leningrad party seizure of power; the alleged ambitions of Abakumov in league with a widespread Jewish conspiracy; and the alleged scheming of Molotov and Mikoyan. Only Stalin could have seen how these pieces might fit together.

As he watched the trial of the Jewish Antifascist Committee become a fiasco, it might well have struck Stalin that he could use Shcherbakov's death as he had used Zhdanov's—to rid himself of another potential rival, in this case, Abakumov—while purging large numbers of cadres in the MGB loyal to Abakumov, as Abakumov had purged hundreds of Leningrad party members. At the same time he may have realized that the scenario of Dr. Etinger's alleged murder of A. S. Shcherbakov would open the question of Jewish revenge against a government leader with well-known anti-Semitic views,[111] thus enabling him to expand the plot to include Jews and their relationships with America and Israel, while at the same time making the case that Kremlin leaders were immediate targets of terrorist plots. Ironically, most Kremlin leaders at the time were under the real threat of physical destruction. The threat, however, did not emanate from terrorists operating within the Kremlin at the direction of American and British intelligence, but from Stalin himself.

In due course Stalin turned not simply against Abakumov and the Jewish cadres allegedly supporting him, but against the MGB as a whole. His constant expression of outrage against the ministry once the interrogation of the doctors began in earnest in 1952 contained the threat of much wider action to come. At this time Stalin told Ignatiev in no uncertain terms that the "Chekists [MGB officers] can see nothing beyond their own noses . . . they are degenerating into ordinary nincompoops, and . . . they don't want to fulfill the directives of the [Central Committee]."[112] At one point, in frustration over the course of the interrogation of the doctors, Stalin demanded to know whether Yegorov (who was not Jewish but would be treated as if he were) had yet been subjected to physical torture.

> "Have you put manacles on him?" When I [Ignatiev] reported that in the MGB manacles were not used, com. STALIN became enraged, cursed me in gutter language I had not heard up to then, called me an idiot, adding, "you are politically blind, bonzes, but not Chekists, you never do this with enemies, and you should not act as you are acting," and demanded that everything he had ordered be

done unquestionably and with all exactitude, and that the fulfillment of his orders should be reported back to him.[113]

In the March 1953 letter to Beria, Ignatiev recalled these portentous words of Stalin from the summer of 1952:

> "I have continually said that RYUMIN is an honorable man and a communist, he helps the [Central Committee] uncover serious crimes in the MGB, but he, the poor fellow, has not found support among you and this is because I appointed him despite your objections. RYUMIN is excellent, and I demand that you listen to him and take him closer to yourself. Keep in mind—I don't trust the old workers in the MGB very much."[114]

Stalin did not "trust the old workers in the MGB very much"; nor did he trust Zhdanov after the Lysenko affair; nor did he trust Kuznetsov and others in the Leningrad party; nor did he trust Molotov, Mikoyan, Voroshilov; nor did he trust Jews. He "trusted" Ryumin, the prototypical New Soviet Man, an "honorable man and a communist." Ryumin was the product of the terror of Bukharin's general purge, a man in whose mind ambition was inseparable from suspicion, whose life was governed by "*everlasting* distrust." Stalin "trusted" him, but as Ryumin would eventually learn to his regret, only up to a point.

"If by some monstrousness," Ryumin wrote in his confession, "I must fall into the hands of the Abakumovshchini and they punish me at the stake, then my last words would be—In 1951, I went to the Central Committee in truth. When I must die, independent of who or whatever circumstances may cause it, my last words will be—I am loyal to the party and its Central Committee."—Ryumin directed these words to Beria whose new MVD (Ministry of Internal Affairs) had taken control of state security after Stalin's death.[115] Ryumin had been loyal to the party and Central Committee and came to it "in truth." Ryumin wrote one last letter to Beria, dated September 24, 1953, attempting to justify himself. Beria had been arrested in June. Stalin's truth gave way to Beria's. After Beria's would come Khrushchev's.

FOUR

THE GRAND PLAN: "NOTHING" COMES FROM NOTHING

LUBYANKA, JULY 1951 — AUGUST 1951

———

Among the doctors there undoubtedly exists a conspiratorial group of individuals, intending through medical treatment to shorten the life of leaders of the party and the government.

—SECRET LETTER FROM THE CENTRAL COMMITTEE, JULY 13, 1951

There was nothing here, absolutely nothing.

—ABAKUMOV TO HIS INTERROGATORS, AUGUST 8, 1951

The question of your guilt is decided by the fact of your arrest, and I do not wish to hear any kind of conversation on this.

—M. D. RYUMIN TO I. B. MAKLYARSKY, NOVEMBER 16, 1951

If the period from Etinger's death, March 2, 1951, to Ryumin's letter of July 2, 1951, can be reconstructed only with difficulty from the limited archival sources, the picture of how the plot against the doctors turned into a plan of action becomes clearer in the days immediately following Ryumin's ostensibly brazen approach to Stalin. Having received Ryumin's July 2 letter, Stalin at once convened the Central Committee to discuss it. On July 4 Abakumov was removed from office, and on July 12 he was arrested.

The July 2 meeting of the Central Committee, the results of which up to now have been known only indirectly, produced the outline of the plan by which Stalin would move against the doctors and the MGB. From that dual basis other segments of Soviet society and government would eventually be implicated.

The top secret *zakrytoe pismo* (secret letter), dated July 13, 1951, of the

Central Committee is the outcome of the July 2 meeting. Made public here for the first time, it contains a "decree" of the Central Committee of the party issued on July 11 and offers a record of the conspiracy's inception. The title of this secret letter is: "On the Unsatisfactory Situation in the Ministry of State Security of the USSR." The letter consists of three and a half single-spaced, typeset pages. According to its headnote, it was distributed throughout all the central committees of all the Communist parties of the Soviet republics, the regional parties, district parties, and the ministries of state security of the Soviet and the autonomous republics, as well as to the regional and district administrations of the MGB. At the bottom of the last page, under a heavy black rule, is the caption: "The present secret letter of the CC VKP(b) must be returned to the *Osobyi sektor* [Special Sector] of the CC VKP(b) after 15 days." The document was both a record of a top secret meeting and a signal to all the elements of the party throughout the Soviet Union that the anticosmopolitan campaign had reached a new stage. After distribution, each copy was returned to the Special Sector of the Central Committee.

According to this document, the Central Committee appointed a commission, consisting of Malenkov, Beria, Shkiryatov,[1] and Ignatiev, to "verify" Ryumin's charges. Their verification, which was concluded by July 4, produced five allegedly inarguable "facts":

1. The Jewish nationalist Etinger was arrested in November 1950, and "without any pressure confessed that during the treatment of com. Shcherbakov, A. S., he had a terrorist intention toward him and took practical measures in order to shorten his life." Consequently, the Central Committee concluded, a "conspiratorial group" undoubtedly existed among the doctors dedicated to murdering Kremlin leaders, which Abakumov refused to investigate.[2]

2. In August 1950 one Salimanov, the former general director of "Vismut," was arrested; and despite the fact that he had betrayed the fatherland, Abakumov hid the case from the Central Committee.[3]

3. In January 1951 the participants of a Jewish anti-Soviet youth organization were arrested. Despite the fact that they had "ter-

rorist plans" against the leaders of the party and the government, Abakumov concealed their terrorist designs from the Central Committee by falsifying the protocols of their interrogation.[4]

4. The MGB "crudely violated" the procedures established by the Central Committee for reporting the interrogations of prisoners by means of composing so-called general protocols based on interrogation notes and draft memoranda, instead of truthfully transmitting all that a prisoner might have said.

5. Abakumov denied these charges and showed no trace of penitence or readiness to admit that he had committed these crimes.[5]

The Central Committee "decree" stated:

In the course of the verification, the Commission interrogated the head of the investigative unit for especially important cases of the MGB, Leonov, his deputies comrades Likhachev and Komarov, the head of the second chief directorate of the MGB com. Shubnyakov, the deputy head of the department of the second chief directorate, com. Tangiev, the assistant head of the investigative unit com. Putintsev, the deputy ministers of state security comrades Ogoltsov and Pitovranov, and in addition heard the explanation of Abakumov.

In view of the fact that in the course of the confirmation of the facts, presented in the statement of com. Ryumin, the CC VKP(b) decided immediately to remove Abakumov from his responsibilities as minister of state security and commissioned the first deputy minister, com. Ogoltsov to execute immediately the responsibilities of minister of state security. This was done on July 4 of this year. . . .

The CC VKP(b) has concluded that the testimony of Etinger deserves serious consideration. *Among the doctors there undoubtedly exists a conspiratorial group of individuals, intending through medical treatment to shorten the life of leaders of the party and the government.* It is impossible to forget the crimes of those well-known doctors, committed not that long ago, such as the crimes of doctor Pletnev and doctor Levin, who poisoned V. V. Kuibyshev and Maxim Gorky at the direction of foreign intelligence agencies. These villains confessed to their crimes at an open trial and Levin was shot, but Pletnev was sentenced to 25 years of prison.

However, the minister of state security, Abakumov, receiving the testimony of Etinger about his terrorist activity, in the presence of investigator Ryumin, deputy head of the investigative unit Likhachev, and in addition in the presence of the criminal Etinger proclaimed that the testimony of Etinger was made up. He declared that the case did not deserve attention and would lead the MGB astray, and he forbade any further investigation on this case. Abakumov scorned the warnings of the MGB doctors and knowingly placed the seriously ill, arrested Etinger in conditions dangerous to his health (in a damp, cold cell) the consequence of which is that on March 2, 1951 Etinger died in jail.

In this way, extinguishing the Etinger case, Abakumov hindered the CC in uncovering *the unquestionably real conspiratorial group of doctors, who were fulfilling the instructions of foreign agents in terrorist activities against the leadership of the party and the government.* In this it is worth noting that Abakumov did not consider it necessary to inform the CC VKP(b) of the confessions of Etinger and in this manner hid this important case from the party and the government. [Emphasis added.]

It took the Central Committee two days to "verify" Ryumin's charges and remove Abakumov from office, the same number of days as to prepare a printed copy of the secret letter for distribution! Given the contradictory and complex testimony at its disposal, including Abakumov's own continued denials, the Commission could not possibly have conducted, let alone completed, a thorough investigation in this time.

There is no evidence that it attempted to do so. This high-level Commission did not examine Ryumin's working notes purportedly sent daily to Putintsev, because Ryumin could never produce any notes; nor did it investigate the reports used later by the MGB procurator's office in its review of the Etinger case, which led it to the opinion that Ryumin had been responsible for Etinger's death. The commission was not interested in investigating *why* Abakumov had ordered Leonov to direct Ryumin temporarily to "shelve" the Etinger case. It confirmed only that he had done so. Nor did it question why Ryumin had never written down, even in draft form, any of Etinger's allegedly terrorist confessions.

The facts were, in fact, not checked at all. Rather, Ryumin was asked to give further testimony that others were compelled to corroborate. At the beginning of July, soon after Ryumin's letter to Stalin, Likhachev and Leonov were summoned by the Commission and were informed of Ryumin's accusation that Abakumov had "smoothed over" Etinger's confessions of terrorism and thereby deceived the Soviet government. Likhachev said that Ryumin's declaration to the Central Committee "shook" him. He could "see no basis for it."

> I laid out [for the Commission] the foundation of everything that had occurred thus far. I attempted to refute RYUMIN's assertion and honestly declared that I did not see and do not see in ABAKUMOV's conduct an attempt at smoothing out this case. I said that RYUMIN was twisting the facts, and that he incorrectly presented them. But RYUMIN insisted on his statement, and, as I very well remember, threw the following phrase at me in the presence of the Commission: "You, LIKHACHEV, left your honor in your office."[6]

Likhachev said he thought Ryumin's accusation that Abakumov had smoothed out the Etinger case simply did not "correspond to reality"; he suggested that Ryumin was probably acting for mercenary reasons. No one listened or cared.

In August 1951 Leonov was again brought in for questioning, but he answered every question evasively—he could not recollect; he wasn't in Moscow when something happened; no one told him; although he had heard something, it couldn't be confirmed . . . Leonov was able positively to confirm only that Etinger had been hostile to Soviet power. When asked about the Shcherbakov confession he stated blandly that in the investigative materials assembled for the case, there was nothing relating to Shcherbakov.[7] He went on to say that the "notes of the bugging" of Etinger's apartment also revealed nothing about Shcherbakov. They pressed him:

"But there were terrorist expressions [in the notes]?"

"There were terrorist expressions, but there were no direct conversations about terror . . . "[8]

Leonov was too dull or too wily to satisfy the Commission. Ryumin turned back to Likhachev, who was arrested in October. He, too, at first resisted unequivocal statements about the murder of Shcherbakov. After his arrest, Likhachev continued to insist that Ryumin had been twisting the facts for his own purposes. At this point, a prosecutor, Ivanov, from the MGB procurator's office, came to Likhachev and told him directly that

> by rejecting RYUMIN's statement concerning the intentional smoothing out of the case of the terrorist ETINGER, who was presumed to be an agent of English intelligence, *I all the more opposed the decision of the Central Committee of the party.*[9] [Emphasis added.]

On July 30, 1951, the procurator's office of the MGB had issued a report clearing Etinger of the charge of terror; scarcely three months later, a representative from this same office broke Likhachev's will to resist by insisting on this charge.

By opposing Ryumin, Likhachev "all the more went against the decision of the Central Committee of the party." The decision of the Central Committee was formed *before* Likhachev ever began to testify. Once it dawned on Likhachev what the situation was, he quickly began to comply. Even so, he was handicapped because he simply "did not know what other confessions to give." What did they want him to say? Ryumin helped him out, "reminding" him of Etinger's confessions, suggesting that Likhachev remember that these were the confessions Etinger gave in his presence to Abakumov. Faced with the overwhelming hopelessness of his situation, Likhachev claimed that he "acted against [his] conscience and deviated from [his] true confessions." He finally gave the Commission what it wanted.

> RYUMIN, in particular, persuaded me that ETINGER, both at my interrogation and at ABAKUMOV'S, presumably gave confessions that, owing to his hostile convictions, he [Etinger] had the goal of killing A. S. SHCHERBAKOV and he realized this evil act by means of incorrect methods of treatment.
> RYUMIN "proved" to me that ETINGER presumably stated at

the interrogation with me and at ABAKUMOV'S that he knowingly prescribed incorrect doses of medicine, that he rejected the correct advice of academician VINOGRADOV and when he [Vinogradov] made a mistake, ETINGER readily supported him.

RYUMIN forced the thought on me that the entire interrogation of ETINGER by ABAKUMOV was conducted in order to smooth out the case and force ETINGER to repudiate his confessions. In this RYUMIN extensively explained what ABAKUMOV'S criminal conduct concretely consisted in.[10]

Resisting Ryumin's logic was useless.

Seeing that it was impossible to convince RYUMIN, that they didn't believe me, but rather believed him, and not wanting to go against the decision of the party, I waved my hand at everything. I saw my complete fate and I agreed with everything RYUMIN thrust on me. I was able to ask RYUMIN only whether he was able to remember everything well. RYUMIN answered that he remembered everything well and before he had made his statement to the Central Committee, he had carefully thought everything out.[11]

To adapt Yegorov's words, *they* believed Ryumin and not some sort of Likhachev. In Bukharin's world of total, *"everlasting* distrust," this was all that mattered. Ryumin remembered "everything well" and "had carefully thought everything out" before he had made his statement to the Central Committee. He wanted Likhachev not merely to confess something he thought was false; he wanted him to remember it, thus falsifying not simply external facts but internal reality as well. Likhachev could not entirely comply.

In connection with this I said to RYUMIN that although I didn't remember that ETINGER had confessed that he had consciously killed A. S. SHCHERBAKOV, I agreed with RYUMIN.

When the prosecutor began to write down Likhachev's confessions, however, Ryumin stopped him. He wasn't satisfied, and he began himself with a pencil to write down "my confessions." At this point, RYUMIN stated to me—let them write it down so that in the course of the investigation it won't be necessary to return to this question.[12]

The notes Ryumin made were then transcribed, as Likhachev recalled, "cleanly into the protocol that I then signed."

> When the protocol was signed by me, RYUMIN declared that now my confessions appeared close to those of LEONOV and that if I had given these confessions to the Central Committee in the first place, they wouldn't have arrested me.
>
> In the course of all the subsequent interrogations up to March 1953, I repeated these "confessions" not seeing the sense of yet again setting out on the path of truth.[13]

The government and its "plenipotentiary . . . to uncover the Jewish nationalist center" were not interested in "the path of truth." They were interested in politically useful statements. When Ryumin produced his description of the testimony of a certain Dr. Yerozalimsky in his July 2 letter to Stalin, he simply invented its contents for the same reason. Ryumin stated that Yerozalimsky had testified that Etinger had hostile intentions toward Shcherbakov. He had written to Stalin that

> . . . [Etinger's] hostile relationship to the Party was incontrovertibly confirmed by materials secretly overheard and by the confessions of his adherent, the arrested YEROZOLIMSKY [sic], who, by the way, in the investigation spoke about the fact that ETINGER expressed hostility concerning comrade SHCHERBAKOV.

This was crucial in establishing Etinger's motive in killing Shcherbakov, a known anti-Semite. However, it appears from all of the recorded documents concerning Yerozalimsky's testimony that Etinger never said this. The MGB procurator's Protocol of Review of the Etinger Case, of July 30, 1951, cited the testimony of Yerozalimsky, but it did not find that Yerozalimsky had said that Etinger intended to harm Shcherbakov, or that Etinger had even expressed hatred of Shcherbakov, in particular. The protocol stated only that on March 12, 1951, Dr. G. V. Yerozalimsky, an intern at the Botkin Hospital, who had been arrested four days after Etinger, testified that Etinger had uttered anti-Soviet expressions to him, had spoken of his criminal ties with other Jewish nationalists, and hated the Soviet leadership because of its anti-Semitic policies.[14] The Procurator's Office concluded:

> From these investigations it is evident that ETINGER acknowledged that he permitted several anti-Soviet expressions; cast aspersions on the national politics of the Soviet government and on the Soviet reality; expressed hatred toward the leaders of the party, disseminated malicious inventions about the leader of the Soviet people; and that he extolled the service of American scholars in the area of medicine, remaining silent about the priority of Russian and Soviet authors.
>
> *His interrogations shed no light on the questions of ETINGER'S terrorist activity.* [Emphasis added.][15]

The ministry found no evidence to support the allegation of terrorism in Etinger's interrogations or the interrogations of others. When Yerozalimsky's 1951 testimony was scrutinized in 1956 during the process of his own rehabilitation, no mention was made of Shcherbakov's alleged "murder."[16] He never made the accusation. Had the Central Committee cared to investigate whether Etinger had told Yerozalimsky that he intended to sabotage Shcherbakov's treatment, it could have consulted the protocol of Yerozalimsky's interrogation attached to the MGB procurator's July 1951 Protocol of Review and would have come to the same conclusion as the procurator. Apparently, it did not do so. The Commission of the Central Committee was not searching for the "truth." It was searching for ways of broadening the charges against Etinger and Abakumov, while widening the scope of the conspiracy.

Abakumov had been accused of *umyshlennoe smazyvaniye*—"the intentional smoothing over" of the Etinger case and of "deceiving the Central Committee."[17] In other words, Abakumov had purposely ignored Etinger's confession of medical sabotage and had guided the prisoner in retracting his confession. Why had he done so? As Grishaev's summer

1952 draft memorandum for the Central Committee stated: "It has been established by the investigation of the case, that ABAKUMOV, in attempting to usurp the highest power of the land, put together a subversive group of several foreigners in the MGB USSR. . . . Thus, the confessions of the Jewish nationalist ETINGER concerning his terrorist activity . . . were kept secret from the Central Committee. The investigative case on ETINGER was smoothed over . . . "[18] According to this document, Abakumov was planning a coup d'etat.

Abakumov denied any wrongdoing. In his interrogation of August 8, 1951, he stated again and again that Etinger's confessions were nonsense. "I said to him that he [Etinger] must speak about how he had treated Shcherbakov. He conducted himself as if he was not guilty of anything and he became anxious to prove to me in detail that Shcherbakov was very ill. He began to explain the nature of his illness, pointing to the region of the heart. He said that Shcherbakov had been doomed."[19]

This sounded plausible to Abakumov, who himself suffered a heart attack in April 1952, which delayed the investigation.[20] In addition, Abakumov knew that Etinger had been only one of several doctors treating Shcherbakov and as such he was not even the attending physician but only a consultant. Abakumov concluded the session with Etinger, saying, "So, did you make all this up in prison?" Etinger answered yes and retracted everything incriminating he had said, or had been forced to say, about his role in Shcherbakov's death. Abakumov recalled that "as Leonov told it, it was entirely clear that he [Etinger] was circling around and speaking falsehoods. . . . No, I think that Likhachev said to me that Etinger was confused and spoke rubbish."

Etinger was brought to Abakumov again in December 1950. Abakumov demanded that Etinger tell them about his "subversive acts." In Abakumov's office, the elderly Jewish doctor needed to sit down. He could not sit down. He needed to sleep. He could not sleep. His knees and ankles were so inflamed that he could barely stand and had to prop himself up against the wall.[21] His entire body was swollen. Dizzy and disoriented from lack of sleep, with an unremitting tightness in his chest from chronic angina pectoris, Etinger supposedly uttered some anti-Soviet, nationalistic sentiments. He spoke of the suppression of the Jews. "Moreover," Abakumov added, "he mumbled and spoke unclearly. I shouted at him

'Speak the truth!'" When Abakumov pressed him on the subject of Shcherbakov, the old man said, "'they pressured me, demanded a confession from me. Therefore I said that I could have treated Shcherbakov better concerning the prescriptions. But all this was untrue. In fact everything was done correctly, there was nothing criminal in what I did.'"

Abakumov was annoyed. The nearly incoherent doctor was barely able to stand in front of him.

"You could have treated him better?"

"Peaceful conditions should have been created for Shcherbakov. But this didn't depend on me because appropriate conditions for Shcherbakov had been created. . . . A. S. Shcherbakov was the kind of patient who would not listen. He did what he wanted."

Abakumov could not disguise his distaste. "Etinger was such a typical Jew and spoke with an accent. . . . I already . . . knew that my presumptions from the first time were confirmed—there was nothing here, absolutely nothing here."[22]

"Nothing will come from nothing," King Lear said. After July 2, 1951, Ryumin took charge of trying to revise this ancient formula. The first *something* this nothing produced would be signed confessions from Likhachev and Leonov acknowledging that Etinger had confessed to murdering Shcherbakov and that Abakumov had in fact conspired with the enemy to conceal the confession. This something that was nothing would grow.

The alleged conspiracy that ultimately united victim and victimizer—Abakumov and Etinger, Abakumov and Kuznetsov—*against* the Soviet government was, in reality, a conspiracy *by* the Soviet government. By July 1951, the plot that began with Stalin's own careful moves in ridding himself of Zhdanov and Kuznetsov had become state policy, legitimated through the Central Committee. Furthermore, through the mechanism of the special commission, leading Politburo members Beria, Malenkov, and Ignatiev had become complicit, though without necessarily understanding Stalin's larger vision.

Ryumin's July 1951 letter to Stalin achieved certain political aims but left much unexplained. Ryumin did not, for instance, suggest a motive for Abakumov's protection of Etinger. Nor did he suggest why it was in Abakumov's interests to protect the anti-Soviet Jewish doctor while, at precisely the same time, Abakumov mercilessly interrogated members of the Jewish Antifascist Committee (including two doctors—Shimeliovich and Shtern). Nor did Ryumin explain why Abakumov might have wished to cover up an alleged six-year-old crime against Shcherbakov,[23] while having Kuznetsov, Voznesensky, Popkov and others in the Leningrad affair condemned to death as enemies of the people only two months previously on charges that they were attempting to usurp the Soviet government. Much remained unexplained and many connections had yet to be made.

The detail that showed which way this nothing would grow was the almost invisible move from Ryumin's limited accusation against a single man, Etinger, to what the secret letter described as "the unquestionably real *conspiratorial group of doctors*, who were fulfilling the instructions of foreign agents in terrorist activities against the leadership of the party and the government." According to Ryumin's testimony, Etinger had acted alone, not part of a "conspiracy" among a "group" of doctors. Two days after his letter was received an "unquestionably real" conspiracy emerged, like one of Lysenko's oblong potato-tomatoes. Only the predisposition to construct this conspiracy could have induced the Central Committee to conclude that it existed. If the conspiracy existed *prior* to the Commission's finding, it did so only in the mind of one man—Stalin. How Stalin's mind worked in constructing this conspiracy is demonstrated in how the decree continued:

> It is impossible to forget the crimes of those well-known doctors, committed not that long ago, such as the crimes of doctor Pletnev and doctor Levin, who poisoned V. V. Kuibyshev and Maxim Gorky at the direction of foreign intelligence agencies. These villains confessed to their crimes at an open trial and Levin was shot, but Pletnev was sentenced to 25 years of prison.

Etinger's testimony may well have deserved "serious consideration," but nothing in Ryumin's letter to Stalin or subsequent testimony provided grounds for concluding that a "conspiratorial group of individuals, intending to shorten the life of leaders of the Party and the government" existed or could be connected to the events of the 1930s. This was pure invention, put into, rather than derived from, a consideration of Etinger's testimony.

Scholars have debated whether Beria, Malenkov, or other Kremlin leaders may have been the motive forces behind the Doctors' Plot.[24] Many of these leaders undoubtedly exploited the conflicts the plot engendered for their own purposes, but none of them could have conceptualized its totality. Beria, Malenkov, or Khrushchev may well have found opportunities to exploit Stalin's increasing paranoia for their own ends and may have benefited from some of its consequences—the elimination of Zhdanov, Abakumov, or Kuznetsov, for instance, may well have benefited Beria, but it is doubtful that Beria would have found it necessary or useful to connect Kuznetsov with a plot against Zhdanov *four years after Zhdanov's death and one year after Kuznetsov's*. The Doctors' Plot was being constructed against a larger political horizon.

Although Malenkov may well have colluded with Ryumin to write the July 2 letter in order to get rid of Abakumov, it was in neither his nor Beria's interests to invent a "conspiratorial group" of doctors dedicated to murdering Kremlin leaders that might cause the purge to spread uncontrollably throughout Soviet society and government. The Central Committee's "decree" puts this alleged conspiracy into that larger context: Doctors had been responsible for the alleged "murders" of Gorky and Kuibyshev in the 1930s; they had been brought to an open trial to confess their guilt and were shot. Now another group of doctors had undertaken the same subversive aim. The invocation of the 1930s could only have terrified the other members of the Central Committee, a body that had been purged as ruthlessly as all the other parts of Soviet society and government during the Great Terror of 1937–1939.

Stalin's name does not appear in the document, but his hand is clearly visible in the insertion of this terrorist "conspiratorial group," which was essential in building the plot into a plan. With this addition, the plot assumed a mass character, and therefore a mass action could be taken. The Doctors' Plot can therefore be dated from this document, composed one and a half years *before* Timashuk's letter was made public in December 1952 as Stalin's evidence of Abakumov's perfidy. Timashuk's letter had nothing to do with inaugurating the plot. It had everything to do, after the fact, with helping to establish the existence of the conspiratorial group.

Though essential, the idea of a "conspiratorial group" was in itself not sufficient to launch the Doctors' Plot. One more vital link was necessary:

the Jewish character of the conspiracy. Why this was so will emerge fully in Chapter Five. The secret letter characterized Etinger as a "Jewish national-ist," but did not describe the conspiratorial group as Jewish. Nevertheless, the idea of a Jewish conspiracy was implanted through this document by means of the trick of associating Etinger with another "plot" against the government that Ryumin had not even mentioned in his letter to Stalin or investigated. The secret letter included as one of its "inarguable facts" the existence of this other, specifically Jewish plot.

> In January 1951, in Moscow participants of a Jewish anti-Soviet youth organization were arrested. During the interrogation, several of the arrested confessed that they had terrorist designs in connection to the leadership of the party and the government. However, in the protocols of the interrogations of the participants of this organization presented to the CC VKP(b), the confessions of the participants concerning ter-rorist designs were deleted at the order of Abakumov. Interrogated by the Commission on this question the head of the Investigative Unit, Leonov, and his deputy, Likhachev, and in addition the deputy minister of state security, com. Ogoltsov, confirmed that the confessions of the prisoners about their terrorist intentions were in fact not included in the protocols of the interrogation. The indicated comrades attempted to explain this criminal falsification of the protocols sent to the CC VKP(b), by asserting that further verification [of the confessions] had been intended. But, despite the importance of the terrorist designs of the participants of the anti-Soviet youth organization, and the fact that a sufficient amount of time had passed—no further information from the MGB was sent to the CC VKP(b).[25]

Just as Abakumov had "smoothed out" the Etinger case, he also had falsi-fied the protocols of the "Jewish anti-Soviet youth organization." Consequently, among the principal tasks assigned to the new administra-tion of the MGB following Abakumov's expulsion was

> to renew the investigation in the case of the terrorist work of Etinger and the Jewish anti-Soviet youth organization.[26]

Although Etinger was never identified as a member of this organization (which he was not), and was identified as a Jew only once in the document, the wording of this order by the Central Committee to the MGB suggests that Etinger and the youth organization were implicitly connected. While Etinger's work as a doctor was gratuitously associated with that of Pletnev and Levin, his identity as a Jew was now gratuitously associated with the "Jewish anti-Soviet youth organization," which sounded portentously like the Jewish Antifascist Committee, whose members were now facing trial.

Stalin's method of widening this conspiracy and discovering unities of intention where none existed depended on associations of this kind. Covering up a specifically Jewish anti-Soviet *conspiracy* was not one of Ryumin's original charges against Abakumov but it was introduced here as though it was, without apparent connection to Etinger, a Jewish doctor. Over time, the two elements would merge into a continuum of alleged subversion and betrayal.

Do the documents provide evidence that Stalin took advantage of an opportunity, or that he invented the opportunity in the first place? Only Stalin could have inserted the element of a conspiracy in the secret letter because only Stalin would have seen where this might lead. The purge of the MGB could have been initiated without it, as could the purge of Mikoyan and Molotov. A Jewish conspiracy was not necessary in the 1930s to decimate the Soviet power elite and galvanize Stalin's supreme control over the country. It was necessary in the 1950s. In the 1930s Stalin was able to connect the alleged betrayals of Bukharin, Trotsky, and others to a supervening threat: the actual Nazi menace. After the war the situation was quite different. Stalin's real internal enemies had been all but eradicated, and the United States, a former ally, had not yet attained the character of an actual enemy. What the Leningrad affair lacked as well as the case of the Jewish Antifascist Committee was this supervening threat. Using each "affair," Stalin could eliminate specific rivals to power, such as Kuznetsov, or troublesome perceived intriguers, such as Mikhoels, but he had not yet found the "full guarantee" for his power in the postwar period that terror and terror alone could give him. Terror could be used against both real and potential enemies. The Jewish element was necessary in this regard because it could be assumed that as the object of widespread discrimination and oppression, Jews would have a real grievance against the government.

Many, such as Etinger, did by 1950. Furthermore, Jews could be found in all strata of Soviet society and throughout the nation. By providing the plot with a Jewish character, Stalin could broaden the conspiracy outside the boundaries of the Kremlin to the international scene in which America and the newly formed state of Israel posed significant, if still potential, threats to Soviet power and prestige. The Jews would become the "canal" through which American subversion flowed directly into the Soviet government, threatening to destroy it. The new menace would have both an external and an internal component.

Nevertheless, the Jewish component of the plot had to be built up carefully. A wild accusation in the secret letter would have been fatal for the plot's credibility. The chief reason for this was that none of the doctors who had treated Zhdanov, with the exception of Sophia Karpai, the EKG specialist, had been Jewish. Likewise, of the five doctors who had treated Shcherbakov, though three were Jewish, by 1951 two of them were dead: Etinger, in prison, and G. F. Lang, who had died of natural causes in 1948. The necessary first step was to associate Etinger with the Jewish anti-Soviet youth organization, but many more links would have to be forged for the plot to germinate and grow.

The first link was the association of Abakumov with Etinger; the second was between Abakumov and the Jewish MGB officer Shvartsman, so that it could be shown how deeply the Jewish threat had penetrated the Soviet state. The decree stated that Abakumov protected Etinger, implying that Abakumov sympathized with the aims of the Jewish anti-Soviet youth organization. Jews hated the Soviet Union because the Soviet Union was unfair to them. Jews loved America because America supported Israel. They had many family connections in America. Therefore, regardless of whether he was an anti-Semite or not, Abakumov became the ally of the Jews.

The accusation against Abakumov contained a significant omission. The secret letter stated that the Central Committee received Ryumin's "statement" (*zayavleniye*) on July 2, without indicating that the statement was a letter addressed personally to Stalin (not to the Central Committee as a whole). The secret letter did not mention this fact or refer to Stalin by name. Stalin may have turned the letter over to the Central Committee on July 2, when he received it, but he may also have simply summarized it without allowing the Central Committee to read it. The secret letter carefully concealed Stalin's role, just as it would be concealed in all public pronouncements on Timashuk's "statement."

Shvartsman reportedly threatened the investigation by claiming that "the threads went higher." Abakumov was said to have tried to "frighten the investigation" by telling his interrogators that his case was "peculiar and delicate," and he didn't wish to speak to them about it because it would "burn up the investigators."[27] Abakumov, too, could have shown that the threads of the conspiracy penetrated the blank omission of the July 13 secret letter and led to Stalin himself. Abakumov could have revealed the truth of the Timashuk letter; he could have revealed much about the conduct of the investigation of the Jewish Antifascist Committee, the murder of Mikhoels, and the origin of the order to let Etinger "sit." Abakumov represented a profound threat to the entire conspiratorial undertaking of Stalin's government. But he was also essential to it. He seems to have understood his role, like many who came before him, protesting his innocence and declaring his loyalty at the same time.[28]

Before his execution in 1940 N. I. Yezhov, the head of the NKVD from 1935 to 1939, asked the court to tell "Stalin that I shall die with his name on my lips," professing unswerving loyalty to the party and the state. As he also pleaded for his life, Bukharin wrote to Stalin that he had "been honestly and sincerely carrying out the *party* line" (emphasis added), and had "learned to cherish and love" Stalin wisely.[29] Abakumov pledged similar loyalty, but he wanted Stalin to know that his loyalty was to him exclusively.

From his cell in Sokolnichesky prison, Abakumov wrote:

> I understand how great is the trust You have placed in me, comrade Stalin, and I am proud of this. I work honorably and have given everything I could of myself, as befits a Bolshevik, to justify Your faith in me. I assure You, comrade Stalin, that whatever assignment You may give me, I am always ready to fulfill it in any circumstances. I have no other life than to struggle for the work of comrade Stalin.—V. Abakumov.[30]

Unlike Yezhov, who thought Stalin might have been deceived by his enemies or that his life was in the hands of the court or Central

Committee, Abakumov had no illusions that anyone but Stalin determined his fate. In his letter, Abakumov did not once refer to the party. He told Stalin that he "has no other life than to struggle for the work of comrade Stalin." By 1951 the work of comrade Stalin subsumed the party and the party line.

Possessing privileged information concerning Stalin's role in the Timashuk case, the Mikhoels assassination, and the Etinger investigation, Abakumov was at the center of a complicated, multidimensional game. A "confession" by him in 1951 would have opened floodgates of repression against the doctors. As it was, these floodgates did not open for another year. There is no record that Abakumov ever produced a suitable confession. He remained loyal to Stalin by not revealing what he knew, while at the same time he refused to comply, even under extreme physical torture, with the investigation's demand to confess his crime of covering up the Etinger case.

The July 11 decree from the Central Committee on the unsatisfactory situation in the Ministry of State Security became the blueprint for a large-scale purge of the MGB, as well as the Doctors' Plot conspiracy. The two actions were tied together at the outset. As soon as Ignatiev took office, he began to implement the principal directives of the July 11 decree, and in November 1952 he summarized his achievements to that date in two lengthy reports addressed to Stalin. One of these is an extraordinary document outlining both the mature contours of the Doctors' Plot and the problems of Soviet counterintelligence at that time in Western Europe. Much of it is given to an analysis of the differential strengths of Soviet and American intelligence forces then operating in Europe, to the need to establish Soviet intelligence capabilities in Israel, and to ways to improve operational efficiency.[31]

Ignatiev characterized the Ministry of State Security as "sluggish, bureaucratic, and swollen." Although a mass action against the Jews did not occur until a year later, the action against the MGB took place almost immediately. From July 1951 to September 1952 some 42,000 individuals were purged from the MGB. Six thousand agents alone were dismissed from agent nets in Ukraine, Belorus, and the Baltics; 3,000 persons were dismissed for breaking the law or violating discipline; 1,583 were dismissed as unfit for service; the rest for other reasons.

Though its underlying purpose may have been to eliminate Abakumov's adherents, the ostensible reason given for this massive reshuffling of personnel and reorganization of procedures was to improve operational capabilities. Ignatiev reported that the Americans and English had up to ninety intelligence organs and spy schools in Western Germany alone. By contrast, the MGB, Ignatiev argued, provided very little training for their agents and therefore could not properly counteract the massive Western intelligence initiatives being undertaken at this time.[32] The "sluggish, bureaucratic, and swollen" security service as a whole, not just Abakumov, had betrayed the interests of the motherland. Both the purge of the MGB and the Doctors' Plot had grown from specific allegations against individuals to mass actions against groups.

The groups kept growing. One week before Ignatiev's November 30 report, a different document was also addressed to Stalin, entitled, "Report on Steps Taken by Security in Accordance with the 11 July 1951 Decree of the Central Committee on the 'Unsatisfactory Situation in the MGB.'" This November 24 document was entirely devoted to the Doctors' Plot.

The November 24 document showed how this process of widening the conspiracy continued. It stated that although the investigation of medical sabotage began with Shcherbakov and general conditions within the Kremlin Hospital, somewhat later, "the MGB received information that the treatment of comrade ZHDANOV, A. A., was conducted as criminally [as the treatment of Shcherbakov]."

Only now did the fates of Dr. Etinger and A. A. Zhdanov become entwined. Once the linkage to Zhdanov's death was made, the mass character of the action against the doctors and the government could assume its penultimate form. The document concluded that "the physicians treating ZHDANOV incorrectly diagnosed his illness, covering up the fact that he had a fresh myocardial infarct, and prescribing a contraindicated regime of treatment for this grave illness which cut short his life."[33]

The source of this information was not revealed; nor was the basis on which a causal, material linkage between the two cases was established. Like Stalin's name in the secret letter, this is a notable absence. At the time little would have suggested a connection between the deaths of these two Kremlin leaders. Shcherbakov, a figure in Stalin's Kremlin of considerably less importance than Zhdanov, had been dead for over six years, and at the

time of his death—one day after the victory over Hitler had been cele-
brated in Moscow on May 9, 1945—nothing suspicious was recorded or
reported concerning it.

The problem for the MGB was that Etinger had not been at Zhdanov's
bedside, and therefore there was no evident conspiracy and no particular
reason for the MGB to begin an independent investigation of Zhdanov's
death. Timashuk's letter was in Stalin's archive. Nothing would have
prompted the security services to connect the two deaths except a predisposition to do so. Not until a Dr. Ryzhikov was arrested in February 1952
could a definite connection, based on an actual confession, be established
through the figure of Dr. Vinogradov, who took part in the treatment of
both Zhdanov and Shcherbakov. There is no reason to think that the
authorities knew of the Vinogradov connection until Ryzhikov's arrest.[34]
In July 1951 there had been no linkage. Furthermore, a memo from
Ignatiev to Stalin, dated September 1952, on the Ryzhikov investigation
was completely silent on the subject of Zhdanov, suggesting that no link
had yet been made.[35] By November, however, Ignatiev would be able to
state unequivocally that "the MGB received information that the treatment
of comrade ZHDANOV, A. A., was conducted as criminally [as the treatment of Shcherbakov]." From September 1952 to November 1952 the
case apparently progressed with extraordinary speed.

Although the November 24 report indirectly referred to the contents of
Timashuk's August 29, 1948, letter to Vlasik and explicitly discussed
Timashuk's conflict with the Kremlin doctors, the letter itself was never
directly mentioned.[36] It contained summaries of confessions from
Yegorov, Vasilenko, Maiorov, and Vinogradov but did not reveal how the
MGB came into possession of the Timashuk letter or why it began to
investigate the death of Zhdanov in the first place. Both the November 30
and the November 24 documents were addressed to Stalin personally. He
not only knew in detail the progress of the investigation; he knew, and
probably controlled, how it was being officially represented.

The November 30 document was signed by Ignatiev, S. Goglidze
(deputy minister of state security), and S. Ogoltsov (deputy minister of
state security).[37] The November 24 document was signed only by
Goglidze, a close associate of Beria, who had taken over the day-to-day
direction of the Doctors' Plot investigation. While acknowledging that

Etinger's alleged confession caused an inquiry into Shcherbakov's death, the November 24 document said nothing about what initiated the investigation of Zhdanov's. This, too, concealed Stalin's role in linking these two elements of the plot.

Stalin's tactic of concealing the role of Timashuk's letter and his own participation in the Doctors' Plot was effective in creating the illusion that the case grew piece by piece as new data were uncovered by the MGB. How effective this has been can be judged from the fact that in a recent study, *Out of the Red Shadows* (Moscow, 1995), despite many new primary documents, the Russian archivist Gennadi Kostyrchenko concluded that Stalin had somehow forgotten about Timashuk's letter of three years before until Poskrebyshev "finally found Timashuk's letter in one of the archive folders and showed it to Stalin."[38] According to Kostyrchenko, Stalin was "bewildered" at seeing his signature on the document, and called a stormy meeting of the Central Committee on December 1, 1952,[39] at which he denounced both Abakumov and Vlasik for concealing Timashuk's letter and Etinger's testimony. In this account, it was not until December 1 that the two sides of the conspiracy fused. The result was that on December 4, the Central Committee issued a decree formulating for the first time the official, government version of the Doctors' Plot.[40]

But this was only an illusion. No evidence substantiates the notion that Stalin was "bewildered," or that Poskrebyshev "finally found" Timashuk's "forgotten" letter in Stalin's archive. The Soviet government's report on the matter at the time stated only that Poskrebyshev "withdrew" the document from the archive.

As comrade POSKREBYSHEV testified in his explanation, in connection with the December 1, 1952 session of the Bureau of the Presidium of the Central Committee, I. V. STALIN stated that the statement of TIMASHUK was concealed from the CC, and com. POSKREBYSHEV withdrew the above mentioned document from the archive and *again* put it before I. V. STALIN. [Emphasis added.][41]

There was implicit acknowledgment that Stalin had seen the letter once before, but nothing was made of this. Stalin blamed Abakumov and Vlasik. In issuing this statement, the government that followed Stalin both revealed and concealed his role. Even after the dictator's death, those who were relieved he was gone remained as loyal to him as Abakumov, whom they eventually executed in 1954.

The documents now in our possession demonstrate Stalin's deliberate management of information in order to obfuscate his direct role in the case, while fostering the illusion of a slow, painstaking accretion of information. The December 1 Central Committee meeting was not precipitated by the unexpected discovery of Timashuk's letter, but rather was the product of many small, unseen, bureaucratic steps taken by Stalin and the MGB over a period of years beginning in July 1951.

The December 4, 1952, decree of the Central Committee, though mentioning Timashuk's diagnosis of Zhdanov's heart attack officially for the first time, did not reveal that Stalin had received Timashuk's August 29, 1948, letter and the accompanying EKG report, even though the November 24, 1952, document referred to this diagnosis. Therefore, up to the eve of the public announcement of the Doctors' Plot in January 1953, knowledge of the existence of Timashuk's letter and what became of it was strictly controlled. The phrasing of the December 4 document shows how this control was managed:

> As early as 1948, the Ministry of State Security had signals at its disposal that obviously spoke to the unsatisfactory situation in the Lechsanupra. Doctor com. Timoshuk [*sic*] turned to the MGB with a statement in which on the basis of electrocardiograms she had confirmed that the diagnosis of the illness of com. Zhdanov, A. A., was incorrectly established and did not correspond to the facts of the inquiry, and that the prescribed treatment for the patient did him harm.[42]

Unidentified "signals" had been at the MGB's disposal, suggesting an intelligence situation still unformed. Dr. Timashuk "turned to the MGB," without Stalin's participation. The MGB officially interrogated Timashuk twice, once in August and once in October 1952, but it had pursued, as we will see, a covert inquiry into the Timashuk denunciation since July 1951 and unofficially for much longer. Dr. Yegorov revealed, "In July 1951, after the removal of Abakumov, Vlasik asked me whether I knew where the statement of Timashuk was. At the same time, Vlasik's deputy, Lynko, demanded all the material from me connected with the investigation of Timashuk's statement. I then sent Lynko the stenogram of the session of 6 September 1948."[43] It is important to note that Yegorov did not state that he sent Timashuk's letter to Vlasik, but only a copy of the stenogram. The letter apparently became available to Vlasik through other means. The entire complex of materials including the Timashuk letter and the stenogram of the September 6, 1948, special session in the Kremlin Hospital was in the hands of the MGB within weeks of Ryumin's July 1951 denunciation of Abakumov.

Vlasik recalled that "the materials connected with the statement of Timashuk were given for investigation to the operational group specially created for this assignment. The former deputy minister of state security, Pitovranov, led the work of this group. . . . On July 17 [1951], an agent report came to the Chief Directorate of the Guard, containing information about the suspicious agitation of Yegorov who appropriated all material having any connection with the treatment of comrade Zhdanov, A. A., and insistently questioned his co-workers to whom a stenogram of the September 6 session of doctors had been sent."[44] It is clear that by July 17, or a mere four days following the secret letter of the Central Committee, the full energies of the security apparatus of the Soviet Union had been engaged. At this point Yegorov understood that the tables now had turned upon him.

The plot was developing simultaneously in multiple and parallel water-tight compartments. One of these compartments contained Sophia Karpai, the first doctor connected to the Zhdanov case to be arrested, one day before Vlasik demanded the materials from Yegorov on July 17. Why she was the first doctor to be arrested in this second branch of the conspiracy remains unclear, as does the role Stalin and the MGB intended for her. If it had been scripted to be a major role, something went amiss. Little is known about her personally, except that Etinger had been her dissertation adviser and that she worked at the Kremlin Hospital as an EKG specialist. She had just completed her thesis, and we know that she had proudly told Zhdanov in Valdai that she was about to get it published. She left Valdai on August 7, 1948, to go on vacation, having delivered her opinion of Zhdanov's condition to the doctors; she had no further connection to the case. Timashuk did not at first directly implicate her in the mistreatment of Zhdanov. When she left Valdai on August 7, Zhdanov was still being treated with strict bed rest. At most she was a minor accessory in an action that had not yet become a case. Nevertheless, she was punished cruelly by the authorities, and we know she demonstrated considerable courage in conditions of terrible psychological and physical suffering.

By July 1951 Etinger was dead. Evidence against him was no longer necessary, and it is doubtful that Karpai was arrested for this purpose. In *Out of the Red Shadows*, Kostyrchenko suggests that Karpai was arrested because she took Shcherbakov's EKGs in 1945, but Kostyrchenko offers no evidence supporting this assertion, or any explanation why this would have mattered. The alleged cause of Shcherbakov's death was not the misdiagnosis of a heart attack, as was the case with Zhdanov, but rather that Etinger and others allowed him to travel by car on the day of victory in May 1945 and had prescribed incorrect doses of medicines.

Furthermore, in all the reports, memoranda, summaries, and protocols of interrogation, Karpai's putative relationship to Shcherbakov was not mentioned. It is in connection with Zhdanov's death, not Shcherbakov's, that she was brought to a "face-to-face confrontation" with Dr. Vinogradov on February 18, 1953.[45] Nothing was said of Shcherbakov. In his interrogation of November 18, 1952, when questioned directly about his role in Shcherbakov's alleged murder, Vinogradov did not mention Karpai.[46]

In his handwritten confession of March 1953, Ryumin wrote, "The doctor KARPAI was the first to be arrested by the 'T' [Terror] department. What served as the grounds for her arrest, I don't remember, but I know, that she deciphered the electrocardiogram of A. A. ZHDANOV."[47]

Why then was Karpai arrested on July 16, 1951, on the heels of Ryumin's letter, the Central Committee's pronouncement on the MGB, and four days after Abakumov's arrest? There may have been several reasons. Karpai had attended Zhdanov in Valdai in 1948—not Shcherbakov in 1945. Karpai was the *only* Jewish doctor in Valdai. Yegorov, Vinogradov, and Maiorov were Russian; Vasilenko was Ukrainian. She was a young woman; the investigators may have assumed that she could be made to confess more readily than a man. The opposite turned out to be the case.

Whoever wanted to connect the alleged confessions of Etinger concerning Shcherbakov to the treatment of Zhdanov also wanted to hold

Jewish doctors responsible. From the outset in July 1951, the conspiracy linking the deaths of Shcherbakov and Zhdanov had an anti-Jewish cast, as the secret letter demonstrates, but the tactical question was how to give this maximum credibility if none of the senior doctors who treated Zhdanov was Jewish. It appears that this was the question with which the MGB and Stalin struggled during this early period. Ryumin's confessions in March 1953 provide some additional light on the subject of Karpai's imprisonment and her value to the government.

> . . . in connection with KARPAI we issued a proposal to the Central Committee that the case should be concluded. After some time, the MGB sent a memorandum that proposed we apply VMN [*vysshaya mera nakazaniya*—the highest measure of punishment, execution] to KARPAI. This was soon followed by the order, as IGNATIEV informed me, from L. P. BERIA, that the investigation in the KARPAI case would continue.
>
> In the interrogations, KARPAI continued to affirm that she had correctly deciphered the electrocardiogram [of Zhdanov] and she rejected all other arguments concerning this question. [48]

The MGB thought she was disposable and proposed she be shot, but Stalin, acting through Beria, ordered that she be kept in prison. It is worth noting that Karpai had been arrested by the "T" Department—the so-called Terror Department of which Ryumin was not a part.[49] Nor did Ryumin connect Karpai with the Shcherbakov case, which he was managing. He connected her with Zhdanov. Otherwise she was nothing more than a shadowy figure of little consequence to him. Ryumin himself saw only part of the plan.

Karpai did not behave as she was supposed to—either as a woman or as a Jew. She was not intimidated by her inquisitors. She never retracted her assertion that her reading of Zhdanov's EKG was properly ambiguous—as indeed it was. She remained unbroken. She was worthless to what the MGB thought was the construction of its new line of inquiry, but Beria ordered that she remain in prison nevertheless, kept under watch, in a damp, refrigerated cell in Lefortovo from which, owing to her asthma, she died not long after her release in March 1953.[50] She would "sit," much as

Etinger sat, waiting for events to dictate how she could best serve the "interests of the state."

Karpai's arrest and Vlasik's request for the Timashuk material from Yegorov occurred simultaneously. Karpai's arrest was not prompted by the documents Yegorov provided. This fact in combination with Ryumin's confession proves that Karpai was being independently interrogated about the Zhdanov case; Beria's order proves that the Central Committee included both the Zhdanov as well as the Shcherbakov case in the Doctors' Plot from the outset, *before* the MGB itself recognized that they were part of the same conspiracy. The MGB thought Karpai's further value to the investigation was nil and recommended she be shot. Stalin thought otherwise. Ryumin himself was trapped in one of the watertight compartments.

The arrest of Karpai produced no tangible result up to the summer of 1952. Though Jewish, she wasn't important enough to continue pursuing; she was too stubborn—too much like Etinger—and therefore this first attempt at forging the link between the Shcherbakov and Zhdanov cases was quietly shelved. What Stalin said to Ignatiev in connection with the attempt on Tito's life —"If this matter is successful, I myself will tell them"—appears to have applied here as well. In due course her usefulness would be revealed, but only once Yegorov, Vinogradov, Vasilenko, and Maiorov were arrested, something that would not happen for some months to come. No one but Stalin could have foreseen their arrests or what those arrests might produce.

Karpai was arrested two weeks prior to the MGB procurator's report of July 30, 1951, which presented Etinger as an isolated instance of an anti-Soviet, Jewish nationalist and specifically charged Ryumin with responsibility for Etinger's death. The report, furthermore, concluded that there were no grounds in any of his statements for concluding that Etinger was a terrorist. Another hermetically sealed compartment in the Soviet bureaucracy.

The arrest of Karpai had led to a dead end. How then did the plot continue to develop? A key to understanding this is provided by the handwritten confession of Vladimir I. Maslennikov, the head of the operational department of the Chief Directorate of Security, working under Vlasik. Maslennikov provided a detailed picture of the activity of yet another side of the security apparatus in the wake of Ryumin's July 2, 1951, letter to Stalin.

Maslennikov revealed that prior to the end of September 1948, no one in the operational department of the Directorate of Security knew of the existence of the Timashuk letter,[51] confirming that Vlasik had not authorized an inquiry into it. Nor was a copy of the letter or even a description of it available in the directorate's own archive. It had never left Stalin's archive. By late summer 1948, however, the operational department did know of the letter. At this point Maslennikov raised the question of verifying Timashuk's statement. Vlasik denied permission.[52] Therefore, despite the absence of an order from Vlasik to investigate the statement and without having actually read it, Maslennikov knew of Timashuk's accusation by other means and knew it deserved attention.

Maslennikov wrote that in September 1948 he was told by two MGB coworkers, Divakov and Rumyantsev, that a secret informant, "Yurina," had covertly reported that Zhdanov had been incorrectly treated in Valdai. He subsequently read agent "Yurina's" report, dated September 8, 1948.[53] He was not told of the Timashuk statement by Vlasik, head of the directorate, or by Lynko, the deputy head. Information penetrated the directorate both through the leadership and at the operational level—from above and from below.

According to Maslennikov, "Yurina" spoke to the operational department through her unnamed handler, while Timashuk wrote letters to Vlasik and other senior government leaders. Vlasik ordered that no action be taken, even as the operational workers were talking among themselves and expressing concern. The system was working against itself. The result was that operational workers in the Directorate of Security were building a case against Vlasik, based on his inaction, even before Ryumin brought his accusation against Abakumov.

Who was "Yurina"? Perhaps one of the nurses. Yet "Yurina's" September 8, 1948, report contained information that the nurses at Valdai could not have known—it contained an account, for instance, of the "unobjective character of the discussion of doctor TIMASHUK's statement at the session held by YEGOROV" in the Kremlin Hospital on September 6, 1948, a week after Zhdanov had died and the nurses had been dismissed. Only the doctors, Timashuk, and the stenographer attended the September 6 session. It is doubtful that "Yurina" could have been the stenographer at that September 6 meeting. Though able to describe the events at this meeting, the stenographer could not have known what had transpired in Valdai except through the disjointed descriptions provided by the doctors on September 6, and she would have been in no position to judge the "objectivity" of the proceedings. Based on the results of this meeting, an otherwise uninformed, nonmedical observer would likely have concluded that the doctors were right and Timashuk had been mistaken.

The solution to "Yurina's" identity may be found in the disclosure that Timashuk herself submitted a secret report to the MGB on the same day, September 8, 1948. "There is a report in the materials of the MGB USSR by Doctor TIMASHUK of 8 September 1948, in which she informed about persecution organized against her by YEGOROV after her statement became known to him, and again she requested a commission be appointed for the investigation into the causes of the death of comrade ZHDANOV, A. A."[54] "Yurina" and Timashuk may have been the same person. To the operational workers in the Chief Directorate of the Security Guard, she may have used the code name "Yurina," while to the leadership of the directorate and the MGB she used her real name.

When the reports of Timashuk and "Yurina" were brought to Vlasik's attention, he said that to undertake any kind of verification at the present time would have "raised undesirable responses around Moscow."[55] Vlasik never clarified what he meant. According to Maslennikov, after Vlasik had indicated that further investigation of the Timashuk accusation would be "inexpedient," "the question of the verification of the signal from TIMASHUK was not raised again before July 1951."[56]

Stalin charged that Vlasik did not verify the Timashuk letter in order to protect Yegorov. Yegorov and Vlasik were drinking buddies, as Maslennikov noted in his testimony,[57] and their friendship may well have gone beyond this. Vlasik, however, had been Stalin's devoted subordinate for thirty-five years, and no evidence exists suggesting that Vlasik might have had a motive to harm his boss. Vlasik's order to Yegorov in 1951 to assemble all materials relating to the Timashuk letter demonstrates that he acted in accordance with directives from above regardless of his personal associations. He did not make such an order in 1948 because, as he said, he received no such directive.

Maslennikov said that during an operational session of the Chief Directorate of Security on July 30, 1951, the day the MGB procurator's office released its report on Etinger, the question was openly raised concerning the verification of Timashuk's charges against Yegorov. At the time, Maslennikov wrote, "it remained unclear whether the contents of TIMASHUK's statement had been brought to the attention of the Central Committee of the party."[58] Nobody knew what anybody else knew. Nobody knew what Stalin knew. At the end of the meeting Vlasik informed the officers that Timashuk's statement was with the Central Committee of the party, which was both true and untrue since only Stalin had possession of it.

The officers at the meeting were quite agitated to learn that nothing had been done. They demanded to know whether Vlasik had seen Timashuk's letter and the statement by "Yurina." He had. Maslennikov wanted to know why no instructions, indicating further action, accompanied the "Yurina" statement. Vlasik answered evasively. Something had impelled the agents to begin searching for the Zhdanov-Shcherbakov link at this time. But no one was told what that was. When they were told that Abakumov and the Central Committee knew about the statements of Timashuk and "Yurina," they were not reassured.

Maslennikov claimed that the July 11, 1951, decree of the Central Committee concerning the "Unsatisfactory Situation in the MGB" was the direct spur to the renewed interest in Timashuk's letter on July 30. He acknowledged, however, that this was not the first sign of interest. As early as sometime in the fall of 1948, the stenogram of the text of the special September 6 session in Yegorov's office had been "seized by the department . . . by covert means."[59] This, too, had been given to Abakumov, who presumably shared its contents with Stalin.[60]

In short, there had not been a time when the Yegorov-Timashuk-Zhdanov affair was not the subject of some form of covert action by the security services. Having both knowledge of the Timashuk statement and the actual stenogram of the September 6 session at its disposal, Maslennikov's department brought the case to Vlasik's attention. He discovered that Vlasik had written on the stenogram that it should be sent with the other statements of Timashuk and "Yurina" to Abakumov. Maslennikov asked the inevitable question:

> With the discovery of this stenogram it became clear that the statement of doctor TIMASHUK and the agent report of the witness

"Yurina" and the stenogram of the session conducted by
YEGOROV were expeditiously put before the leader of the Chief
Directorate of Security and ABAKUMOV. But all the more incomprehensible is the question—why a careful investigation was not
conducted concerning these signals.[61]

Incomprehensible, indeed. Maslennikov reached the only explanation he
could imagine: "the exposed facts of the criminal activity of YEGOROV,
VINOGRADOV, MAIOROV, RYZHIKOV and others reflect a serious
collapse in Chekist work of the former operational department of the
Chief Directorate of Security. As the former head of the department, I
assume responsibility for this collapse. But responsibility for this collapse
should also be assumed by the leadership of the Chief Directorate of
Security."[62] Only a "collapse in Chekist work" could have been responsible for this. It was incomprehensible to him that there could have been a
deliberate choice not to investigate based on Stalin's decision. Maslennikov
did not know, or could not imagine, that the threads went higher. The
truth, as Abakumov suggested, would have "burned up" the investigation.

After the July 1951 secret letter and the decree on the "Unsatisfactory
Situation in the MGB," the investigation began in earnest. Vlasik requested
the material from Yegorov, who became agitated and suspicious. Other
doctors were covertly recruited. Agents "Lvov," a radiologist, and
"Vladimir," a surgeon, provided much "compromising" material on their
colleagues.[63] Deficiencies in the entire Kremlin Hospital system were
exposed, numerous doctors were implicated, and in November 1951 a
draft memorandum of Maslennikov's findings was sent to the Central
Committee.

The ground had now been prepared for purging Vlasik, in addition to
Abakumov. Both could be held accountable for not ordering an investigation. A wholesale purge of the entire security apparatus of the state was at
hand. Both Vlasik and his deputy V. S. Lynko were arrested within a month
after Maslennikov's testimony in December 1952.

When Vlasik was asked why he did nothing with the Timashuk statement, he said that he had believed the results of the 1948 autopsy of
Zhdanov and the conclusions reached by Yegorov's September 6, 1948,
special session.[64] Like Abakumov, he did not say that he assumed that Stalin

believed this, too. After all, every piece of information had been reported to Stalin; Poskrebyshev, Stalin's secretary, had authorized the hurried autopsy in Valdai; and Abakumov had remained silent. All the "signals" that reached Vlasik at that time told him the opposite of what the MGB concluded in November 1952, that "the treatment of comrade ZHDANOV, A. A., was conducted as criminally [as the treatment of Shcherbakov]." He believed that an investigation was not necessary or desired. But now, as Ryumin put it, "political circumstances in the country" dictated something other than what Vlasik had understood in 1948.

Maslennikov denounced his superiors, Lynko and Vlasik, because that was the way the system worked. Timashuk had denounced the doctors; Ryumin denounced Abakumov. Like Timashuk and Ryumin, Maslennikov attributed the inaction of those above him to lack of party spirit, with the difference that Maslennikov did not accuse Vlasik of plotting a coup or knowingly conspiring to assassinate Kremlin leaders. It made no difference. The essence of the accusation was the same: betrayal of the motherland, a charge ultimately made against Vlasik in January 1953:

> **You and Abakumov worked as one, intentionally muffling the signals of incorrect treatment of leaders of the party and the Soviet government.**

"No," Vlasik pleaded. "This was not the case. I don't know why Abakumov did not give an order to verify the statement of doctor TIMASHUK. I am guilty only as a result of my political blindness and negligence."[65]

In the end, Vlasik's captors agreed. Those who succeeded Stalin eventually took up Vlasik's case and in 1954 deemed him too scatterbrained to foment a coup d'etat, too self-indulgent to conspire to murder Kremlin leaders, too sexually degenerate and systematically drunk to construct a plan.[66] He was sent to the gulag to rot.

The question remains: What was the spur that first put the idea of Shcherbakov's murder into Ryumin's mind? In his 1953 confession Ryumin admitted that he had tortured the confession of murder out of Etinger—"pressure in the form of long, nightly interrogations preceded the obtaining of Etinger's confessions."[67] Etinger was not guilty of Shcherbakov's death and did not consider himself guilty. It would have been uncharacteristic of him spontaneously to offer this confession, even after hours of nightly interrogation. Most likely Ryumin suggested the idea, and he was directed to confess to it. Who suggested it to Ryumin?

There is no hard evidence proving this was Stalin. Nevertheless, the charge of medical sabotage against Etinger was essential to constructing the second part of the Doctors' Plot, having to do with Zhdanov's alleged murder in Valdai. Only if Jewish nationalism could be united with the assassination of Kremlin leaders could the plot develop beyond the investigation of a single, localized crime.

The arrest of Karpai on July 16, 1951, suggests that steps were being taken to join both essential elements of the conspiracy at a date far earlier than has been recognized up to now. Neither side of the plot could exist independent of the other. The Zhdanov side was necessary because the Timashuk letter constituted "proof" of medical sabotage. The Etinger side was necessary because none of the doctors with the exception of Karpai was Jewish.

By itself, however, the Shcherbakov case was also not adequate to launch the plot. His death was now over five years old; the circumstances of his alleged assassination were dubious at best; the two Jewish doctors who treated him—Etinger and Lang—were dead; and Shcherbakov, though a member of the Central Committee, was not a national figure, such as Zhdanov or Dimitrov. A retrospective account of his death would not provoke the kind of general panic that could justify imposing a new Terror. Zhdanov, by contrast, was Stalin's close associate and an extremely powerful Kremlin leader with national and international prestige.

Once Jewish nationalism was fixed as the *cause* for the alleged murders, then the case against Yegorov could go forward. Karpai turned out to be a false start but one that revealed the direction the investigation would take. Although the basic contours of the Doctors' Plot had been sketched out in Stalin's mind as early as December 1950 or January 1951 when he ordered Abakumov to "shelve" the investigation of the old doctor, it remained unclear how precisely the plot would unfold, or whether he might even be able to connect it concretely to Zhdanov's death.

But the subject of medical assassination had been on Stalin's mind before then. According to an interview published in 1988, the minister of public health of the USSR, Yefim Smirnov, was visiting Stalin in Sochi,

Stalin's favorite Black Sea resort, probably shortly after the death of Georgi Dimitrov in the winter of 1949. As they strolled around Stalin's garden, Stalin was in an amiable mood and wanted to show Smirnov his beautiful fruit trees—lemons and oranges—speaking in a friendly way about how to take care of them. Then he abruptly changed the conversation. Stalin asked who had treated Zhdanov and Dimitrov just before they died. Smirnov recollected that it was the same doctor. When Stalin heard that he said in a quiet voice, "Isn't it strange? One doctor treated them and they both died." Smirnov protested that the doctor could not have been guilty of such a thing. "How come, 'not guilty'?" Stalin asked.[68] Smirnov did not reveal in the interview which doctor had treated both Dimitrov and Zhdanov. It turns out that it was not just one, but all the doctors who had treated Zhdanov in 1948—Yegorov, Vinogradov, Vasilenko, and Maiorov—also treated Dimitrov in 1949. What is interesting in this anecdote is not that Stalin was thinking about connecting the deaths of Dimitrov and Zhdanov, but that he was thinking about medical sabotage at all—a full two years before allegedly discovering the treachery of the Kremlin doctors. Showing off the fruit trees in his garden, Stalin was considering the fruit of a garden his visitor could not see.

While Stalin relaxed in his garden, Ryumin worked assiduously to achieve complete harmony between the assertions contained in his July 2 letter to Stalin and Likhachev's confessions. "I was interested that Likhachev's confessions confirm my statement concerning the Etinger case," Ryumin testified, and "to achieve this goal," he routinely "corrected" the protocols of interrogation of Likhachev, overemphasizing those confessions of Etinger concerning Shcherbakov. When Likhachev did not say precisely what Ryumin wanted, Ryumin simply changed the testimony, and Likhachev eventually signed the altered text. Thus they worked together to manufacture the necessary proof, fulfilling the portent of Stalin's words to Smirnov.[69] In Ryumin's own dead language: "In this way, and in these circumstances, I manifested a lack of objectivity in the investigative work."[70]

On July 22, 1952, Likhachev and Abakumov were brought face-to-face for a "confrontation"—a time-honored practice of Soviet security ostensibly to get at, but in fact to agree upon, the "truth." In this confrontation Likhachev accused Abakumov of preventing Etinger from confessing to what he now termed "the killing of Shcherbakov."[71]

ETINGER confessed that he incorrectly treated SHCHERBAKOV and nevertheless said that "as it seemed to him" or "as he thought," it might have been possible to extend the life of the patient. Seizing on this, ABAKUMOV knowingly asked ETINGER the incorrect question: "Did you think this up in prison?" and ETINGER had nothing else to do except say: "yes, in prison" and in this way fully retracted his confession concerning his participation in the killing of SHCHERBAKOV.[72]

Ryumin's own eventual confession that he himself did not actually believe Etinger's confessions about Shcherbakov put this seemingly powerful accusation in perspective. The "explanation of Etinger . . . appeared implausible and could not but arouse doubts for the following reasons: In the first place, Etinger was not alone in treating A. S. Shcherbakov. He worked together with other doctors, by virtue of which his activities were not uncontrolled. In the second place, pressure in the form of long, nightly interrogations preceded the obtaining of Etinger's confessions. Finally, both from the account of the illness and from the confessions of Etinger it was clear that for a prolonged period of time A. S. Shcherbakov suffered from a grave cardiac illness and might have died at any time even from a correct course of treatment."[73] Rather than exculpating him, this confession only damned Ryumin further in the eyes of his judges in 1954. Paradoxically, it demonstrates that the origin of the Doctors' Plot had significantly more to do with Stalin's wintry musings about Dimitrov and Zhdanov than with Ryumin's own small-minded machinations.

The Zhdanov and the Shcherbakov cases were vastly different in terms of the medical treatment provided, the conduct of the attending physicians, and the circumstances in which they occurred. Shcherbakov died following the long-awaited, delirious celebration of victory over Hitler. Virtually all citizens of the Soviet Union were jubilant.[74] Stalin had raised the Soviet Union into the ranks of a great world power. Jews throughout the world were grateful to the Vozhd of the Soviet Union whose armies had liberated Auschwitz.

Dr. Etinger, who cared deeply for his people and had adopted a child orphaned in Minsk during the Nazi occupation, would have been no exception. Etinger's anger toward the Soviet state did not develop until later—in 1947 and 1948—with the onset and intensification of the anti-cosmopolitan campaign and the purge of Jews from institutions of higher education, medicine, politics, and the arts. He had no motive in 1945 to murder Shcherbakov on the day of victory over Hitler, nor was such a motive, predating the anti-Jewish campaign, ever attributed to him by the MGB.

If Etinger was so valuable an asset, why was he interrogated so ruthlessly, contravening the warnings of the Lefortovo medical staff, and allowed to die in prison? At the center of this highly calculated game, human egoism, accident, and chance still prevailed. Ryumin was careless. The warnings of the Lefortovo medical staff had been explicit, but Ryumin pursued his own narrow, egoistic ends. Later he recognized the significance of Etinger's death and raised it as an important count against Abakumov, who he claimed had not only "smoothed" over the confession of this Jewish doctor, but allowed the old man to die in a cold, wet cell. Furthermore, Etinger's death vastly complicated the situation because Ryumin had written nothing down and now would have to invent wholly fictitious interrogation protocols, which Etinger would not have been able to sign. Even Ryumin could not carry the falsification to this extreme. Eventually Ryumin had to admit that because of his death, Etinger's "terrorist activity remained uninvestigated."[75] This remained the weakest link in the chain: At the heart of the case against Etinger there were only hearsay accounts of alleged oral confessions. Much would have to be done to overcome this. There was nothing left to investigate. The work of the MGB would now be all in the direction of inventing, not investigating.

fIVE

RECOGNIZING THE ENEMY

LUBYANKA, AUGUST 1951 –NOVEMBER 1951

---

*If you do not expose the terrorists, the American agents among the
doctors, then you will be where ABAKUMOV now is. . . . I am not a
supplicant to the MGB. I can demand and give it to you in the face, if
you don't fulfill my demands . . . We will drive you like sheep.*
—STALIN TO IGNATIEV, NOVEMBER 1951

*Here, look at you—blind men, kittens, you don't see the enemy;
what will you do without me—the country will perish
because you are not able to recognize the enemy.*
—STALIN TO THE CENTRAL COMMITTEE, DECEMBER 1952

*After all it is still a moot question: What precisely
does a man have—a snout or a face?*
—MIKHAIL BULGAKOV, *THE MASTER AND MARGARITA*

I n October 1951 S. D. Ignatiev had been minister of state security for
some four months, succeeding Abakumov. As he often did, Stalin called
him to discuss ministry affairs. In the course of their conversation he asked,
"How is the work on exposing the enemy group among the doctors
going?" Ignatiev had little to report. A few MGB officers had been arrested
in the wake of Ryumin's July 1951 letter; Karpai was in prison but she was
not forthcoming. They had few leads. Nobody was yet producing the confessions Stalin needed and was convinced he would get.

Stalin exploded, accusing the MGB of being filled with Chekists who
could "see nothing beyond their own noses," agents who "are degenerating into ordinary nincompoops and . . . don't want to fulfill the directives

of the [Central Committee]." Stalin demanded decisive measures. He insisted that Ignatiev immediately "expose the group of doctor-terrorists, of whose existence . . . he had long been convinced."[1] But it was unclear to Ignatiev what to do. On what basis would arrests be made? What was the goal?

N. I. Yezhov did not ask these questions in the 1930s. But times had changed. The carefulness of the MGB in the face of Stalin's rage and exasperation is an important indication that a significant, if small, shift in the country had occurred. While there had been no fundamental change in Stalin's underlying system of government, sectors of the governmental superstructure searched for legal legitimacy. Much of the revolutionary fervor that impelled the nation in the 1920s and 1930s was spent, and the quest for stability and order began to be felt throughout Soviet society. Stalin did not recognize this need for legalistic legitimacy in the postwar years, but it appears to have gradually become part of the mental inclinations of the vast Soviet bureaucracy. Despite his machinations and warnings, Stalin found it difficult to change these conditions. His perception of the growing inertia of the postwar bureaucracy may have been yet one more element in his threat to launch "'a country-wide purge' of grandees, idlers, degenerates,"[2] not unlike what occurred in the 1930s.

The Soviet regime had always needed legitimacy. Ideological rhetoric and terror made up for legality in the early years of the Soviet regime, but after the war, recovering from the German occupation and restoring normal life became paramount national goals. Paradoxically, a mood of optimism about the future, despite the painful deprivations of the immediate postwar years, based on the knowledge that Hitler had been defeated and the Soviet Union was now a great, world power, coincided with Stalin's perception that he was less rather than more secure.

Stalin expressed his understanding of the unity of the Soviet state in the famous celebratory toast he gave on November 7, 1937, following the parade commemorating the Revolution of 1917:

> The Russian tsars did a great deal that was bad. . . . But they did one thing that was good—they amassed an enormous state, all the way to Kamchatka. We have inherited that state . . . We have united the

state in such a way that if any part were isolated from the common socialist state, it would not only inflict harm on the latter, but would be unable to exist independently and would inevitably fall under foreign subjugation. Therefore, whoever attempts to destroy that unity of the socialist state, whoever seeks the separation of any of its parts or nationalities—that man is an enemy, a sworn enemy of the state and of the peoples of the U.S.S.R. And we will destroy each and every such enemy, even if he was an old Bolshevik; we will destroy all his kin, his family. We will mercilessly destroy anyone who, by his deeds or his thoughts—yes, his thoughts—threatens the unity of the socialist state. To the complete destruction of all enemies, themselves and their kin! (Approving exclamations: To the Great Stalin!)[3]

Anyone guilty of threatening the unity of the socialist state by his deeds *or thoughts* would be exterminated. Who was not made vulnerable by such a declaration? The ultimate enemy was within the minds of every Soviet citizen.

Following WWII, the threats to the unity of the state multiplied in part because of contact with the West. A powerful America had become the Soviet Union's chief external rival; while younger, more vigorous, internal political rivals had also begun to appear after the war, like Kuznetsov or Voznesensky, along with potential rivals among the old guard, like Molotov or Mikoyan, whom Stalin felt did not understand the threats of the postwar world. Times had changed but the system had not. Repressions continued unabated. Special sections of MGB still existed that could sentence anyone to death without trial. Millions of people remained in concentration camps.

In this context the unprecedented trial of the Jewish Antifascist Committee may have shown Stalin just how dangerous the situation was. Despite the painstaking collection and falsification of evidence, the merciless interrogations, physical torture, and threats of death, the government could not extract the necessary confessions from the defendants. This had never been a problem in the past, but now the trial became hopelessly bogged down. The problem lay not only with uncooperative defendants.

In the 1930s the judges paid little attention to the recantation of confessions or the contradictions in testimony. The dreaded troikas paid attention only to the quotas of death they had to fulfill. In 1937 Yezhov gave a directive to the NKVD, following, as he said, Stalin's telegram to the Politburo, to the effect that the security service had many inadequacies and had lost much time in fulfilling Comrade Stalin's order to rid the land of Trotskyists. "Therefore," Yezhov ordered, "our main goal . . . must be quickly to make up for lost time both in our agent work and in our complete and total crushing of the enemy."[4] This they did. According to government statistics, in 1937 nearly one million people were arrested *from among members and candidates for membership in the Communist Party alone.* Of these, 353,074 were shot outright. The rest were sent to the gulag.[5] Millions perished in the Yezhovshchina. Many had no trials at all or were condemned by the so-called troikas that generally took no more than ten to fifteen minutes to reach a decision.

In 1951 Stalin's recourse to total terror—Bukharin's formula for an "everlasting guarantee"—began to appear less feasible. Some of the problems Stalin now faced are indicated in the letter from Justice Lieutenant General A. Cheptsov, who was one of three judges presiding over the trial of the Jewish Antifascist Committee. He wrote this letter in 1957 to Marshal G. K. Zhukov, minister of defense of the USSR. Cheptsov described his doubts concerning the trial and what he did about it. He wrote that beginning in 1935:

[T]he Military Collegium often carried out sentences that did not correspond with the materials obtained in the court. The judges did not bring their doubts to the attention of the CC [Central Committee], either out of fear or out of trust in the infallibility of Comrade Stalin's decisions, even though the judges could have seen in a number of cases that the information had not been reported objectively to the directive organs.

However, I must note here, that during my time as Chairman of the Military Collegium, from the beginning of 1949 through 1956 and in a number of cases, which were the subject of pretrial discussion in the directive organs, whenever I disagreed with the preliminary decision, I reported my point of view to the CC, which is what happened in the Lozovsky case . . . [6]

Self-serving as it is, Cheptsov described his attempts at achieving some sort of judicial fairness and objectivity in the trial of the Jewish Antifascist Committee. Cheptsov's concerns caused delay in the proceedings. In July 1952 Cheptsov brought his concerns to Malenkov.

[Malenkov] listened to my conclusions and then gave the floor to Rumin [*sic*], who began accusing me of liberalism toward enemies of the people and said that I was intentionally dragging out the hearing for over two months and thereby orienting the defendants to deny their statements made in the investigation, accused me of slandering the organs of the MGB, and denied using physical means of influencing Lozovsky and others. I once again stated that Rumin was behaving illegally. However, Malenkov, after asking me a few questions about the work of the Military Collegium, stated the following, almost literally: "What do you want to do, put us on our knees before those criminals, after all the sentence in this case has the approbation of the people, the Politburo of the CC took up this case three times, execute the decision of the Politburo."[7]

After this interview with Malenkov, the trial went forward and the defendants were executed. Cheptsov did not explain what emboldened him to delay the execution of the sentence.[8] His letter demonstrates that the new "liberalism toward enemies of the people" delayed but could not change the predetermined end. The underlying reality of Stalin's system prevailed. At the same time this letter exposes intentionally or unintentionally the outrageous falsifications of the MGB and objectively attests to the groundlessness of the indictment.

It may be that when Stalin encountered what he took to be similar opposition in the MGB, he recognized that the enemy was not simply individuals with more energy than he possessed or the United States, but was in the bureaucracy of the government itself. As a result, the action that might have begun against individuals such as Zhdanov, Kuznetsov, Molotov, Mikoyan, or Abakumov, or against individual Jewish nationalists, such as Etinger, would take on a mass character.

Many signs suggest that Stalin consciously wished to return the country to the 1930s. One of the most telling is that he wished to link the alleged medical murders of the late 1940s with the still unsolved deaths of Gorky, Gorky's son, Kuibyshev, and Menzhinsky[9] in the 1930s who were said to have been murdered by Dr. Pletnev (who was not Jewish) and Dr. Levin (who was).[10] When an interrogator pointed out to Ryumin that one of the doctors under interrogation appeared unreliable and might repudiate his confessions later on, Ryumin portentously reminded him that "doctor Pletnev had repudiated and then acknowledged his confessions several times, but nevertheless was shot, that this had no significance—repudiate or not."[11]

The new arrests that resumed in the summer and fall of 1951, after the purge of Abakumov and the decree of the Central Committee concerning the conspiratorial group of doctors, reflected the widening character of the Doctors' Plot. Most of the arrests were not of doctors at all, but of officers in the security services, primarily though not exclusively Jews. Ryumin admitted that these arrests were the result of further denunciations.

> The arrests began and continued after the declaration submitted by me to the Central Committee of the party on 2 July 1951. At the beginning a group of leading workers was arrested, at the head of which were ABAKUMOV and SHVARTSMAN, and later in October or November 1951, PITOVRANOV, SHUBNYAKOV and others were arrested.

Ryumin had written a second letter to the Central Committee outlining the charges against these MGB officers. This second letter, written in late October or early November 1951, did not make general observations about the state of the MGB but denounced Ryumin's colleagues by name. Ryumin eventually confessed that this letter, which has not come to light in available KGB files, was written in consultation with Ignatiev himself, who was about to visit Stalin in the south. Ignatiev suggested that Ryumin set out his views on the state of the ministry and what had been done in the wake of the July 1951 resolution of the Central Committee, leaving Ignatiev's name out of it.[12]

> I wrote a letter, not a declaration. . . . The first part of the letter spoke about [what] was done in the area of investigation from the

moment the decision of the CC of the Party was issued on July 11, 1951.

Further, I pointed out that the unsatisfactory situation [in the MGB] was observable not only in investigational, but also in operational work, in particular, in the area of the struggle with agent, foreign intelligence organs.

Moreover, in this sector were to be found concrete guilty parties who had allowed one or another crime and failure in their work. I recall that in the list of those individuals mentioned by me were the names of RAIKHMAN, SHUBNYAKOV, MYAKOTNIKH, STROKOV, KARTASHOV, and in addition the deputy ministers PITOVRANOV, SELIVANOVSKY, KOROLEV, and OGOLTSOV.

In the letter, I remarked that in the MGB there continues to exist the criminal practice of arrests without detailed agent work-ups and with insufficient materials.

In the letter, in support of this, I provided an example from the work established under direction of PITOVRANOV of an operational group whose aim was the uncovering of the terrorist activity of doctors. [Emphasis added.][13]

Ryumin was cleaning out the security service with his denunciations. The MGB officers who were now being arrested included some non-Jews at a very high rank, such as S. I. Ogoltsov and Y. P. Pitovranov,[14] both of whom were deputy ministers of state security under Abakumov; as well as F. G. Shubnyakov, the head of the Second Chief Directorate of the MGB who had signed Etinger's arrest decree on November 18, 1950.

However, by the late fall of 1951, almost all his victims were Jewish. According to Sudoplatov, "Stalin ordered the arrest of all Jewish colonels and generals in the Ministry of Security. A total of some fifty senior officers and generals were arrested."[15] Among these were L. F. Raikhman, the Jewish deputy head of the Second Chief Directorate; Lev R. Sheinin, a writer, dramaturg, and investigator in the general procurator's office, arrested on October 19; and Isidor B. Maklyarsky, a native of Odessa, MGB officer, and a writer twice decorated with the Stalin prize, arrested on November 6.

The pattern of the arrests, beginning with Abakumov and Shvartsman in July, shows how the plot against the doctors would be wound together with the purge of the security services. Pitovranov's case is illuminating in this respect. Ryumin specified that he had been arrested because he had been part of "an operational group whose aim was the uncovering of the terrorist activity of doctors." By the summer of 1952 Abakumov's arrest would be directly tied to Shvartsman's, and they would be linked together in what the MGB labeled the "Abakumov-Shvartsman case," the absurd premise of which was that Abakumov had joined forces with Jewish nationalists to prepare a seizure of "supreme power" in the country.[16] By summer 1952, three distinct foci had been established for the Doctors' Plot: an alleged political-military attack on Kremlin leadership from the MGB, the physical murder of Kremlin leaders by Jewish doctors, and the pervasive, international threat to the USSR from America. By November 1952 these three points would be fused into a single center, but in November 1951 it was still not entirely clear to Stalin how this would be managed.

In November 1951 Ryumin was promoted to the position of deputy head of state security, replacing Ogoltsov, who was now in prison. His assignment was to establish this center. We can see how he went about this in his November 1951 interrogation of the former NKVD officer Maklyarsky. Ryumin demanded Maklyarsky not give the investigation "crumbs" but immediately confess what his Jewish nationalist friends had said about Shcherbakov. Shcherbakov? Maklyarsky was puzzled. Why Shcherbakov? "Your associates killed him" was the answer. "Why don't you speak about it?"[17]

Maklyarsky had nothing to say except that this was slander. All the newspapers, he pointed out, had reported at the time that Shcherbakov had died of natural causes. What was going on? When Maklyarsky showed his incomprehension, Ryumin took another tack around to the question of Jewish nationalism, this time coming directly to Zhemchuzhina, Molotov's wife. He told Maklyarsky bluntly that Zhemchuzhina was a "leading, bourgeois, Jewish nationalist" and an Israeli spy. "Everything is finished with her," Ryumin claimed. But when he began to say something about Molotov, he checked himself portentously in mid-sentence. Maklyarsky could not tell why. It is tempting to speculate that at this point Molotov was being fashioned into one of the central figures of the plot and that Ryumin did not want to get too far ahead of himself.

Maklyarsky gave his interrogators in 1954 another piece of crucial information:

> Moreover, RYUMIN often said that all Jews—it is a spying nation, that he was already finished with Jewish assistants in the MGB and now he was authorized by the government to uncover all Jewish nationalistic undergrounds in the Soviet Union. He spoke further, in order to jog my memory and so that I would give detailed confessions about Jews—responsible workers of the party apparatus.[18]

All of Stalin's major plots required a generic character: kulak or bourgeois social origins in the 20s, Trotskyism in the 1930s, and now Jewish nationalism.[19] The generic link was essential because it provided an a priori disposition—you are guilty before the fact, regardless of actual thoughts, plots, or acts. The potential to commit the crime was sufficient to convict you. As Ryumin pointed out to his stunned victim, who continued to protest his innocence: "The question of whether you are guilty is decided by the fact of your arrest, and I do not wish to hear any kind of conversation about it. For a long time, you and your accomplices have already been declared outside the law. You are arrested at the order of the government and if you do not wish to acknowledge it, you will hardly remain alive." All of this, Maklyarsky said, was "accompanied by a stream of foul language."[20]

In Ryumin's words to Maklyarsky, we can see the practical implications of Stalin's admonition to the Central Committee to "recognize the enemy." Khrushchev has given the following account of Stalin's warning to his colleagues:

> There was no need, in [Stalin's] mind, for an investigation [of the doctors' plot].... He spoke and they were arrested ... Stalin said to Ignatiev: if you don't obtain the confessions from these people, you'll lose your head. He himself called the investigators, himself instructed them, himself ordered the method of the investigation— and there was only one method—beat them. And then the protocol was prepared that we all read. Stalin said: "Here, look at you—blind men, kittens, you don't see the enemy; what will you do without

> me?—the country will perish because you are not able to recognize the enemy"[21]

Slightly modified in *Khrushchev Remembers*, Stalin spoke these words at the fateful Central Committee meeting of December 1, 1952, which issued its proclamation about the Doctors' Plot; but in Ryumin's threat to Maklyarsky it is possible to see how the spirit of these words had already permeated the Soviet security apparatus. The Jews were "a spying nation," and Stalin's postwar initiative was to open the eyes of the country to the new enemy: America and the Jews.

In Stalin's new plot the Jews were inextricably bound up with America. Whoever was Jewish was for America, and whoever was for America was Jewish or had become influenced by Jews. In this regard Dr. Yegorov could not contain his exasperation when he wrote, "I absolutely refuse to understand—why I, a Russian man, must be put in a group of Jewish nationalists."[22] For the purposes of the conspiracy, Yegorov and many others would be treated as if they were Jewish.

Stalin's words had something of the biblical portentousness of Hosea's lament: "My people are destroyed for lack of knowledge."[23] For Stalin now the enemy was America, the Jews; but the *goal*, as Bukharin rightly understood, was not simply to destroy those external enemies; it was to retain and fortify the power of the party—the power of Stalin himself—and thus the unity of the state. Stalin's power was directly proportional to the extent of the enemy threat.

Ryumin pushed Maklyarsky again and again to reveal the enveloping, underground web of Jewish conspiracy. The second time they met was in February 1952. Maklyarsky had been moved to the dreaded Lefortovo. Ryumin was "courteous," almost delicate in his approach. He praised Maklyarsky for his previous confessions. The MGB, he said, could make a great deal out of them. He assured Maklyarsky that they were "extremely important and valuable." Ryumin told him that he would "never forget" this service to the country. Furthermore, Ryumin went on, "I rarely give anyone such a promise, but if you continue in this spirit, you will leave this case with the smallest possible losses." What a guarantee of *his* safety!

Only one small service remained for Maklyarsky to perform. "In your confessions there is a real inadequacy," Ryumin told him. "There are no 'figures' in them." Figures? Ryumin explained. "The people you have named are nationalists, but not major ones. Why," he asked with some exasperation, "do you always avoid naming the main names? Why not Ehrenburg, for instance?"[24] Ryumin was looking for the center, the "figure" around whom the plot could be constructed at an open trial.

Maklyarsky didn't know Ehrenburg, had never met him, was not in any way acquainted with him. This was unfortunate. How about Zhemchuzhina, then? Maklyarsky, Ryumin presumed, had to know her, having worked in the special department of the *komandatura* of the Kremlin. Ryumin demanded to know "precisely which rooms of the leaders of the party and government, living in the Kremlin, Zhemchuzhina visited." In particular, "how often she was in Kaganovich's room."

Maklyarsky couldn't say. Although he had worked three months in the Kremlin, he had never seen Zhemchuzhina on the premises and certainly didn't know in what rooms of the Kremlin the various government leaders lived.

Looking for a central figure, Ryumin was also widening the net. The mention of the last remaining Jewish Politburo member, Lazar Kaganovich, showed that he was hunting for enemies not only in the preserves of the MGB or in the Kremlin Hospital, but in the Kremlin leadership itself. He seemed intent on enmeshing Kaganovich in a new plot with Zhemchuzhina involving more than simply political subterfuge. Molotov would be humiliated, implicated, and then denounced; others would follow. It would make sense for Kaganovich, Stalin's court Jew, to be shown to be both personally dissolute and politically dangerous, intimately involved with the Jewish wife of the treacherous Molotov. Kaganovich, too, would be held up as a "grandee, an idler, a degenerate." Kaganovich along with Molotov may well have become candidates for the "figures" Ryumin was looking for.

But Maklyarsky could not provide this service. "Fine," Ryumin said, "but don't speak about yourself. Give confessions about others." Here Ryumin ordered the investigator, Yazev, to write down all of Maklyarsky's confessions about as many other people as he could think of. Ryumin insisted on important names.

What to do? Being a writer of some accomplishment and coming from Odessa, Maklyarsky decided he could subsequently demonstrate the groundlessness of his arrest and the illegal means whereby confessions were concocted by Ryumin if, in addition to the names of living individuals, he provided names of long-dead writers as well. So he named Isaac Babel and Eduard Bagritsky, both of whom were Jewish writers from Odessa, both dead before the war. Ryumin apparently did not know who they were. When Ryumin asked about Bagritsky, Maklyarsky said only that he was a Soviet poet from Odessa. "I must say that with these people—neither with Bagritsky nor with Babel—did I have any personal acquaintance whatever, if you don't count my having read their works," Maklyarsky told his interrogators in 1954.[25]

Just before he left the interrogation, Ryumin warned that Maklyarsky would have to give additional confessions so that he could take "prophylactic measures." When Maklyarsky asked what these measures might be, Ryumin said simply:

> "In Moscow there live more than a million and a half Jews. They have seized the medical posts, the legal profession, the union of composers and the union of writers. I'm not even speaking of the trade networks. Meanwhile of these Jews only a handful are useful to the state, all the rest—are potential enemies of the state. Especially if you consider that in Moscow are to be found all the foreign embassies, foreign correspondents, etc." Here he gave YAZEV the order to fix this in the future Stenogram. This YAZEV did. If you pick up the Stenogram copybook of the March interrogation, my words will be fully confirmed.

Maklyarsky noted, however, that the official, as opposed to the copybook stenogram of the interrogation, to be found in the case file, did not include this remark by Ryumin. When he subsequently asked Yazev why this explicit anti-Semitic comment did not appear in the formal protocol, Yazev told him, "At the present time, they decided that it was not possible to write this." Later, Ryumin told Maklyarsky "he intended to put the question to the government about the expulsion of the Jews from Moscow."[26]

At the present time, Yazev explained, it would not be possible to write so overt an anti-Semitic statement into an official protocol. But soon the time would come when they could because Ryumin was preparing to propose the expulsion of all the Jews from Moscow. If from Moscow, then from all the major cities. If expulsion, then what else? But Stalin was not ready to take such drastic steps in the fall of 1951 or even the spring of 1952. Malenkov indicated the nature of the necessary preparation when he ordered Cheptsov to finish off the trial of the Jewish Antifascist Committee and execute the prisoners: "after all the sentence in this case has the approbation of the people, [and] the Politburo of the CC . . . " Much work was still necessary in order to obtain the "approbation of the people" for the kind of mass action Stalin envisioned. More details would have to be connected, more plots uncovered, more confessions obtained, more spheres of government brought into the conspiracy.

Knowing his rights as a prisoner, Maklyarsky wished to lodge a formal protest to Ryumin concerning the relentlessly provocational nature of the interrogation to which he was subjected. We do not have copies of those interrogations, but Maklyarsky described them as a series of "compromising questions about members of the Politburo, ministers, members of the Central Committee of the Party." Despite his prior assurances, it seems that Ryumin had not given up on this line of questioning. When he heard Maklyarsky out, Ryumin responded with a kind of exaggerated calm. "Well," he said, "some of them *were* ministers, several *were* members of the Central Committee." Then he left the room with the implication hanging in the air that the investigation had already achieved many of its aims and had already shown that the enemies of the state had penetrated to the highest levels of government. Though this was not in fact the case—no ministers except for Abakumov had yet been purged—Ryumin's demeanor shows where the case was headed. Soon after Stalin's death, Ryumin was asked to explain why he sought compromising information about the leadership of the party and the Soviet government. At first Ryumin angrily denied this; then admitted he had done so but gave no explanation. Ryumin never suggested, as Shvartsman had, that the "threads went higher." But there is little doubt they did.

Timashuk may unwittingly have given Stalin the first sign that Zhdanov's death could have benefits beyond getting rid of a potentially dangerous rival. This must have become gradually clearer to Stalin once Dr. Etinger was arrested and Jewish nationalism could be directly associated with medical malpractice. In July and August 1948 this was not as clear as it was in November and December 1950. Israel had declared its statehood in May 1948—three months before Zhdanov's death—and Stalin initially supported the Jewish state. In May 1948, therefore, the Jews and Israel were not yet the enemies of the Soviet state they soon became. In 1951 it was clearer still.

Although the anticosmopolitan campaign attacked Jews as Jews, it became virulent only after Israel became a state fully identified with the Western camp. The shift is registered in Ilya Ehrenburg's September 21, 1948, essay in *Pravda*, "The Union of the Snub-nosed," in which he argued that that as a bourgeois state, Israel could not decide the Jewish question. Nor could it unite Jews around the world. In addition, Ehrenburg wrote, the charges of anti-Semitism, discrimination, and the "squeezing" of the rights of Jews in the USSR was nothing more than malicious slander brought by the enemies of the Soviet order.[27]

We will never know what went on in Stalin's mind, but on December 1, 1952, at the dramatic meeting of the Presidium of the Central Committee, Stalin declared that "every Jew is a potential spy for the United States."[28] The entire Jewish people had thus become a target. There is telling symmetry between Ryumin's language to Maklyarsky in November 1951— "the Jews—it is a spying nation"—and Stalin's to the Central Committee a year later, suggesting that what Stalin said openly in December 1952 had been uttered privately many times before.

Though arrested earlier than Maklyarsky, Lev Sheinin was not interrogated until February 26, 1952. At first Ryumin was courteous, gentle, almost affectionate when he said, "Really, you, Lev Romanovich, a man with great experience in investigative work. Who if not you should understand that confessions in the MGB must be pointed and politically sharp, that you must consider the political circumstances in the country and the international position."[29] Ryumin praised Maklyarsky's "excellent confessions about the Jewish nationalist underground with ties to America, with the center, and everything as it should be." But then as he had done previously with Maklyarsky, Ryumin's tone became reproachful. Why had Sheinin not named anyone important?—perhaps Ehrenburg. Why had he not gone into "the heart of the matter"? He urged Sheinin to read Maklyarsky's confessions "who much more correctly approached the goal. And," Ryumin added, "it wouldn't be suitable for you to lag behind him."[30] He went on, "If you . . . provide politically sharp confessions according to the plan I will give you, then I'll declare to the Central Committee that Sheinin was disarmed and that he divulged the underground center, and he may be forgiven and released. But if not, then there may be a different variant." What that different variant was, Ryumin immediately dramatized by putting out his index finger and pulling an invisible trigger.[31]

Ryumin continued in this vein, telling Sheinin about the armed Jewish nationalists who had been captured the very same day as Sheinin himself in the offices of the MGB—apparently, according to Ryumin, about to take over the government. "The Jewish nationalists were armed even in the MGB, and we captured them on the same night as you. They did much here even in 1940 . . . " Ryumin did not say what he meant by this last, provocative remark, but it suggests some connection between the present situation and the prewar situation in the country. Ryumin may also have wanted to suggest that the Jews were somehow responsible for the onset of the Great Patriotic War itself, but he did not elaborate. Instead, "catching himself abruptly . . . he blurted out superfluously . . . 'Of course, anti-Semitism is also a bad trick.' "[32] He never explained what he meant.

Perhaps Ryumin wanted to let Sheinin know that he didn't hate Jews as such but only as elements in the current political situation. He may have been signaling to Sheinin that he, Ryumin, was above this petty Jewish conspiracy and knew well enough that Sheinin, though Jewish, was really not a bad fellow after all. It was a game they both would have to play. Sheinin had to play along. If he did, he would be saved. If not, he would be shot.

The interrogations of Sheinin and Maklyarsky further reveal that at this point Ryumin still knew and was developing only half of the eventual plot. The half concerning Etinger and Shcherbakov. Ryumin insisted that Sheinin confess he knew the aged, terrorist doctor prior to 1951—"maybe in 1946, perhaps when he took a vacation somewhere in the Baltics and you might have met with [Etinger] there. For you," Ryumin added, "this has no particular significance, but to us this is important."[33]

Sheinin couldn't understand what was going on. He continued to say, honestly, that he did not know Etinger before he had been arrested. Ryumin grew impatient and declared that Sheinin was to be transferred back to Lefortovo. Sheinin begged not to be taken back a third time.

Ryumin laughed. "I understand why you're afraid of Lefortovo—you're afraid they will break you. Well, that can't be excluded."[34]

Both Sheinin and Maklyarsky ultimately cooperated as much as they could. Upon being returned to Lefortovo, Sheinin was deprived of free time, books, additional food. He was questioned late at night and deprived of sleep. They applied manacles to him in November and December 1952, refused him warm clothing, and eventually threatened him with physical torture.

With Maklyarsky another tactic was used. "Don't stand on ceremony with him," Ryumin instructed Maklyarsky's interrogator after one session at which Ryumin was present. Then to Maklyarsky he added, as he left the room, "keep in mind if there is no confession about nationalism by morning, your wife and son will be arrested tomorrow."[35] After an hour, Maklyarsky signed a declaration, dictated by the investigator and addressed to Ryumin, that he would provide confessions about the alleged nationalist underground operating in Moscow.

Despite this assurance, little of material value came from the interrogations of Sheinin and Maklyarsky that could amplify the larger scenario. Nor was much progress being made in developing the medical side of the plot. Stalin demanded faster results.

Ignatiev revealed that immediately following the abusive telephone conversation with Stalin in October 1951, cited earlier, he formed a special operational group in the Second Chief Directorate consisting of MGB officers Goglidze, Ogoltsov, and Pitovranov, of which the latter two would be denounced by Ryumin and arrested within a month.[36] Their job was to study all the material available on the medical workers and in conjunction with corresponding departments in the MGB to prepare "the necessary operational measures for further development." Ignatiev wrote:

> In the whole time of the existence of this group no results were ever achieved and to the end of January 1952 in almost every conversation with com. STALIN, I heard not only his sharp abuse but also threats of approximately this character: "If you don't expose the terrorists, the American agents among the doctors, then you will be where ABAKUMOV now is," "I am not a supplicant to the MGB. I

can demand and give it to you in the face, if you don't fulfill my demands," "We will drive you like sheep," etc. (By the way, the abuse and threats, as is known, continued up to recently.)[37]

At some point, probably the autumn of 1951, Stalin realized that it would be impossible to make the Jewish Antifascist Committee trial into a public show trial. The defendants proved too stubborn. They recanted their confessions and contradicted one another endlessly. Despite all the cruel efforts of the MGB, no hard and compelling evidence could be adduced to prove the charges against them. Furthermore, as Cheptsov's letter to Zhukov illustrated, the Ministry of Justice itself was dragging its feet. Stalin was becoming increasingly frustrated. He needed to find the enemies that would win "public approbation" for the actions he hoped to take. The Jewish Antifascist Committee would not serve this purpose. But the doctors in combination with American plots against him and the government, in combination with subversion in the MGB and the party leadership itself, might provide his "full guarantee."

Though no arrests against doctors had occurred between Karpai's arrest in July 1951 and the MGB purge in November, the medical side of the plot continued to develop as well. What was apparent from Ryumin's interrogations of Sheinin and Maklyarsky now became evident in the MGB's effort to widen the scope of medical malpractice beyond Zhdanov and Shcherbakov. In November 1951 a draft memorandum was prepared for the Central Committee concerning the allegedly "suspicious" deaths and illnesses of various high-ranking party members, including Shcherbakov, Zhdanov, Georgi Dimitrov, and Andrei Andreyev.[38]

This November 1951 memorandum was not prepared by Ryumin but by Maslennikov, who worked under Vlasik in the Directorate of the Guards. On the one hand, this is further evidence that all the intricate sides of the plot had not yet fused; on the other, it demonstrates that all sides of the plot were being coordinated *from above*. It was not by accident that the medical and the political sides of the plot were expanding in the same way toward greater and greater inclusiveness: the plotters were potentially all Jews, and all Jews were potential enemies of the state. Jewish doctors were the first target, but Ryumin's recurrent interest in having Maklyarsky and Sheinin implicate Ilya Ehrenburg, the internationally known Soviet journalist and novelist, suggests that he was looking for a way of including all Jewish intellectuals, not only doctors, in the plot.

"I am not a supplicant to the MGB. I can demand and give it to you in the face, if you don't fulfill my demands," Stalin warned Ignatiev. "If you do not expose the terrorists, the American agents among the doctors, then you will be where ABAKUMOV now is." But Stalin was not content with simply investigating the deaths of Shcherbakov and Zhdanov. He insisted that the MGB expose those who murdered comrade Dimitrov and those "who led com. Andreyev astray." Subsequently Stalin insisted that Thorez and Tokuda also had been mistreated by Soviet doctors.[39]

Stalin said this to Ignatiev in January 1952, but we know that Maslennikov had sent a report to the Central Committee concerning Thorez and Tokuda in November 1951 that had been under preparation since at least August 1951.[40] In his testimony Maslennikov revealed that in August 1951 he placed all the agent material concerning Timashuk and the death of Zhdanov before Yevgeny Pitovranov, then in charge of a special operational group responsible for uncovering medical wrecking in the Kremlin Hospital. At the time, Maslennikov stated, Pitovranov told him directly that "the department, using proven agents from among specialists of the Lechsanupra of the Kremlin, would study the history of the illnesses of comrades _____, _____, _____, and _____ with the aim of possibly uncovering facts deserving attention in connection with incorrect methods of treatment and identifying those who were guilty. Somewhat later this same work was assigned in connection with the study of the history of the illness of comrade _____ ."[41] The spaces left blank, as security protocol demanded, were clearly the names of Dimitrov, Andreyev, Shcherbakov, Thorez, and Tokuda. Zhdanov's name would not have been among them because up to now Pitovranov was exclusively concerned with the alleged plot of Etinger against Shcherbakov. Therefore at the same time that the major elements of the plots—the deaths of Zhdanov and Shcherbakov—were being covertly interlinked, the death of Shcherbakov was also being developed to include a much wider circle of important names. This shows us that the initiative for expanding the operation had to have come through Ignatiev to Pitovranov. Pitovranov on his own would not have—could not have—understood the vast dimensions this operation was assuming. And if the initiative came from Ignatiev, who up to now had shown so little initiative in dealing with the doctors, it could have emanated only from Stalin himself. Up to that time, there was not the slightest reason for anyone in the security services, Ryumin included, to have connected the names of Dimitrov, Andreyev, Thorez, and Tokuda with Shcherbakov, let alone with Zhdanov.

Some of the alleged victims of this epidemic of terror had been dead since the mid-thirties—Maxim Gorky and his son, for instance; others in jeopardy were still living, such as Andreyev, Thorez, and Tokuda, as well as Soviet army personnel. Dimitrov, the president of Bulgaria, had died in 1949. Certainly the Vozhd himself was the principal target, but this would not be publicly disclosed until the very end. This great criminal undertaking would soon encompass the whole of Soviet society and politics. But in November 1951 the two sides—Zhdanov and Shcherbakov—had not yet grown seamlessly together. Nor would they for several months to come.

SIX

WAITERS IN WHITE GLOVES

LUBYANKA, NOVEMBER 1951 — NOVEMBER 1952

"Beat them!"—[Stalin] demanded from us, declaring: "what are you?
Do you want to be more humanistic than LENIN, who ordered
DZERZHINSKY to throw SAVINKOV out a window? . . . you
work like waiters in white gloves. If you want to be Chekists, take off
your gloves. Chekist work—this is for peasants and not for barons."

—STALIN TO IGNATIEV, NOV. 1952

We will beat you every day, we will tear out your arms and legs,
but we will all the same learn everything down to the
last detail about the life of A. A. Zhdanov, and all the truth.

—INTERROGATOR SOKOLOV TO YEGOROV, NOV. 1952

Com. STALIN as a rule spoke with great anger, continually expressing
dissatisfaction with the course of the investigation, he cursed,
threatened and, as a rule, demanded that the prisoners be beaten:
"Beat them, beat them, beat them with death blows."

—GOGLIDZE TO BERIA, MARCH 1953

Tell us everything, and we ourselves will decide
what is true and what is a lie.

—RYUMIN TO MAKLYARSKY

The year 1952 was terrible—a praeludium.

The Soviet Union and the United States had become locked in a global conflict that would last another forty years. "We must not lose time because we never know whether the Russians will attack first" was a view expressed by more than one U.S. general.[1] The global struggle was reflected in the mounting tensions within the Soviet Union and its satellites as well. In March 1952 the secret trial of the Jewish Antifascist Committee began. In August the defendants were executed in Lubyanka. In November, Rudolph Slansky, general secretary of the Czechoslovak Communist Party and a Jew, was tried, at Stalin's instigation, on the charge of being a Zionist conspirator. With ten others, he was publicly hanged in Prague on December 3, one day before the Central Committee decree on the Doctors' Plot. Significantly, Slansky and his doomed colleagues were accused of "taking active steps to cut short the life of the republic's president, Klement Gottwald" by means of employing "medical doctors from enemy circles and of dubious backgrounds . . . establishing close ties with them, planning to use them for his hostile plans."[2] Their bodies were burned and their ashes scattered in the wind. Later that month Stalin would denounce Molotov and Mikoyan as spies.

While Jews in the Soviet Union were accused en masse of being a "spying nation," in the United States, Ethel, who was innocent, and Julius Rosenberg, who was guilty, were in prison sentenced to death for betraying the United States to the Soviets. As McCarthyism spread recklessly throughout American society, confusing authentic patriotism with demagoguery, it would have been inconceivable for the Rosenbergs or their many Jewish friends, relatives, and supporters to have imagined that they were sacrificing their lives for a country in which in August 1952, thirteen innocent Jewish intellectuals would be executed, a country whose leaders soon thereafter would turn against the entire Jewish population of the Soviet Union and, at the highest governmental levels, was seriously considering the idea of the detention and deportation of hundreds of thousands, if not millions, of innocent people. Such bitter ironies paralyze the imagination.

Throughout 1952 Stalin's daily show of exasperation and frustration suggested that almost all the initiatives in building up the case against the Kremlin doctors and the MGB had stalled. In April 1952 Ignatiev wrote to Stalin that the investigation of Karpai had been unproductive. Therefore the MGB was having difficulty in establishing the key linkage of medical terrorism to Jewish nationalism without which the larger plot could not go forward. On the security services front, another wave of arrests of Jewish MGB workers had taken place in December 1951. Raikhman, Navlovsky, Sverdlov, and others were thrown in jail, demonstrating the truth of Ryumin's remark to Maklyarsky that he was "already finished with Jewish assistants in the MGB." But these arrests apparently also yielded little of value.

Stalin was cautious. But he was not infinitely patient. He had become increasingly frustrated with the slow pace of events. He needed results. Perhaps Stalin knew that his own time was running out. He had been ailing for many years with high blood pressure, sclerosis, and exhaustion. According to Volkogonov, in 1952 he had been advised Dr. Vinogradov not to continue with his duties as head of state.[3] Stalin met this suggestion with contempt, vowing never to see a doctor again. From that point, up to his fatal stroke in March 1953, the head of one of the two superpowers of the world, in possession of a vast military machine with nuclear weapons, was without professional medical assistance.

Perhaps Stalin thought that this was the moment to strike a blow that would firmly return Soviet society to his grip. "We will drive you like sheep," he had threatened Ignatiev toward the end of January 1952. And so the investigation returned to the Jewish Dr. Etinger's treatment of Shcherbakov. On February 26, 1952, Dr. Roman Isaevich Ryzhikov was arrested, and by midsummer the MGB had what it hoped was its first real breakthrough on the medical front.

Ryzhikov had been the former deputy director of the medical unit of the government health sanitarium Barvikha, where in 1945 Shcherbakov had been treated. From December 1944 to May 1945 he was Shcherbakov's daily attending physician. Not nearly as brave as Etinger, he gave confessions practically from the beginning. By July 24 he had admitted that he had "facilitated the terrorist Etinger in conducting enemy actions."[4]

The alleged crime was that the doctors had recommended that Shcherbakov be allowed to travel by car to witness the victory day celebrations in Moscow; in addition, they were alleged to have prescribed incorrect doses of some medicines. According to his signed confessions, Ryzhikov went along with this terrorist scheme. But it remained difficult to pin him down because Ryzhikov also, from time to time, would retract in whole or in part that confession, admitting only that he might have made an incorrect judgment in how to treat Shcherbakov's aneurysm.

"You consciously shortened the life of comrade SHCHERBAKOV, A. S. Testify about this."

"I had no desire to shorten the life of A. S. SHCHERBAKOV."

"Don't lie, Ryzhikov," they warned him. "You knew back then—and not only now—that the treatment of comrade Shcherbakov, A. S. was criminally incorrect. You were one of the direct executors of the terrorist plan. . . . We demand that you tell the truth."[5]

But when Ryzhikov tried to tell the truth, he was threatened with torture and death until he told *their* truth. In the end, according to the official protocols of interrogation, Ryzhikov placed most of the blame for Shcherbakov's death on Etinger and a Dr. G. F. Lang.[6] About Vinogradov he said only that Etinger had led him astray by exerting undue influence over him. "Therefore," he stated, "I think that it was precisely both ETINGER and then LANG who, insisting on an active, movement regimen for the patient, destroyed the life of A. S. SHCHERBAKOV."[7]

On October 29, 1952, Ignatiev wrote to Stalin that a special medical examination of Shcherbakov's heart had been completed in the summer of 1952 by an elite group of medical examiners, which confirmed the charges against the doctors. The internal organs of all deceased Soviet government leaders were customarily preserved in the Kremlin, and Shcherbakov's heart was among them, kept in a jar of formaldehyde during the previous seven years.

This expert commission secretly "investigated the heart, studied the history of the illness and the analysis and protocols of the autopsy of the body of comrade SHCHERBAKOV, A. S." In addition to this, another group of medical experts had been appointed in April 1952 to verify the drug prescriptions ordered by the doctors who had cared for Shcherbakov. Both commissions came to the same conclusion that the medical treatment of comrade Shcherbakov had been "criminal."

As it was reported to You No. 6367/I of 25 September 1952, concerning the case of the arrested RIZHIKOV, R. I.,—the former deputy director for the medical unit of the sanitarium "Barvikha"— a qualified medical examination was planned to investigate the heart of comrade SHCHERBAKOV, A. S., and secretly to interrogate as a witness Professor RUSAKOV, and the chairman of the Mosgorzdravotdel, doctor PRIDANNIKOV, who were present during the autopsy of the body of comrade SHCHERBAKOV, A. S.

At the present time the expert commission has finished its work. . . . The commission investigated the heart, studied the history of the illness and the analysis and protocols of the autopsy of the body of comrade SHCHERBAKOV, A. S.

Simultaneously a second, expert medical commission was created. . . . They have also completed their work.

The conclusions of both commissions fundamentally coincide; they correspond to the conclusions of the previously conducted medical examination and confirm the criminal treatment of comrade SHCHERBAKOV, A. S. [8]

Both commissions diagnosed the cause of death as a myocardial infarction; they concluded that the doctors had provided improper doses of prescription drugs and never should have allowed him to undergo the ordeal of traveling by car from Barvikha sanitarium to Moscow to watch the victory day celebration.

Yet when Ignatiev, who has been described by those who knew him as an orderly, well-trained party man, wrote his October 29 memorandum to Stalin, the MGB already possessed information that should have made him think twice. On July 10, 1952, Major General D. Kitaev, deputy senior military procurator in the MGB, wrote an assessment of the legal standing of the two medical examinations Ignatiev mentioned. Kitaev had been assigned to review the medical examinations—a third verification. It may be that it was this document that caused the delay from July to October in reporting the results of Ryzhikov's interrogation. At the top of the Kitaev report in Ryumin's clear hand is the notation that this finding was to be distributed to comrades Sokolov and Garkusha, both of whom were the senior investigators working under Ryumin responsible for conducting the Ryzhikov investigation.

Kitaev stated that the work of both commissions was, from the outset, flawed, sloppy, tendentious, and largely illegal. "These deficiencies" in the work of the commissions, Kitaev concluded, must be "immediately eliminated."[9] There is no evidence that they were. Of the April examination, Kitaev wrote: "Although the examination is considered finished, it is impossible to think that it was satisfactory; that is, it was conducted in violation of the demands of the law."[10] He had a similar view of the second examination begun in June. Kitaev pointed out that there were numerous violations of medical practice, legal procedure, and so on.

The most glaring impropriety, Kitaev thought, was that Lidia F. Timashuk was included in it. Her name is conspicuously absent in Ignatiev's October 29 report to Stalin, listing the doctors who prepared the "expert examination" of Shcherbakov's heart. Kitaev insisted that she should never have been allowed to participate.

> **TIMASHUK should have been removed from the staff of the exam-ination because the latter, as is evident from the testimony of the accused RYZHIKOV of April 21, 1952, being formerly a doctor in the Kremlin hospital at the time when one of the leaders of the party was ill, took an electrocardiogram of the patient and insisted to academician VINOGRADOV that giving digitalis to the patient should be prohibited because it might disrupt the functioning of the heart and consequently also the data of the electrocardiogram.**[11]

How the tables had turned for Lidia Timashuk! Now she sat in judg-ment of her former bosses who had once judged her. Apparently, in his April 21, 1952, interrogation, a copy of which we do not possess, Ryzhikov had testified that he knew something about Timashuk's earlier dispute with Vinogradov concerning Timashuk's demand not to adminis-ter digitalis to "one of the leaders of the party"—no doubt Zhdanov, though Ryzhikov apparently did not mention his name.[12]

Along with Ryzhikov and Etinger, Vinogradov had also treated Shcherbakov in 1945. Learning through Ryzhikov's April 21, 1952, inter-rogation about the earlier dispute between Vinogradov and Timashuk, Kitaev correctly concluded that Timashuk had a direct conflict of interest in the conclusions of the "expert" commission. Two of the seven questions posed to the "expert medical commission," Kitaev pointed out, bore directly on Timashuk's field of work:

a. **What conclusions about an illness could be made on the basis of an electrocardiogram and the history of an illness?**

b. **Was it allowable to administer digitalis while performing an EKG and what effect would this have on establishing a diagnosis?**

"Under these circumstances," Kitaev concluded, "Timashuk absolutely should have been interrogated as a witness about [the episode with Vinogradov], and according to statutes 43 and 48 UPK RSFSR should not have been an expert in this case."[13]

Timashuk could not be expected to be impartial. So much for Timashuk's profession of "objectivity" when she protested the September 1948 session held in Yegorov's office. "I would like this question to be decided by some third party and not only between the two of us—you and me," she had said at the time.[14] Whatever might otherwise be said about Timashuk's role in the Doctors' Plot, she, too, was part of the same system that made Ryumin temporarily successful.

Kitaev concluded his memorandum to Ryumin unambiguously: "RYZHIKOV confessed that, being antagonistic to Soviet power, he allowed anti-Soviet expressions and slanders among his circle. Despite his confession concerning this point of the accusation, there are *no other objective proofs* in the case." (emphasis added)[15] It is not known whether Ignatiev saw this report, but it is likely he did. The delay from July to October when he sent his account to Stalin may have had to do with his realization that he needed more than the findings of these "expert" commissions, which might be called into question by the same judicial, bureaucratic body that had caused endless delays in the trial of the Jewish Antifascist Committee.

On October 29, 1952, the day he sent his memorandum to Stalin, the MGB managed to extract a full confession from A. A. Busalov, the former chairman of the department of surgery of Yaroslavsky Medical Institute and former Head of the Kremlin Hospital before Yegorov. He, too, had participated in the treatment of Shcherbakov and had been responsible for recruiting Etinger for the case. Busalov confessed that

LANG behaved criminally, he did not attribute serious significance to the illness of A. S. SHCHERBAKOV and did not want to think about the danger threatening the patient . . . As a rule, LANG supported ETINGER . . .

Prescribing an active regime to A. S. SHCHERBAKOV did not accord with the functional condition of the patient's heart, ETIN-

GER, LANG, and VINOGRADOV shortened the life of A. S. SHCHERBAKOV.

I committed a grave evil, and stamped everything with criminal irresponsibility, in that [I allowed] LANG, ETINGER, and VINO-GRADOV to direct me. This occurred throughout the entire period of treatment.

And the greatness of my guilt consists in the fact that after the death of A. S. SHCHERBAKOV I acted as an associate of LANG, ETINGER, VINOGRADOV and continued up to the present to conceal from the government the actual causes of A. S. SHCHER-BAKOV'S death.[16]

This was what the MGB needed to counteract Kitaev's petty concerns about due process and the absence of "objective proofs."

Whether Ryumin forwarded Kitaev's report to Ignatiev, he definitely sent it to Sokolov and Garkusha, who were directly in charge of Ryzhikov's interrogation. When asked about this report during an interrogation in June 1953, Ryumin professed to know nothing about it.[17] Ironically, one of the MGB officers interrogating Ryumin after his own arrest was the same Sokolov to whom a copy of the report had been given the previous year.

In any event, whatever Ryumin knew, he did nothing about it, except to substitute "confessions," such as Busalov's, for facts. Ryumin was looking for *truth,* not evidence, and he would not be deterred by facts, logic, mere empirical reality, or due process. Such concerns, to quote Lysenko in the field of biology, were nothing more than disguised counterrevolutionary, bourgeois idealism, on the order of Kantian metaphysics. Stalin's truth had nothing to do with such so-called facts or concerns, but rather, as Zhdanov had observed in connection with Akhmatova's poetry, served the interests of the party and the state. Truth in Stalin's system was always subject to change with these interests.

In yet another turn of fate, it was this same Major General Kitaev who on March 16, 1953, sanctioned the arrest of Ryumin on the grounds that he "was the organizer of falsifications and perversions, permitting in the course of the investigation in the case of the arrest of leading scholars of the country, academics VINOGRADOV, GRINSHTEIN, professors YEGOROV, VASILENKO, PREOBRAZHENSKY and others, unconscionable accusations of terrorist activity against the leadership of the party and the government and of having ties with foreign intelligence organizations."[18] In the summer of 1952, however, Kitaev could do little to redirect the investigation of the doctors. Stalin remained dissatisfied with all the MGB had thus far produced.[19] He demanded more.

There is yet one more ironic dimension, in the borderland between Marx Brothers absurdity and Shakespearean tragedy, to the medical examination of Shcherbakov's heart: One panel was formed to determine whether Shcherbakov had been given proper medicines; the second to determine whether he had died of a myocardial infarction—the same cause of death attributed to Zhdanov. In fact, no one had ever disputed that Shcherbakov had died of a heart attack. No one had questioned it at the time. There had been no disagreement among the doctors or technicians. The autopsy recorded that Shcherbakov had died of an infarct. Why then an examination seven years later to prove that he did?

If the doctors had covered up a heart attack in the case of Shcherbakov, there would be symmetry with the Zhdanov case in which a heart attack also had been covered up. The MBG was now preparing to make the criminal treatment of Shcherbakov and Zhdanov conform to the same pattern. According to this pattern, Shcherbakov, who had been treated by the Jewish doctors Etinger and Lang along with Vinogradov and others, had been murdered because the doctors had not properly treated his heart attack; Zhdanov, who had been treated by Vinogradov and others, had been murdered in the same way. Vinogradov became the conduit whereby Etinger's criminal influence entered into the treatment of Zhdanov. *This* was the way these murderous doctors routinely acted disguising their crimes, using guile and deceit, typical of Jews, not daring to act in a forthright manner.

We do not know who put Timashuk on the commission to investigate the charge against Shcherbakov's doctors. As a covert MGB agent, she could well have been appointed at the recommendation of her MGB handler Suranov, who already knew of her feelings against Vinogradov. There is no evidence of Suranov's intervention, but there can be no doubt about the reasons for choosing her. Timashuk would serve as another link between the Zhdanov and Shcherbakov cases. The hero of one would help adjudicate the other. Vinogradov connected them at the level of medical treatment; Timashuk would connect them at the level of dissenting diagnoses. Timashuk's presence on the commission, despite Kitaev's concerns, guaranteed a level of public acceptance, once her role was made public, as it would be on January 20, 1953, when she received the Order of Lenin for unmasking the enemy doctors in the Zhdanov case. Her credibility would be assured. Once the Doctors' Plot became public knowledge in January 1953, Timashuk was presented as a great hero, she was treated, according to Yakov Rapoport, as "the Mother of God or, at least, Joan of Arc."

> The Soviet press went into raptures about the perspicacity and courage of this paragon of virtue. Poems were dedicated to her, she was styled "a great daughter of the Russian people," and near-religious veneration was accorded her. She was specifically compared to Joan of Arc.[20]

Her presence on the commission gave it, to use Ryumin's words, the sharpness and political point it needed.[21] Almost before our eyes we see how this case against the doctors was growing together like crystals on a string.

But it was Ryzhikov, not Vinogradov, who was arrested. Why? The delay of perhaps a month between Stalin's threatening telephone call to Ignatiev in January 1952 and the arrest of Ryzhikov in February, though possibly the result of bureaucratic sluggishness, suggests some hesitation. Ryzhikov was to Shcherbakov what Maiorov was to Zhdanov—the on duty, daily attending physician, by no means the most prominent or most important member of Shcherbakov's medical entourage.

Ryumin claimed at his trial that Ryzhikov was not arrested because of involvement in Shcherbakov's death, but rather because he was charged with having poisoned two members of the Soviet army.[22] This was certainly not true but was another ploy to increase the verisimilitude of the investigative process. It is contradicted by Ryumin's confession that he initially instructed Ryzhikov's interrogator to concentrate on extracting a confession of guilt in Shcherbakov's death and that "in general the RYZHIKOV case from the beginning was headed in this direction."[23] The MGB, like Stalin, took care to conceal its tracks. The intent was clearly to "unmask" Ryzhikov, so that the investigation could claim that it had discovered the Doctors' Plot through the course of a careful examination of some serious but unrelated charges, rather than that it had been constructing the Doctors' Plot piece by piece. Ryzhikov, like Karpai, was a minor character who followed the instructions of superiors. He was arrested to establish Etinger's guilt, and from this starting point a whole network of "Zionist" agents could be exposed. Through this network, including Jews and non-Jews alike, the death of Zhdanov could be explained as an assassination to a horrified population.

Each of the actions against the doctors followed the same pattern with the arrest of relatively inconspicuous subalterns, such as Karpai and Ryzhikov. The groundwork of conspiracy and confession that began at the bottom and gradually extended farther and farther throughout the medical and political hierarchy ensured that more eminent figures would eventually be compromised and engulfed in a tangle of denunciations and "evidence" from which they could never free themselves. This procedure was in marked contrast with the one begun at the top with the arrest of Abakumov and followed by the arrest of his subordinates. Stalin employed different strategies in different circumstances, depending on whether the consequences would be made public, as in the case of the doctors, or could be concealed from view, as in the case of Abakumov.

Ryumin himself suggested the logic for this in his discussion of the "expert analysis" of Zhdanov's heart. He noted that it was done covertly and that Zhdanov's name was carefully removed from all documents relating to it, because "to make an open examination of the preparation of A. A. ZHDANOV's heart might stir up a lot of unnecessary noise, threatening unpleasant consequences (not excluding the possibility of suicide among the doctors who were mixed up in the incorrect treatment)."[24]

Stalin's brutality was not merely the result of a paranoid or fanatic mind. It was the combination of cruelty with cautiousness, of paranoia with shrewdness, of detached inhumanity with an extraordinarily realistic grasp of human nature that made Stalin so dangerous and so mesmerizing to his followers and enemies alike. His plots possessed an almost geometrical precision. Vinogradov would have been the obvious choice if Stalin wished to publicize the operation under way, but he did not. If Vinogradov had been arrested first, the case would have received considerable notoriety and the plot, such as it was, could not have been portrayed as a Jewish-Zionist invention.

Ryzhikov, however, was not the only possible choice. The MGB had at its disposal a Dr. Kadzharduzov, of Armenian descent, who also attended Shcherbakov in 1945. But Stalin wanted a plot in which Jews, pursuing anti-Soviet, Zionist aims, murdered Kremlin leaders at the behest of American intelligence. Russians, Armenians, Georgians, Ukrainians, and Byelorussians could be brought into the picture later. Eventually Kadzharduzov would also be arrested, but at the outset the Armenian, Kadzharduzov, did not fit properly into the structure the MGB was shaping.

Ryzhikov's biblical, Jewish-sounding patronymic Isaevich (son of Isaiah), though not an uncommon Russian name of a certain generation, may have helped the MGB associate him with Jews. Perhaps the MGB counted on the resonance of the name to suggest, if only subliminally, to the public that Ryzhikov's father, at least, might have been Jewish. It may also be that the MGB itself, at first glance, thought they had the doctor they had been looking for—a Jew directly tied to the death of a Kremlin leader. But if this was so, they quickly found out that poor Ryzhikov wasn't Jewish at all, but rather Byelorussian.

Stalin's interest in names was not new. During the anti-cosmopolitan campaign artists and writers of Jewish origin who had adopted Russian pseudonyms were routinely unmasked as Jews in the press. K. M. Simonov, then editor-in-chief of *Literaturnaiia gazeta*, a major organ for literary affairs, was present at a Politburo meeting on February 26, 1952, not long after the arrest of Ryzhikov, when the lists of candidates for the Stalin prizes were presented for consideration to Stalin. Next to certain names, Simonov recounted, were other names in brackets. In each case the bracketed name had a noticeably Jewish appearance such as "Rovinsky."

Stalin was irritated. "Why does it say Maltsev, but in brackets 'Rovinsky'? For what purpose? For how long will this continue? . . . " he demanded. "Why is this being done? Why are two names being written?" Stalin appeared offended. He proceeded to instruct his amazed audience: *"If a man chose a literary pseudonym—that's his right. We're not speaking about anything other than elementary decency . . . But apparently someone thought to under-*

line the fact that this man had a double name, to underline that he was a Jew. Why would you underline this? Why would you do this? For what purpose instill anti-Semitism? Who needs this?"[25] In fact, Orest M. Maltsev, the author in question, was Russian, a native of the Kursk district. He had chosen the Jewish sounding pseudonym Rovinsky when he wrote his collection of stories *Vengerskaya rapsodiya* (*Hungarian Rhapsody*). Simonov corrected Stalin about Maltsev's nationality, and an awkward moment of silence ensued.

According to Simonov, Stalin's comment astounded the important literary and government figures who attended the February meeting and had an impact on Soviet literary society. News of this incident spread by word of mouth in upper echelon cultural circles, the effect of which was to distance Stalin himself from the crude anti-Semitic campaign still under way that was the cause of the parentheses in the first place. It made people think twice about their own accusations. How could ordinary citizens expose Jews as Jews if Stalin himself could not support such invidious considerations?[26]

In reality, the incident related by Simonov was but another instance in which Stalin cynically obfuscated his true intentions. By cooling down for the moment the boiling pot of state-sponsored anti-Semitism, the coming revelations of villainous Jewish doctors would be made to seem even more explosive and credible because the people would know that Stalin could not be accused of anti-Semitism.

At the same time that Stalin "smoothed over" (to borrow Ryumin's locution) the anti-Semitic policies of his regime while choosing the Stalin prize recipients, Dr. Ryzhikov received a different Stalin prize. This one was awarded in Lefortovo prison, where Ryzhikov was subjected to vile, degrading, and brutalizing conditions to compel his confessions about the Jews whose civil rights Stalin seemed intent on protecting. Even so, Ryzhikov's confessions had to be retouched.

> . . . I was transferred to investigator GARKUSHA, who created extremely hard conditions for my interrogation. I was transferred to Lefortovo Prison where the interrogations were conducted at night. Lieutenant GARKUSHA threatened to subject me to physical torture, cursed me, shouted that they would flay me, that he would keep me naked in a cell, that I would sing my testimony, that he would kill not only me but my children as well. Being a gravely ill

person . . . and not feeling myself guilty, I quickly fell into depression; that is, I wept a great deal, I did not react to anything that happened, and so on.

During the interrogations, investigator GARKUSHA, as a rule, sat in an arm chair or on a divan, slept two to three hours, but then, assuming some strength, composed the testimony he needed and forced me to sign it.[27]

Investigator Garkusha, to whom a copy of Kitaev's July 10 memorandum was sent, was in charge of Ryzhikov's interrogation from March 22, 1952, to July 28, 1952.[28] It was not at all unusual for the interrogation sessions to last up to seven hours, from, say, ten o'clock in the evening to five o'clock in the morning. One session lasted from 10:20 P.M. to 6:45 the next morning.[29]

One can imagine the small, bare interrogation office, with a picture of a visionary Lenin, arm outstretched, gazing into the future, and another picture of Stalin glaring down paternally from the stained, yellowy walls— the Plato and Aristotle of this modern Renaissance. There would have been a desk, an armchair or divan, a hard wooden chair for the prisoner, who was made to stand or sat bewildered and terrified while his interrogator most often snored in a drunken stupor. Ryzhikov had not been a well man. Painful headaches had twice forced him to undergo the extremely dangerous operation of cranial trepanation; he had also had his gall bladder removed, and suffered from kidney stones, sclerosis, hypertension, and chronic vertigo. His tormentor, Garkusha, was also not healthy. No doubt medicated with vodka, he slept through much of the proceedings. After two or three hours he might rouse himself, write out the confession for Ryzhikov to sign, show Ryzhikov his fist, growl out some crude threat; and Ryzhikov would sign.

Investigator Garkusha illustrates the system in its most elementary form. When asked in 1954 why he had selected Garkusha for this important assignment, Ryumin answered that he knew Garkusha to be "not very bright and insufficiently qualified, but 'zealous' . . . " Ryumin explained, "When we had worked together in the same position, I had had a good personal relationship with him. He believed me and, as I knew, aspired to serve."[30]

Ryumin knew that the slow-witted Garkusha would not bridle at his order to "extract" from Ryzhikov a confession of his "criminal ties with Etinger, because Etinger was a terrorist and Ryzhikov was one of the doctors who treated A. S. Shcherbakov [along with Etinger]."[31] The logic of this was perfectly simple.

But everything did not go simply. Just as Kitaev had discovered improprieties in the "expert" commission's analysis of Shcherbakov's heart, in September 1952 similar improprieties were discovered in the interrogation of Ryzhikov. While Garkusha was on vacation, his temporary replacement, investigator Ivanov, in preparing a memorandum on the case, reported to his supervisor, N. M. Konyakhin, deputy head of the investigative unit working under Ryumin, that he had discovered "facts" of falsification in the protocols of interrogation compiled by Garkusha.[32] Ivanov told Konyakhin that he had found these facts in Garkusha's working notes and drafts of protocols in which he had composed false testimony that Ryzhikov was forced to sign. Konyakhin called Ryzhikov to his office, interrogated him, and found this to be true.

Konyakhin immediately brought the situation to Ryumin's attention. But Ryumin refused to look into the charge because, as he put it, he "didn't see anything unclear in it. Similar situations in investigative work were not uncommon."[33] Konyakhin explained that Garkusha had falsified a confession of murder by not reporting that Ryzhikov either had not said what was recorded or had retracted testimony clearly obtained through intense pressure. It made no difference.

Ryumin patiently tried to explain. If Garkusha were punished in this matter, other investigators would be adversely affected. It would have a chilling effect. They might well lose their "sharpness," as he put it.[34] Their effectiveness would be undermined. In the end, when Konyakhin pressed the matter, Ryumin told him flatly: "We won't do this. We have no such confession," referring to Ryzhikov's retraction.[35]

In June 1953, during his own interrogation, Ryumin was accused of having "stopped the exposure of the falsification of the confessions of the prisoner [Ryzhikov]." But he fiercely denied this.

"Judging by what KONYAKHIN has testified," Ryumin said, "I gave him a criminal order. In fact, nothing of the sort happened. KONYAKHIN came to me and said that it had become known to him what GARKUSHA'S behavior had been and there and then gave me the report by GARKUSHA himself which requested that he be released from investigative work on the grounds of his health."[36] Garkusha subsequently was released from the MGB, but not as Konyakhin had wished, punished "according to party procedure."[37] That is, he was released from service but never actually punished.

Garkusha was nothing more than Ryumin's tool in building the plot against the doctors, just as Ryumin was Stalin's tool. He did what he was told and suffered no great fate. Ryumin told him to familiarize himself with the "confessions" of Karpai and with the testimony of Likhachev in advance of interrogating Dr. Ryzhikov, despite the fact, as Ryumin eventually admitted, that "there was no evidence of the association of RYZHIKOV to terrorist work or of his criminal ties in this direction with Professor ETINGER."

Garkusha was instructed to "interrogate RYZHIKOV about his possible terroristic activity with ETINGER."[38] He did so. Subsequently, Ryumin admitted that in giving Garkusha these instructions he did not also tell him that Etinger had testified that Shcherbakov's illness was untreatable and that he, Ryumin, had his "own doubts about the truth of ETINGER's confessions."[39] It hardly mattered. Ryumin knew that Garkusha would "not be particularly perplexed at [his] orders," no matter what they might be.[40]

Just as Stalin never revealed the whole of his intention to any one of his subordinates, Ryumin did not reveal to his chief interrogator a crucial element in the case. The system continually replicated itself. Garkusha's job was not merely to torture Ryzhikov, but to invent a confession. This "not very bright and insufficiently qualified, but 'zealous' " Garkusha, bored and perpetually half asleep, became an "engineer of human souls," like the Soviet poets and novelists admonished by Zhdanov in 1934, in this case, Dr. Ryzhikov's.

Dr. Ryzhikov had treated Shcherbakov in 1945 and in that capacity had worked with Etinger, Vinogradov, and several other doctors, Jewish and non-Jewish alike, in the days before Shcherbakov's death. Their names, particularly the Jewish ones, would shortly be attached to the investigation: Busalov (Russian), Lang (Jewish), and Kadzharduzov (Armenian).

Since Ryzhikov himself wasn't Jewish, even Garkusha realized that it was impossible for him to confess credibly that he took part in a Jewish nationalist conspiracy, and the most his interrogators could wring out of him or invent was that his superiors had committed crimes to which he was a witting accessory. Ryzhikov languished in Lefortovo throughout the spring and summer of 1952.

When he was asked in October 1952, "For what purpose were you cooking up this traitorous affair?" Ryzhikov could provide no concrete answer, quickly deflecting all responsibility for these larger concerns from himself to the "council" consisting of Etinger, Vinogradov, and Lang. Despite the disheartening fact that Ryzhikov was not a Jewish nationalist conspiring to kill a Kremlin leader, Ignatiev seized upon the admissions in the faked protocols provided to him by Ryumin as proof that Etinger had these intentions. And if Etinger did, then so did the others. Once Etinger was definitely incriminated, the case could progress to the next level.

This it rapidly began to do. If the MGB had acted slowly and carefully in the initial stages of the plot, it now acted with considerable speed and force. Once Ryzhikov confessed in July to having abetted the killers of Shcherbakov, three things happened in relatively quick succession. On August 11 Timashuk was formally interrogated for the first time by a senior MGB investigator. On August 12 the members of the Jewish Antifascist Committee were shot. The third development was that sometime in early August investigator Inspector Yeliseyev, who was interrogating Timashuk, suggested to Ryumin that Zhdanov's heart be examined by the same kind of "expert" commission as had examined Shcherbakov's. Overriding Ignatiev's concerns, Ryumin approved the initiative, and by the end of August took full credit for it. As a reward for his faithful service Stalin admonished Ignatiev that he must "take Ryumin closer to himself."

On one Sunday (in the evening), at the end of August 1952, com. STALIN summoned me to *Blizhnyaya* [Stalin's dacha close to Moscow] and after a very sharp conversation about the fact that the Chekists had forgotten how to work, had grown fat, had become dissolute and had forgotten the traditions of the ChK times of DZERZHINSKY,[41] that they had torn themselves from the party, that they wanted to put themselves above the party, he grasped in his hand the memorandum that was the result of the expert medical examination of the preparation of com. ZHDANOV's heart, and he asked who showed this initiative. And to my answer that this was done by RYUMIN with his workers, com. STALIN said: "I have always said that RYUMIN is an honorable man and communist, he helps the CC [Central Committee] uncover serious crimes in the MGB, but he, poor fellow, has not found support among you and this is because I appointed him despite your objections. RYUMIN is excellent, and I demand that you listen to him and take him closer to yourself. Keep in mind—I don't trust the old workers in the MGB very much." This time I was not able to object.[42]

Many of the "old workers" were by now in jail. The rest of the fat, dissolute "grandees" of the MGB would soon be there if they didn't do what Stalin wanted.

Did Yeliseyev conceive the idea of an examination of Zhdanov's heart? Or did Ryumin suggest it to him, as Ignatiev had suggested that Ryumin write a letter in October 1951 denouncing his colleagues in the MGB? We will never know for sure, but we do know that the point at which Ryumin took control of the investigation of Zhdanov's death is the point at which both the Shcherbakov case and the Zhdanov case fused into a single case.

On August 11, 1952, Yeliseyev interrogated Timashuk and formalized her version of Zhdanov's death in a lengthy protocol. Two months later, on October 17, Yeliseyev, Ryumin, and one other investigator interviewed Timashuk for a second and final time. Between August 11 and October 17 the case had assumed its final form. The way A. A. Kuznetsov appeared in the October 17 protocol, having been absent in the August 11 protocol, demonstrates what had happened. Earlier in the summer, in the draft report on the "Abakumov-Shvartsman case," Kuznetsov became associated with Abakumov's alleged attempted usurpation of the Soviet government at the direction of American or British intelligence. "Abakumov," it alleged, "made use of the help of the former Secretary of the Central Committee, Kuznetsov . . . [to] safeguard the placing of cadres advantageous to himself in the Ministry and positions in the Chekist organs."[43]

But now, in Timashuk's October 17 interrogation, Kuznetsov became for the first time implicated in Zhdanov's murder as well, thus connecting the murder of Zhdanov with Abakumov's plot, the traitorous doctors with the "unsatisfactory condition in the MGB." Furthermore, by connecting Kuznetsov with Zhdanov's death, the MGB created a truly Shakespearean conspiracy in which the Brutus-like Kuznetsov turned on his former friend, mentor, and leader, Zhdanov, as if to foreshadow the way Stalin would accuse Molotov and Mikoyan of turning on him.

At the October 18 interrogation Timashuk revealed that she had written to Kuznetsov on September 15, 1948, about Zhdanov's medical treatment but had received no reply. She said that she had called the secretariat but could not get through; Kuznetsov did not return her call, although his assistant told Timashuk that he had indeed received her letter. In early 1949 she wrote once more to Kuznetsov. Again she received no reply. The implication was obvious: Kuznetsov took no interest in saving Zhdanov's life.

Superficially, this might seem to have been the case, but Kuznetsov's silence had other causes. On the one hand, party discipline prevented him. Formally he was a party supervisor of MGB but only formally. As an experienced member of the party apparatus, aware that Stalin personally guided the matter of Timashuk's letter, he could not handle this without Stalin's approval. After Zhdanov's death, when Malenkov took over Zhdanov's position as the number two figure in the Soviet government, Kuznetsov couldn't intervene without Malenkov's permission. He might well have shared Timashuk's letter with Malenkov, who would thereby have been brought into the Timashuk matter early on.

A further explanation may be that by September 1948 Kuznetsov knew he was doomed and there was nothing he could do to save either Zhdanov or himself. Having been sent by Stalin to Valdai upon Zhdanov's death on August 31, he met Voznesensky, who was already present at Zhdanov's bedside—as if in anticipation of the event. He, too, would have read Yuri Zhdanov's letter in *Pravda* and would have drawn the obvious conclusion.

Whatever the case, Timashuk's testimony put a powerful new weapon in the hands of the MGB: Kuznetsov allegedly was behind Zhdanov's death. His silence *proved* it. It is important to see that this element developed only during the interview with Ryumin in October, and as we will soon see it became an explicit charge in the final version of the Doctors' Plot presented to Stalin on November 24, 1952, by S. Goglidze.

At the same time the confession of the arrested YEGOROV showed him to be the direct organizer of ZHDANOV'S death; that he acted at the instruction of the enemy of the people, A. A. KUZNETSOV,

who in connection with his own hostile plans was interested in the removal of comrade ZHDANOV.

Behind KUZNETSOV were the Americans and the English.[44]

Showing Ignatiev the report on Zhdanov's heart, Stalin showed he *did not need Ignatiev to get things done.* Little wonder that on September 25, 1952, a day after Goglidze's report to Stalin, Ignatiev wrote Stalin that

> . . . the testimony of RYZHIKOV about the incorrect methods of treatment of comrade SHCHERBAKOV, A. S., was confirmed in the conclusions of the medical experts. . . . Therefore, the material now collected on the case gives us reason to think that the attending doctors of comrade SHCHERBAKOV, A. S., despite having the clinical facts, established an imprecise diagnosis of his illness and correspondingly prescribed an incorrect regimen and treatment for the patient, the fulfillment of which in the last analysis led to the premature death of comrade SHCHERBAKOV, A. S.[45]

Ignatiev concluded that "the RYZHIKOV case . . . is being conducted with the intention of exposing the doctors who conducted enemy work against the leadership of the party and the government."[46] In other words, as Ryumin said to Maklyarsky, "the question of whether you are guilty is decided by the fact of your arrest."

The investigation was not simply intended to frame innocent people while letting the guilty go free. The investigation of the so-called Doctors' Plot was not an investigation at all. It did not begin with evidence and end in a theory, however distorted by passion, prejudice, or ideology. It began with a political goal and ended with the concoction of "evidence" with which to achieve it.

Despite his July confession, Ryzhikov was interrogated relentlessly throughout July and August 1952. In order that he provide, as Ignatiev said to Stalin, further proof against "the doctors who conducted enemy work against the leadership of the party and the government," he was subjected to twenty-four harsh sessions conducted during this period, often lasting up to six hours. The interrogations steadily intensified as time went on and, with occasional interruptions, continued up to the day of Stalin's death on

March 5, 1953, when he was interrogated twice. On the day Stalin died, Ryzhikov was interrogated first for three hours and forty-five minutes and then later for two hours and twenty-five minutes.[47]

September and October 1952 were very busy months for the Ministry of State Security. In addition to the covert examination of Zhdanov's heart, and the continuing interrogations of Ryzhikov and Abakumov, Yegorov's predecessor as head of the Lechsanupra of the Kremlin, A. A. Busalov, was arrested on October 1. Timashuk was interrogated again on October 17. On October 18 Yegorov was rearrested after having been arrested once already but then dispatched to a hospital for treatment because he suffered a heart attack of his own. Zhdanov's bodyguard, Belov, was interrogated for the first time on October 18 as well. The tide of arrests grew and in a matter of months swept the country. The Shcherbakov and Zhdanov cases were conjoined; Fefer had recruited Etinger; Etinger had recruited Vinogradov; Vinogradov had recruited Yegorov; Kuznetsov had sold out to the Americans and had abetted Abakumov, who was preparing for a seizure of supreme power. Within a month all the principals would be rounded up.

The Jewish issue was now fully merged with Abakumov's alleged betrayal and his attempted coup d'etat. But the new element of Kuznetsov's treachery had been added. Soon Molotov and Mikoyan would be added. The network of conspiracy and betrayal now extended from Jews, doctors, and the MGB to the highest reaches of the Soviet government itself.

Yegorov's interrogator, Levshin, put the matter succinctly during one of his most brutal interrogations of the former head of the Kremlin Hospital. Levshin shouted at Yegorov to confess that Kuznetsov had recruited him to murder Zhdanov: "hating Soviet power, you sold out to KUZNETSOV for a mess of pottage."[48] Like Esau, Yegorov had sold his birthright. Kuznetsov now stood for Jacob, the clever usurper, who defrauded his older brother. Jews, usurpation, and political assassination merged in a single image. Yegorov told Levshin that he didn't know what he was talking about, but his protests meant nothing.

Yegorov's arrest followed the "analysis" of Zhdanov's heart in mid-September, not Timashuk's denunciation in August 1952: a significant, if minor detail of timing. Timashuk was first formally interrogated on August

11, 1952. At least one month elapsed between Timashuk's interrogation and
Yegorov's arrest. At her August 11 interview, Timashuk stated again what
she had revealed to her MGB handler Suranov in 1948 and again in 1951—
that Yegorov and the other doctors had incorrectly diagnosed Zhdanov's illness, prescribed treatment contraindicated by the EKG she had taken,
allowed negligent nursing care, and demonstrated criminal indifference to
their patient.[49] Abakumov was purged two days after Ryumin's denunciation in July 1950, but now Stalin waited, as he had after the first Central
Committee meeting on the subject of Etinger's arrest. The timing of
Yegorov's arrest did not, however, simply reflect Stalin's belief that
Timashuk's testimony was too insubstantial for the wide-scale, public operation he was now contemplating. Stalin waited because each "expert analysis" and confession aided Stalin in denying his role in conceiving and directing the operation. He also wanted the initial arrests to be as concealed from
public view as possible.

From Ignatiev we learn how the arrest of Yegorov and other doctors
was handled.

> [Stalin] gave the order to replace YEGOROV, to send him to the
> provinces, and on the way or upon his arrival at his destination to
> arrest him and put him in manacles. Here it was proposed to create
> a commission in the CC [Central Committee] for the verification of
> the work of the Lechsanupra of the Kremlin, to replace professor
> VASILENKO, who was then in China with com. TOKUDA, and
> upon his arrival on the territory of the USSR to arrest him, put him
> in manacles, and place him on an airplane for Moscow. On this
> evening I received the order for the arrest of doctor MAIOROV
> and the wife of YEGOROV.[50]

Yegorov would be sent on a journey and arrested "*at his destination*," not in
Moscow, where it might cause undue alarm. Vasilenko would be put in
manacles the second he returned to Soviet territory from China, far from
Moscow. Maiorov and Yegorov's wife, minor characters, would be arrested
as well to seal one part of the case shut. For the time being, the most eminent of the doctors, Vinogradov, would be left untouched.

Having suffered a minor stroke as well as a heart attack after he was apprehended, Yegorov had to be admitted to an MGB hospital, where he remained until October 18, when he could be brought back to Lubyanka and formally rearrested. As for Dr. Vasilenko, Ignatiev was forced to admit that when they went to arrest him, the MGB agents mistakenly arrested the wrong man![51] Stalin was furious. He called Ignatiev and his MGB a pack of "hippopotamuses, people incapable of quickly and conscientiously fulfilling the orders of the [Central Committee]."[52]

Of the four principals who had treated Zhdanov—Yegorov, Vinogradov, Vasilenko, and Maiorov—Vinogradov was the last to be arrested, on November 4. Other prominent doctors quickly followed: Miron Vovsi and Borukh B. Kogan were arrested on November 12 for the treatment of Dimitrov, who died in 1949 in a Kremlin hospital. Eventually the MGB would extract damning confessions from Yegorov, Vinogradov, Vasilenko, and Maiorov that they intentionally made use "of medical wrecking to lead comrade Andreyev, A. A., astray, to shorten the life of comrade Dimitrov, G. M.; and did much harm to the health of comrades Thorez and Tokuda."[53]

Resolutions and arrests followed in quick succession. Trusted heroes such as Kuznetsov were soon exposed as seditious monsters; trusted physicians like Vinogradov and the "wise Meyer," as Miron Vovsi was called, would be revealed to be murderous, plotting enemies of mankind. Vovsi's arrest, in particular, showed another direction the case might begin to take. A cousin of the great actor and head of the Jewish Antifascist Committee, Solomon Mikhoels, murdered in 1948, Vovsi had been the chief therapist of the Red Army. His confessions would soon lead the investigation to a consideration of the military, thus potentially widening the net of terror and purges. Because of his connection to Mikhoels, who had gone to America on his famous trip in 1943, Vovsi was particularly useful in demonstrating that American intelligence directed the actions of the entire conspiracy. On November 25, 1952, Vovsi told his interrogators that he wanted to make a full confession and "set out everything in order."

> Yes, we had a terrorist group. In this group were doctors including a number of Jewish nationalists—KOGAN, BORUKH BORISO-VICH, and TEMKIN, YAKOV SOLOMONOVICH. . . .
>
> I, VOVSI, was the inspiration of the group. The terrorist group from the group of Kremlin doctors TEMKIN, YA. S., KOGAN, B.B.,

and myself was nationalistic, formed as an enemy to Soviet power, from the first; that is, before the participants had decided to use in their struggle against Soviet power these extreme means such as terror by means of criminal methods of treatment with the aim of destroying the health of leading workers. . . .

I earlier confessed that SHIMELIOVICH gave the first directions, approximately in 1948. From that time, without question, we set out on subversive work, intending to shorten the lives of specific leading workers.

The interrogators wanted to know for whom Shimeliovich worked.

I am convinced that all the nationalistic work of the ringleader of the Jewish Antifascist Committee, hiding under this name its subversive work against the Soviet state, was directed by Anglo-American imperialistic circles.

But this was not enough. For whom did Vovsi himself work?

The Anglo-American bourgeoisie, which was actually our boss. Such is the logic.[54]

Such was the logic in Stalin's world.

On December 8, the investigators wanted to know more about Vovsi's relationship to Mikhoels.

MIKHOELS—my cousin, was by his convictions a Jewish bourgeois nationalist. MIKHOELS seriously attempted to prove that world culture without Jews looked dim. . . . MIKHOELS' nationalism achieved especial strength after his visit to America in 1943. . . .

Having gotten the false idea that he himself was the leader of the Jewish nation, MIKHOELS set forth, as the first part of the plan, the fact that as a result of his visit to the U.S. he succeeded in making ties with leading Jews of America who promised to show every sort of support to the Jews continuing to live in the Soviet Union.

Painting the American way of life in joyful colors, in every way glorifying his mission to that country, MIKHOELS said smugly with great pompousness that when they had met him in the U.S., "Balconies with Jews fell to the ground. Jews who wanted to see us and hear our voice"—said MIKHOELS. . . .

Vovsi mentioned that Mikhoels often showed people, "with especial warmth," the photograph of himself and the American pro-Soviet journalist B. Z. Goldberg posing at the grave of Sholem Aleichem. Goldberg was married to Sholem Aleichem's daughter.

"What kind of instructions did MIKHOELS bring from the U.S. for the deployment of enemy work in the Soviet Union?" the interrogators asked.

The situation was thus, MIKHOELS said, that for us, Jews, it was necessary at all costs to maintain key positions in science, art, literature and in pedagogical institutes and only then, he underlined, would we be able to unite the uncoordinated strengths and preserve the unity of the Jewish population.

MIKHOELS instructed that in every way we had to extol Jews occupying important posts in science, art, literature and education, to strengthen old positions and expand them by means of insinuating into various other spheres Jews from among young people. . . .

In this way, it was then already clear to me that the name "Jewish Antifascist Committee" was only a smoke screen under which Jewish nationalists realized their anti-Soviet, nationalistic goals; and this fact that the Committee directed to the U.S. various kinds of information about the Soviet Union, directly shows that he essentially served the interests of Zionist circles in the U.S.

Did Mikhoels give specific terrorist instructions?

I did not receive detailed orders from MIKHOELS. I already showed the investigation what the direct arrangements for terror were that SHIMELIOVICH gave me soon after the death of MIKHOELS.

I must remark that SHIMELIOVICH kept me abreast of the

work of the Committee and in several conversations shared with me that the ruling bosses of the Jewish Antifascist Committee, of which he himself was a part, representing the sphere of medicine, supported the ties with the U.S.

He, like MIKHOELS, declared that the interests of Jews, living in the Soviet Union and the USA, coincided and therefore we had to work together.

Informing me of the activity of the Jewish Antifascist Committee, SHIMELIOVICH often said that Jewish nationalists in the Soviet Union had to orient themselves toward America—toward a united strength that will save the Jewish nation.

SHIMELIOVICH gave me clearly to understand that the criminal work of specific leaders of the Jewish Antifascist Committee was directed by imperialist circles in the U.S.

The best proof of this declaration might be the information that SHIMELIOVICH collected about the Soviet Union. . . . I can't deny the fact that I provided SHIMELIOVICH with espionage information.

The interrogators wanted Vovsi to elaborate and began to aim toward the military.

For myself, naturally, the most important information for SHIMELIOVICH was information about the activity of the Kremlin Polyclinic in which he appeared particularly interested.

Beginning in 1946, SHIMELIOVICH . . . asked me in detail about the character of the illnesses . . . of the patients Marshal Tolbuzhin, Marshal Konev, concerning the nature of the treatment of the patient DIMITROV, G. M., and about the course of his treatment, and in addition about the course of the illness of GRANATKIN. . . .

In addition, I informed SHIMELIOVICH about the situation of medical affairs in the polyclinic of the Kremlin hospital.

I informed about applied methods of treatment, about new therapies, and about the techniques for treating leading Soviet and party

workers, about the character of the ties that exist between the polyclinic and the hospital. I enumerated the inadequacies of the consultant work being conducted. . . .

Simultaneously, I informed SHIMELIOVICH in detail about the
personal qualities, experience and degree of preparedness of the
doctors of the Lechsanupra of the Kremlin who had some connection to the treatment of leading Soviet and party workers; . . . about
the use of new medical means, especially antibiotics, that at this time
had begun to be widely employed.

I also informed SHIMELIOVICH about several questions of treatment in the Soviet Army, in so far as I was there for a long time as chief
therapist. Thus, SHIMELIOVICH showed the greatest interest in the
questions of methods of treatment used in the military-treatment institutions, the structure and organizations of the military-medical services during the Patriotic War.

He knew in addition from my words about the level of preparedness of the doctors conducting therapy during the post-war
period in the provinces. I informed SHIMELIOVICH also about
other questions interesting to him.[55]

Vovsi's confessions about terror produced little more than mundane information such as the nature of the treatments used in the Red Army. It hardly
mattered. Like those of Etinger, Ryzhikov, and the other doctors, his statements, compelled by threats, physical torture, sleep deprivation, and sheer
mental and physical exhaustion, could eventually be shaped to immediate
political needs. Much of what he alleged was absurd, such as that he, Vovsi,
was the inspiration for the terrorist group. Nevertheless, this "confession"
was necessary for the investigation because the Jewish nature of the conspiracy had to be beyond question.

Behind Vovsi's confessions lay a specific policy decision by Stalin.
According to Ignatiev, Stalin demanded that decisive measures be taken to
expose the group of doctor-terrorists, of whose existence, as he said, he
had "long been convinced." Stalin *knew* that the doctors were guilty and
became apoplectic when the investigation did not expose their guilt
quickly enough. According to Goglidze, who took over the investigation
of the doctors in November 1952 after Ryumin's expulsion:

Almost every day, Com. STALIN showed interest in the course of the investigation concerning the case of the doctors and the case of ABAKUMOV-SHVARTSMAN, speaking with me on the telephone and sometimes calling me to his office. Com. STALIN as a rule spoke with great anger, continually expressing dissatisfaction with the course of the investigation. He cursed, threatened and, as a rule, demanded that the prisoners be beaten: "Beat them, beat them, beat them with death blows."[56]

In November and December 1952 Stalin, though every bit as brutal and powerful as he had been in the 1930s, was perhaps less confident. V. A. Malyshev detected a sign of this in 1947 when, on a visit to Stalin, he heard the Vozhd remark in connection with a Politburo decision, "If I'm still alive after a year, I will raise the question." The phrase stuck Malyshev, who had never heard Stalin speak in that manner before. He did not know how much longer he would live and all around him he sensed incompetence and bureaucratic delay.[57] Stalin was not simply paranoid. He did not see enemies everywhere—he *invented* them because he needed them. Stalin knew the doctors were guilty, but at the same time he knew that they were not guilty. The contradiction did not matter. What mattered was the supervening political imperative.

Stalin changed his "truths" continually in relation to political and social circumstances. In 1943 the Jews were useful to him, and he sent Mikhoels and others to America to raise money and goodwill for the Soviet war effort; in 1947, he supported the establishment of the state of Israel and allowed, if unwillingly, his daughter to marry Grigory Morozov, a Jew. By 1951 the Jews had become a nation of spies, and by 1952 those he had sent to America in 1943 were either assassinated or faced the firing squad.

Stalin is Godot, absent from an empty landscape. We wait, we guess, we attribute motives, we receive incomprehensible communications, but in the end he will not reveal himself, and there is no direct way toward understanding him as a "person." When asked whether Stalin ever appeared in his dreams, Molotov answered, "Sometimes. In extraordinary situations. . . . In a destroyed city . . . I can't find a way out, and I meet him. In a word, very strange, confusing dreams."[58] It is in a destroyed city of man from which no one can find an exit that Stalin appears.

On October 29, 1952, ignoring the Kitaev report, Ignatiev told Stalin that the medical commissions devoted to the question of Shcherbakov's death produced conclusions that "fundamentally coincide . . . and confirm the criminal treatment of comrade SHCHERBAKOV."[59] Ignatiev also told Stalin that doctors Busalov, the former head of the Kremlin Hospital system, and Ryzhikov continued "to hide from the Government the actual causes of the death of comrade SHCHERBAKOV, A. S."[60]

As could be expected, Stalin blew up, unable to contain his impatience and rage at what he took to be further evidence of obstructionism and equivocation. The result was that Stalin summoned Ignatiev, Goglidze, Ryumin, and other MGB officers to the Kremlin on November 2 and gave the direct order to obtain the confessions by any means necessary, including physical torture.

Comrade Stalin "declared to all of us," Goglidze wrote to Beria in March 1953, "the extremely serious nature of the claims in the investigation of the doctors of the Lechsanupra of the Kremlin." But Stalin said more than this. He told them that "the investigators worked without spirit, that they clumsily used contradictions and slips of the tongue of the prisoners in exposing them; they clumsily posed questions, not seizing like hooks on every possibility, even the tiniest, in order to clutch, in order to grasp the prisoners in their hands, etc., etc."[61] Stalin ordered them to beat the prisoners with "death blows." He told them to put prisoners in manacles and chains. Though never written down, this order has been confirmed from many sources. Goglidze said that he had "personally heard [it] from the lips of com. STALIN."[62]

Ignatiev gave a slightly different account:

Beginning with the end of October 1952, com. STALIN more and more often in categorical form demanded from me, com. GOGLIDZE and the investigators to apply physical torture to the arrested doctors who had not acknowledged their enemy activity: "Beat them!"—he demanded from us, declaring: "what are you? Do you want to be more humanistic than LENIN, who ordered DZERZHINSKY to throw SAVINKOV out a window?" DZERZHINSKY had for this job special people—Letts who fulfilled this commission. DZERZHINSKY was no match for you, but he didn't shirk the dirty work. You work like waiters in white gloves. If you want to be Chekists, take off your gloves. Chekist work—this is for peasants and not for barons."[63]

Ignatiev, however, did not immediately comply. Though the direct order was given on November 2, it was not fulfilled until November 12. Ignatiev, it appears, held back. We will never know fully why. More bureaucratic obstructionism from those "old workers of the MGB" and elsewhere whom Stalin no longer trusted? It is doubtful that this delay represented a principled disagreement with the Vozhd.

Though Ryumin claimed that the order was not fulfilled because the MGB feared for the life of the doctors, it is also possible that the delay was caused by some combination of Ignatiev's own mental and moral condition at the time and the growing alarm felt by other members of the Central Committee over the accelerating course events were now taking. Khrushchev hinted at this in his memoirs, where he wrote:

In those days anything could have happened to any one of us. Everything depended on what Stalin happened to be thinking when he glanced in your direction. Sometimes he would glare at you and say, "Why don't you look me in the eye today? Why are you averting your eyes from mine?" Or some other such stupidity. Without warning he would turn on you with real viciousness. . . . Bulganin once described very well the experience we all had to live with in those days. We were leaving Stalin's after dinner one night and he said, "You come to Stalin's table as a friend, but you never know if you'll go home by yourself or if you'll be given a ride—to prison!"[64]

This air of crisis and instability intensified as the Doctors' Plot got fully under way. In Khrushchev's words:

The interrogations began. I heard myself how Stalin talked to S. D. Ignatiev. . . . Stalin used to berate him viciously over the phone in our presence. Stalin was crazy with rage, yelling at Ignatiev and threatening him, demanding that he throw the doctors in chains, beat them to a pulp, and grind them into powder. It was no surprise when almost all the doctors confessed to their crimes . . .

That's how the so-called Doctors' Plot arose. It was a shameful business. . . . The doctors' case was a cruel and contemptible thing.[65]

While Stalin may have been "crazy with rage," his political objectives were highly logical. Though the Doctors' Plot was a "cruel and contemptible thing," it served a well-thought-out plan. Not until Beria's, Malenkov's, and Khrushchev's roles in the intricate political maneuverings surrounding the Doctors' Plot are disclosed will we know the extent to which members of the Central Committee were prepared to sacrifice this plan because of the degree of Stalin's irrationality.

Throughout the fall of 1952 Ignatiev suffered from extreme physical and mental exhaustion brought about by Stalin's daily threats. When Goglidze inquired about his state of mind, Ignatiev would refer simply to the "slow development of the case" against the doctors. He told Goglidze that he received warnings every day from Stalin, but there was little to be done. Even worse, the case against Abakumov and Shvartsman was also terribly delayed. The "prisoners had not given the necessary testimony," Ignatiev informed Goglidze. Stalin was very critical.[66]

On November 14, 1952, two days after Stalin's order was implemented, Ignatiev collapsed. According to Khrushchev, he suffered a "near-fatal heart attack." Khrushchev recalled that Ignatiev, a bureaucrat at heart, not a security chief, "was mild, considerate, and well liked. We all knew what sort of physical condition he was in." [67] Ignatiev, who had good reason to fear for his life, did not return to active work until January 1953. He had seen what happened to Abakumov. In the past when the state demanded confessions, the state got them, usually by beating the prisoners to a pulp, as Khrushchev put it. Ignatiev, who supervised this process beginning in 1951, did not shrink from the task.[68] But now, in 1952, the strain caused him to break down. Ignatiev described the situation to Beria.

After his surrender in the middle of September, YEGOROV became ill and was transferred to the hospital where he stayed until the first half of October. In connection with this, on several occasions, *com.*

*STALIN accused us, me in particular, of minimizing YEGOROV's respon-
sibility, saying that YEGOROV's illness was an invention of the MGB.
More than once, I was told that I would pay with my head for protecting
YEGOROV.* [Emphasis added.][69]

Abakumov had alledgedly protected Etinger, and now Stalin was
preparing the way for the accusation that Ignatiev was protecting Yegorov.
The wheel would turn yet again. Beyond this, Stalin was preparing to punish the entire MGB. Though Ignatiev appeared to have realized this, he
could find no means to oppose it. Rather, his comments to Beria reflect a
state of desperate malaise, something far different from Yezhov's celerity in
fulfilling Stalin's orders in 1937. Ignatiev's malaise and exhaustion, however, did not prevent him from slavish obedience. No organized opposition
to Stalin appears to have existed within the MGB or elsewhere. His power
remained real, as the beatings of the doctors and as the fantastic developments related in the next chapter make clear. What Stalin took to be opposition may have been nothing more than a bureaucracy filled with
Garkushas on the one hand and venomously divisive figures like Ryumin
on the other. The degree of his rage may have had as much to do with his
knowledge of his own approaching death as with real conditions in his
government. Nevertheless, the confusion and depression into which
Ignatiev had sunk signaled another decisive, if small, shift in sensibility
within the upper stratum of Kremlin leadership that would eventually
make Khrushchev's secret speech possible three years later.

Despite Ignatiev's hesitation, the beating and torture of the prisoners
commenced on November 12, 1952. According to Ryumin, Ignatiev
received an order from Malenkov in the Central Committee to apply physical torture to the prisoners immediately. If these dates are correct, then
there was a delay of approximately two weeks from Stalin's order to apply
physical torture to its implementation. Ryumin testified:

The Minister called me on 12 November. He spoke to me in an agitated way about a conversation over the telephone, which he had
with comrade MALENKOV who gave him the order to beat the doctors with death blows. Soon I went down to the prison to [investiga-
tor] MIRONOV where they beat VASILENKO, VINOGRADOV (I

don't remember—perhaps not), and they frightened YEGOROV in
MIRONOV'S office. [70]

Ryumin claimed that he warned all the prisoners that if they gave false
confessions, they would be beaten to death. Even before morning on
November 13, Yegorov and Vasilenko were ready to confess their guilt.

But November 13 held another surprise, this time for Ryumin. In the
middle of the interrogation-beating of Vasilenko, Ignatiev called Ryumin
to his office to show him the written decision of the Central Committee
that he was to be removed from his position in the MGB on the grounds
that he was "unequal to the task."[71] Very little else is known about the
decision to remove Ryumin at this crucial moment in the investigation.
The fact that the decision came from the Central Committee signifies
Stalin's consent. Only two months before, Stalin had instructed Ignatiev to
"listen to [Ryumin] and take him closer to" himself. He had been deputy
minister of state security of the USSR for one year.

Ryumin was removed ostensibly because he turned out to be ineffec-
tive. Stalin knew he had botched Etinger's investigation and was little more
than a small-minded, utterly venal man who knew too much. On
November 12 Ryumin had sent Stalin a report that Vasilenko had been a
Trotskyist in 1925, but Stalin didn't want revelations from 1925; he wanted
revelations about Vasilenko's espionage in 1948. The *shibsdik* could not
deliver.

There may have been other reasons as well. Ryumin's colleagues in the
MGB hated him openly, in part because he boasted about his relationship
to Stalin and spoke freely about the importance of his assignments.[72]
Goglidze alluded to this when he told Beria that Ryumin "boasted about
the complexity of the assignment given to the Investigative Unit and the
application of a series of operational combinations . . . "[73] Ryumin could
not be trusted not to boast too much. Ryumin, the "honorable man and
communist" who helped "the Central Committee uncover serious crimes
in the MGB," was also expendable. He had nearly blurted something out
about Molotov to Maklyarsky at one of their sessions. What else had he
said indiscreetly that may have gotten back to Stalin? He, too, had become
someone Stalin could not trust.[74] Everyone was expendable.

Ryumin made one last attempt at subverting his enemies and restoring his place in the MGB. In a desperate, undated, almost incoherent note, but certainly sent on November 13 after he learned of his dismissal, Ryumin summarized for Stalin his accomplishments in obtaining confessions from Vasilenko and Shvartsman, and in pushing the case of the doctors toward its conclusion. It was a most unusual communication, based on a previously drafted directive from Ignatiev in early November to provide Stalin with information on the case. The document had been prepared and sent to Ignatiev, and according to Ryumin, it was returned as unsatisfactory. Nevertheless, on November 13 Ryumin apparently reworked the "draft" summary and sent it to Stalin, bypassing Ignatiev. During his interrogation he was asked why he took this very irregular step.

> I had been charged with uncovering the core of subversive and espionage ties of individuals who had been accused of the indicated crimes. To do this was impossible because the charges put forward by us, as I have already confessed, were based on falsified investigative materials. In this way I became bankrupt and was removed from my work as incompetent, and I became a victim of my own falsifications.
>
> *At this time, knowing about the application of physical torture to the prisoners, I understood, that future, invented confessions about espionage and terror might be obtained from them, and I feared that IGNATIEV would do to me what I had in my time done to ABAKUMOV, that is, accuse me of intentionally minimizing the case.*
>
> Thinking to protect myself from the possible intrigues of IGNATIEV, I wrote I. V. STALIN a letter, in which together with the acknowledgment of my own guilt in the unsatisfactory results of the investigation, pointed to my "service" in the unmasking of the "doctor-terrorists" and the "sabotage" in the organs of the MGB. [Emphasis added.][75]

Now Ryumin feared that he, too, would be judged, caught in the same web of lies he had spun for Abakumov. And so he would, but not until after Stalin himself was dead.

Meanwhile Lieutenant Sokolov replaced Ryumin as the acting head of the Investigative Unit for Especially Important Cases. Following what was apparently another caustic rebuke by Stalin at a meeting of the Central Committee, on November 13, shortly after Ryumin had been dismissed, Ignatiev informed the MGB interrogators that the government had "deemed the investigation thus far unsatisfactory," and that they had to "take off their white gloves," and "observing caution," make use of torture. Ignatiev let everyone know that this order came "from above."[76] The next day Ignatiev had a heart attack requiring hospitalization.

A special torture chamber was set up in Lefortovo prison, fitted out with special screens to deaden shrieks of pain; and a special detail of "helpers" was recruited from the Directorate of the Guards, because Ignatiev did not want MGB investigators to take part in the actual physical torture for fear of demoralizing them. "I did not think it was possible not to fulfill the demands of com. STALIN," Ignatiev told Beria. He ordered that two workers of the prison guard be selected for implementing the physical torture. Ignatiev claimed that he was never "even once informed . . . that anyone among the investigators took part personally in the application of physical torture to the prisoners or violated the decreed regime for the prisoners. I received no complaints from the arrested doctors. On my own initiative, I did not allow the application of these measures to a single prisoner. This was done only after the very severe demands of com. STALIN."[77]

The MGB was careful to keep its hands clean. But Ignatiev's letter to Beria of March 27, 1953, is not surprisingly self-serving and untrue. Although much of it is confirmed from other sources, we know that the doctors suffered actual physical torture, in the form of sleep deprivation and manacles, as well as immense psychological torture in order to compel their confessions, even *before* November 13, 1952.

Two weeks after his formal arrest in October Yegorov was brought to see Ignatiev in the presence of Ryumin, Sokolov, and senior investigator Levshin. "Yegorov," Ignatiev said, "your crimes have been factually proven. Your accomplices in everything have acknowledged their crimes. Your silence and stubbornness is incomprehensible. What are you calculating here? You weren't simply arrested and put in prison," Ignatiev assured him, telling him that his case had been discussed twice in the Central Committee. Now finally it was clear who believed what and who trusted whom. It would be better for Yegorov, Ignatiev advised, if he complied quickly and fully in divulging his crimes and his associates.

Now, YEGOROV, you understand that for you to wait for some kind of help is senseless. If you are indeed a Soviet—then correct yourself and continue. But if you aren't a Soviet then at the least you are a Russian man, a former soldier, so in the interest of our Motherland you must fully divulge the conspiracy. We know that you played the leading role.[78]

Ignatiev gave Yegorov his personal word of honor that if he confessed truthfully, "his life would be saved and he might even be allowed to work in his specialization" again in the Kremlin Hospital. But if Yegorov continued to be stubborn, Ignatiev admonished, he would authorize sanctions. What kind of sanctions? Yegorov explained that he was still a sick man and would appreciate help from the investigators to sort things out "within the limits of the truth." Ignatiev reassured him. "Don't worry," Ignatiev said, "the investigators will help you."[79] So they did.

Yegorov related that Ryumin had "created unbearable conditions" for him.[80] He signed his confession because he had "lost his human aspect," as Yegorov put it, having fallen into a condition of deep "spiritual depression," owing to the fact that he had been a very sick man prior to his arrest. Sokolov and Levshin, who directly interrogated Yegorov, threatened him with severe beating; they cursed him endlessly, calling him a "murderer in a white coat," a "bandit," a "wolf hound." They told him that his wife, who had also been arrested, was a common whore; that she wanted to be the wife of a general. They told him that if they were to show him on the street and say to passersby, "Here's Yegorov, the murderer of Zhdanov," the people would immediately tear him to pieces out of righteous fury.

Inspector Sokolov, soon to replace Ryumin, made a momentous declaration. He assured Yegorov that he had been officially delegated by I. V. Stalin to explain that if he confessed truly and fully, telling the investigation who his associates were and for whom they were working, the government, with Stalin's own personal guarantee, would preserve his life. "You understand," Sokolov added, "that all the world knows He always fulfills his promises."

Yegorov understood. Only too well.

"Then," Sokolov continued, "you must realize that this is the only chance you have for saving your life. Otherwise, you would be hanged.

They wouldn't shoot you," Sokolov explained. "In the case of stubbornness it would be precisely only hanging for you."[81]

In the course of further interrogations, Yegorov's new interrogator, Levshin, would often remind him of this unique promise, saying that though he had worked as a senior investigator in the MGB for many years, this was the first time he had ever heard that the Vozhd himself had made a promise of this kind. However, if Yegorov remained stubborn, Levshin could assure him that they would use all the "catalysts" at their disposal to make him talk.

The list of things they wanted Yegorov to admit was long but relatively simple:

1. He, Vinogradov, Vasilenko, Maiorov, and Karpai had a premeditated plan to murder A. A. Zhdanov.

2. He and the others murdered Zhdanov on a direct order from Kuznetsov. When Yegorov protested, saying that there had never been such an order from Kuznetsov, Sokolov said cynically and, as Yegorov later recalled, "without any hesitation: 'If it wasn't Kuznetsov—then let it be foreign spies.'"[82]

3. He and other doctors had knowingly murdered Georgi Dimitrov and had systematically mistreated Andreyev, Thorez, and Tokuda.

4. He and other doctors had murdered A. I. Yefremov.

5. He and other doctors planned to murder marshals of the Soviet Union, A. M. Vasilevsky, A. L. Govorov, I. S. Konev, and General S. M. Shtemenko.

6. They had also attempted to get Yegorov to admit to the "wrecking treatment" of Politburo member A. N. Bulganin, as well as a failed attempt at fatally poisoning Svetlana Alliluyeva during her recent pregnancy.[83]

7. Levshin attempted to link Yegorov to international espionage
   through the United Nations, and to "expose" Yegorov's association with "unknown Jewish nationalists." This last accusation
   utterly nonplussed Yegorov. Levshin thought "as if seriously,"
   Yegorov told Beria, that he—Yegorov—was "a particular
   guardian of Jewish nationalists."[84]

Yegorov could not understand, but Levshin did. If it was a Jewish plot,
everyone became Jewish or was controlled by Jews. When they finally
arrived at the question of what Yegorov and his associates hoped to gain
from changing Soviet policy toward America and from murdering the
leadership of the party, Yegorov was stymied.

"Write down whatever you want," he told Levshin.

No. Levshin demanded a "confession." "So? What was it?" he insisted.
Yegorov stumbled. He didn't know.

"That the Soviet Union was to become the headquarters of a bourgeois-democratic republic?" he offered.

Levshin thought that was fine. The stenographer wrote it down. "But
it's not right to put that in," Yegorov protested. "You know," he said,
"you've already unjustly thrown two or three nooses around my neck.
Don't also throw a fourth." But Lieutenant Colonel Levshin only smiled.

"No," he said. "We'll write down just that."[85]

Yegorov complied with the investigation. He gave them everything
they wanted, in particular, the confession that he, Vinogradov, Vasilenko,
and Maiorov murdered Zhdanov in Valdai. But in order that his interrogators could give him the "help he needed," he and all the others were tortured. The torture consisted of systematic beatings with rubber truncheons, fists, and lashes. Manacles were applied to both the feet and the
hands of the prisoners. The slightest movement caused them to tighten,
producing unbearable pain to the elderly and generally ailing physicians.
The doctors were deprived of sleep; made to stand for hours; and held in
cold, dark, wet cells. In the end they all signed "invented, fabricated, and
knowingly false confessions produced by the investigators."[86] The MGB
interrogators were no longer waiters in white gloves.

According to the decree of March 31, 1953, terminating the action against the doctors and chronicling key events of the case, soon after the call from Malenkov to Ignatiev, Dr. Vasilenko was taken to one of the special rooms in Lefortovo with many doors leading in and out. Ryumin was there. Would he speak the truth? Ryumin wanted to know. Vasilenko said he would. Ryumin cursed him horribly and hit him in the face with his fist. Three "helpers" bound the doctor, gagged him, and beat him with rubber truncheons until he bled, facedown on the floor. He lost consciousness. They revived him. Ryumin stood over him. Would he confess the truth or wouldn't he? Ryumin insisted on nothing but the truth. He warned all the prisoners again and again against making false confessions. The truth was what he wanted. The truth was what Comrade Stalin wanted. Vasilenko again said he would speak only the truth. Again Ryumin hit the doctor in the mouth with his fist. Again the "helpers" threw him to the floor, beating him with truncheons until he lost consciousness. After the third beating, barely conscious, Vasilenko agreed to tell everything. But this was not enough. They beat him still more, lashing him with a flat whip. Bleeding profusely and barely conscious, he "lost his entire human aspect and fell to his knees begging for mercy." They beat him until the skin of his entire backside turned black. He lost his will. He would sign anything. His interrogator, Yazev, threatened to kill his entire family, to obliterate Vasilenko's name from the face of the earth. Then they brought the protocols of interrogation for signing. He signed, stating unequivocally that he had fully participated in the premeditated murder of A. A. Zhdanov.

Vinogradov was interrogated for the first time on November 13. As with the others, he initially denied any and all criminal activity or intent. But he, too, was brought to the specially padded isolation chamber in Lefortovo. After three days of continuous beating, he suffered a grave heart attack. They kept him in manacles around the clock while he recovered. Soon, not able to bear the suffering, he confessed to everything.

Maiorov, too, was taken to the little room. They beat him with truncheons. At the first interrogation, the investigator, Sedov, hit him in the face with his fist. He was forced to name Vinogradov as a spy and a murderer.

Ryumin personally forced Yegorov to confess that Kuznetsov was behind the scheme to kill Zhdanov. They cursed him, threatened him, told him he would be burned with a red-hot iron. Even after he signed the confession, they beat him in the special room, calling him a prostitute, a doctor-murderer, an "enemy snout," and "the head of a gang of doctor-murderers and bandits."[87]

In the March 31, 1953, decree that described these beatings and confessions, the investigators noted that there was not a single *written* order to falsify confessions and to beat and torture the prisoners until they signed them. All orders in these cases were oral.[88] What happened in that specially fitted room on the third or fourth floor of Lefortovo prison was instigated through a chain of oral commands, telephone calls, meetings at which no stenographer was present. Such meetings occurred in Stalin's Blizhnyaya dacha and must have formed the background of Beria's order to Ignatiev about Karpai's investigation; Poskrebyshev's telephone call with Yegorov about Zhdanov's autopsy; Yegorov's certainty in 1948 that "they" believed him and not some sort of Timashuk; and Poskrebyshev's denunciation of Molotov and Mikoyan in 1948. Much remains unseen in the tragedy of these doctors.

Yegorov's own words help us understand more of what we cannot see.

Lieutenant Sokolov came up to me, put his fist to my face, threatening to beat me. Then he added, "By the way, it's disgusting to beat you, it's disgusting even to slap that enemy mug of yours." Lieutenant colonel Levshin repeated this. "See my fist? If I move it, good luck—then you'll find out"—everything was approximately in this spirit. Lieutenant SOKOLOV with the aim of intimidation and with a sadistic smile on his face took a rubber truncheon out of a drawer of his table. He gave it to me to feel and said that with the help of this "little thing" criminals become more talkative. I asked, "but how much do you do with it?" And he answered, "Precisely as much as is necessary to get the confession."

In general, I signed under these unbearable conditions, and as soon as I did, my relationship with Ryumin sharply changed. If he earlier had threatened me with tortures . . . if earlier he had repeated the whole time, "and this type, think about it, was the head of the Lechsanupra—what a shame," "it's even disgusting to talk to this type"—now (after I signed) his face quite literally beamed with joy, and now he even came to look at me with approval, in a friendly way, and no longer threatened me with torture, such as red-hot iron. By the way, it turned out that on the very next day they beat me again. But he didn't do it; others did it to me in his presence—a detail, but with special significance, it seems, for pushing my psyche finally to understand the "order" necessary to the investigation. They beat me in a separate room (if I'm not mistaken, in the office of the head of the prison), specially fitted out, it seems, with screens for the purpose of intimidation. At the entrance of the room, 3 men jumped on me with shouts. They threw me onto the divan and began to beat me with rubber truncheons, all the time demanding (Ryumin himself demanded) that I speak "truth." After, it seems, two such whippings, Ryumin asked—"Will you, you traitor, speak everything? No, you won't. I see it in your eyes. You won't tell us the whole truth." I expected more torture, but it didn't come. There was only the sanction for manacles that I wore day and night for approximately 2 months. I slept with manacles.[89]

Lieutenant Sokolov told Yegorov, "we will beat you every day, we will tear out your arms and legs, but we will all the same learn everything down to the last detail about the life of A. A. Zhdanov, and all the truth." Yegorov wrote to Beria after Stalin's death, "Oh, if only this case went according to the real truth!"[90]

Yegorov could not understand how or why he was being charged with such nonsense. "I would have to have been an utter idiot," he wrote to Beria, "to concoct a plan together with Kuznetsov for a bourgeois-democratic government in our country. This is total nonsense, but I could not get through to the investigators. In truth, they probably didn't want this to get through. On the contrary, they cynically argued, 'It pays for you to be upset about your anti-Soviet feelings. The more you say everything bad about yourself now, the better things will be for you in the future.'"[91]

When he tried to tell his investigators that he had no anti-Soviet feelings, that he was perfectly content in the Soviet Union, that he was pleased with the hard-line policy toward Anglo-American imperialism, he got nowhere. He was "a voice howling in the wilderness." He had "sold out to Kuznetsov for a mess of pottage"[92] as his interrogator explained.

All the other doctors sold out for a "mess of pottage" as well—each one was duped by the "clever Jew" to sell his birthright. Once the deaths of Zhdanov and Shcherbakov were irrevocably linked, the investigation galloped on. Vinogradov testified in December 1952:

> In the past interrogations I fully confessed my guilt both in spying and in terrorist activity against the leadership of the party and the Soviet government, and in addition I spoke about all the facts touching the disorganization of medical work in the Lechsanupra system accomplished by me, YEGOROV, BUSALOV and others of our associates.

But the interrogator was not satisfied. "You are still far from a full, sincere confession, and you must testify about a whole series of facts of criminal methods applied by you and your associates in the treatment of patients," Vinogradov was told. What facts? "You are still hiding much. Tell us: in the course of what period of time did you treat comrade M. I. KALININ?" Old, ineffectual Kalinin, too, it seems, was to have been the object of a murderous plot. But once the investigator was apprised of the fact that Kalinin had died of cancer in 1946, his name was quietly dropped from further protocols.[93]

On November 18, 1952, Vinogradov was still able to deny a premeditated plot to kill Zhdanov: "I allowed a mistake in the diagnosis that led to grave consequences and then to [Zhdanov's] death. There was no evil plan in my action. . . . I want only to repeat that at the basis of this crime, its original source, was medical error that I allowed as a consultant, leading the treatment of A. A. Zhdanov."

But Sokolov, who was interrogating Vinogradov, shot back: "We will unmask you further. . . . Who is your boss?"

"I have no boss. I did not murder A. A. Zhdanov or A. S. Shcherbakov intentionally. In what I have already testified to there was no influence of anti-Soviet spirit or ties with people who were enemies of the Soviet power."

Sokolov was not satisfied. He reversed directions, tying Vinogradov together with other known medical saboteurs of the past. Eventually Vinogradov admitted to having enemy ties with the dead Pletnev, one of the alleged murderers of Gorky; and Drs. Etinger and Pevzner, who had recently died.

They knew that Vinogradov liked to live well and collected valuables. "I don't deny that my anti-Soviet convictions, my tie with Etinger and other enemies of Soviet power, whom I have already named, speaks to my treatment of the leaders of the party and the Soviet government. I did not show concern for their health and this question didn't bother me. I lived in my own world and my own interests: collecting valuable pictures, buying up diamonds. I have a passion for money."

"Especially for dollars and pounds sterling?" they asked.

But Vinogradov wouldn't take that bait. "I didn't have that passion," he told them. The interrogation found itself unable to get him to admit what they needed: an unequivocal statement that he had been working for the Americans and British to murder Zhdanov and others. They gave him one last chance:

DECLARATION OF THE INVESTIGATION

We have been charged by the leadership to transmit to you that for the crimes you have committed you already may be hanged, but that you may yet preserve your life and receive the possibility to work, if you truthfully say what was the root of your crimes, toward whom were you oriented, who was your boss and associates.

We are also charged to tell you that, if you wish to repent to the end, you may set out your testimony in a letter to our Leader who promises to preserve your life in the event of an open confession by you of all your crimes and a full exposure of your associates.

It is known to the entire world that our Leader always fulfills his promises.

After hearing this declaration, Vinogradov was speechless.

"Why are you quiet?" they wanted to know.

"I find myself in a tragic position," Vinogradov replied, "and have nothing to say. I did not serve foreigners; no one directed me; and I myself drew no one into crimes."[94]

Vinogradov could not tell the "truth" without physical torture. Therefore, the torture was necessary. So they beat Vinogradov more. By mid-December the "truth" came spilling out about his ties to numerous living doctors, most importantly Miron Vovsi, who by this time was freely providing the MGB whatever confessions it required. The link between Kuznetsov and Abakumov extended the plot vertically into the Central Committee and the Ministry of State Security. The link between Vovsi and Vinogradov extended the plot horizontally into the entire network of medicine and Jewish intellectual life in the Soviet Union, and ultimately, through Mikhoels and the Jewish Antifascist Committee of which Mikhoels had been the head, across the world to America.[95]

The tragic situation of Vinogradov was the same for Yegorov, Vasilenko, and Maiorov. In Valdai in 1948 they may well have *thought* they were acting on behalf of the state and that their treatment of Zhdanov was sanctioned from above. Why else would Poskrebyshev have authorized the autopsy in Zhdanov's bathroom in Valdai? Why else would Timashuk's statement, which Yegorov knew had been passed on to Abakumov, have produced not a single ripple of reaction from the Kremlin? Why else could Yegorov fire Timashuk summarily in September 1948 with no repercussion? What was the meaning of the disfavor into which Zhdanov had fallen, Yuri's letter in *Pravda*? Did not all this signal Stalin's disposition in the case and show them—because they were vigilant Soviet citizens—what to do? Yet now Stalin was telling them to confess things utterly beyond their comprehension—that they could have been guilty of betraying the Soviet state and comrade Stalin at the very moment they believed they were being most loyal. This is the contradiction that rendered Vinogradov speechless. It was truly beyond words. Everything Vinogradov could say was true and not true at the same time.

By the end of November 1952 the picture was complete. On November 20 Pitovranov was released from Lefortovo in order to commence work on reorganizing foreign intelligence, and on November 24 the entire investigation of the Doctors' Plot up to that point was summarized in the memorandum sent to Stalin. This memorandum provided the "truth" Ryumin had constructed about the deaths of Shcherbakov, Zhdanov, and other leaders, including a tantalizing detail that one of the doctors had even imagined at some point making an attempt on the life of the now powerful Grigory Malenkov, Zhdanov's rival in the Central Committee.[96] It related the present plot to the alleged medical wrecking of Pletnev and Levin that took place some fifteen years previous in the deaths of Gorky, Menzhinsky, Kuibishev, and Gorky's son. It tied together the evil plotting of the doctors with Abakumov's conspiracy through the figure of Kuznetsov and through Kuznetsov with American and English imperialism. It named all the doctors and noted that "Vovsi and Kogan were Jewish nationalists actively working against the USSR under the cover of the Jewish Antifascist Committee." It exposed the existence of an extensive "Jewish nationalist underground with ties to foreign anti-Soviet centers" and concluded by saying that "the investigation proceeds. It is not yet known who concretely controls the underground and how it is organized."[97]

Stalin was searching for a unifying structure.

SEVEN

AN INTELLIGENCE PHANTASMAGORIA: THE PLAN OF THE INTERNAL BLOW

1951 — 1953

Ryumin again rejected my proposal, saying that I was not a child and must understand that in politics there is no role for the confirming of this or that fact, but that it is important to achieve the established goal.

—INSPECTOR PAVEL GRISHAEV, MAY 25, 1953

There is no better chicken in the world than pork Kolbasa.

—N. S. KHRUSHCHEV

Stalin's answer to the question of who controlled the underground and how it was organized formed the basis for one of the more bizarre, if hitherto unknown, episodes of the Cold War, known as the Varfolomeyev case or the "Plan of the Internal Blow." The only evidence for this case comes from the Soviet archives, and much of what the documents contain was shaped to fit the purposes of state security. They provide a picture of a plot so incredible that the Soviets themselves attempted to erase all trace of it. All signs of the existence of Ivan Ivanovich Varfolomeyev vanished soon after his execution on September 10, 1953, six months after Stalin's death. Tried before the Military Collegium of the Supreme Court of the Soviet Union, he was found guilty of "criminal acts against the Soviet Union . . . according to statutes 58–1, 58–4, 58–6 (ch.1) & 18.11" of the criminal code of the Russian Federation.[1]

Varfolomeyev's case was carefully scrutinized, investigated, and prepared over a period of some two years, going back to December 25, 1950, when allegedly he had been arrested as a spy by the Chinese in the city of Tien-Tsin and turned over to the Soviets.[2] Varfolomeyev came into custody at the same time that Dr. Etinger was being interrogated in Lefortovo.

Varfolomeyev was not Jewish. He was a Russian whose parents had immigrated to Japan in 1919, fleeing the revolution. He stated that in 1933 he became a paid agent of Japanese intelligence operating in China. Later he became an American agent working in China, collecting intelligence about Soviet economic and political policies. Little else is known of his personal story. He had nothing in common with the doctors under investigation by the MGB, except for the fact that Ignatiev ultimately transferred his case to the Investigative Unit for Especially Important Cases, where M. D. Ryumin took it over.

At the same time that Etinger was made "to sit," Varfolomeyev divulged a story that began to attract ever more attention from key MGB personnel, the Central Committee, and Stalin. From the evidence discovered in the KGB archive it appears that, initially at least, Varfolomeyev was going to be brought to an open trial at approximately the same time as the trial of the doctors. The revelations of his trial would have rocked Soviet society and stunned world opinion. It would have dramatically answered the question posed by Ignatiev in his November 24, 1952, memo to Stalin about the organization and intent of the underground operating in the Soviet Union. In combination with the Doctors' Plot, it might well have ignited a new world crisis.

If the outline of the plot about to be sketched seems fantastic—it is. How much was believed by anyone, including Varfolomeyev, is not clear. Though never publicly divulged or explained, the Varfolomeyev case illustrates the absurd extent to which the Soviets were willing to falsify facts to suit the "political realities" of the day, to make, in Khrushchev's colorful words, a pork Kolbasa into a chicken. According to eyewitness testimony, when Stalin heard that some of what Varfolomeyev told investigators was "difficult to believe," he responded that "one could expect anything from the Americans, and therefore if the confessions of Varfolomeyev are unconvincing, then . . . the MGB must make them convincing for the trial."[3]

The MGB claimed to have credible information, including Varfolomeyev's own confession, that Varfolomeyev was a spy who had worked for the Japanese prior to the Great Patriotic War and for the Americans beginning in 1948.[4] His job, he allegedly told them, was to obtain information about the kind of aid provided to China by the Soviet Union and anything else he could learn of an economic and political character.

When the Korean War began in June 1950, Varfolomeyev's work assumed a specific character. He allegedly received an assignment concerning supposed American infiltration of North Korea and was partly responsible for a group of agents with a radio station directed against the North. The alleged purpose of this assignment was to collect information on the dislocation and movement of Korean troops, the mood of the local Korean population, and the role of the Soviet Union in the North Korean war effort, in particular the sale of arms and ammunition.[5] According to the Soviet documents, he also appears to have been responsible for providing a certain amount of disinformation about the war to the local Chinese population.

Confessing in connection with espionage work for the Americans, VARFOLOMEYEV declared that at the end of March 1949, the American espionage *residentura*, into which he had entered, received the assignment, connected with the preparation by the American aggressors . . . of the armed invasion of North Korea.

According to VARFOLOMEYEV, the Americans charged the leader of this *residentura*, YU DZUN-BIN, to detail a group of agents with a radio station and direct it at North Korea with the purpose of collecting information on the dislocation and movement of North Korean troops, on the mood of the local Korean population, and on the facts of arms and ammunition aid on the part of the Soviet Union provided to the North Koreans.

According to the further confession of VARFOLOMEYEV other agents of the *residentura* of YU DZUN-BIN, left in China, in 1949 had the assignment from the Americans to spread among the Chinese population provocational fantasies about the situation in North Korea, and after the attack on North Korea to spread propaganda against giving aid to North Korea.

As an American spy, VARFOLOMEYEV associated in China closely with the American intelligence officer ROGALSKY, P.A., with whom he was tied for many years in hostile activity against the Soviet Union and in the service of foreign intelligence.[6]

Varfolomeyev's story took on a lurid twist with the mention of Peter Arsenievich Rogalsky, another anti-Soviet Russian. It has been impossible independently to confirm the existence of Rogalsky. The best we can offer is Varfolomeyev's own account of their relationship, as contained in the protocols of his interrogation. According to Varfolomeyev, Rogalsky told him that he had joined the pro-tsarist White Guard at the time of the revolution and actively participated in the armed struggle against the Soviets. Rogalsky eventually fled to France, where he was recruited by French intelligence. As with Varfolomeyev, Rogalsky participated in various White émigré anti-Soviet organizations.

In 1940 Rogalsky left France for the United States, where he married the daughter of a man named Hartman, who, according to Varfolomeyev, was employed as a senior secretary of Pierre Dupont. Because of this family connection, Rogalsky became employed by Dupont and gained access to both business and military circles in the United States. At some point, he claimed, he was recruited by American intelligence.

Varfolomeyev told his interrogators he became acquainted with Rogalsky in Tien-Tsin in 1950 when Rogalsky was on an assignment to set up a "reliable financial base" in the territory of Northern China.[7] Rogalsky's elaborate and improbable scheme involved filtering American money into the region through a number of houses purchased from the mission of the French Catholic Church located in Northern China. According to Varfolomeyev's story, the Catholic Church owned various private houses for its missionary work throughout the country, and Rogalsky was supposed to buy these houses for the Americans, rent them to locals, and finance American intelligence efforts in the region using Chinese currency.[8]

Rogalsky reassured his friend that the American setbacks in Korea were purely temporary because of the extraordinary military strength America would soon bring to bear on the region that would "not only destroy North Korea but also put Chiang Kai-shek back into power in China." This was only the first step. Rogalsky allegedly also revealed a plot, as improbable as it might have been, to "realize a plan for the military invasion of Soviet territory" across the China-Russia frontier.[9]

Varfolomeyev told his captors that Rogalsky had informed him that his association with Dupont made it possible for him to learn much about the American military-industrial complex, and he revealed that Dupont, and his brother Lammot,[10] were part of an organization called "the financial center," the purpose of which was to finance and facilitate the USA's military arms programs in preparation for war with the Soviet Union.[11]

In the end, the composition of "the center" was to have included Owen Young, chairman of the board of General Electric, and D. S. Abrams, president of the Standard Oil Company. These titans of American capitalism were working closely, Rogalsky allegedly told Varfolomeyev, with the American military to speed production of advanced nuclear weapons. In particular, they worked with General LeMay, General Twining, a Lieutenant MacPherson, General Vandenberg, and General Omar Bradley, chairman of the Joint Chiefs. President Harry S. Truman was said to support their initiatives. The most important of these initiatives was "The Plan of the Internal Blow."[12]

The "Plan of the Internal Blow" consisted of a plot to fire five nuclear devices using new "noiseless" ejectors at the Kremlin. These fantastic bombs (obviously science fictional prototypes of tactical nuclear weapons) developed tremendous heat during explosion that would "destroy the entire Kremlin and burn up everything living" in the vicinity.[13] The whole Soviet central government would go up in smoke, as would most of Moscow. According to the scenario Varfolomeyev related, the weapons would be fired from the windows of the American embassy in Moscow. The implementation of the plan was initially entrusted to a Lieutenant Colonel O'Daniel, the American military attaché allegedly in Moscow at that time.[14]

According to Rogalsky, as related by Varfolomeyev, the necessary missile apparatus was scheduled to land on the Baltic coast of the Soviet Union by the end of 1951. Initially the date set for incinerating the Kremlin was March 1952, though this date supplied by Varfolomeyev was not included in the April 1952 memorandum eventually sent to Stalin by Ignatiev, for obvious reasons. According to the original scenario, in March 1952 the United States of America would declare war on the Soviet Union. The ghosts of June 22, 1941, the date of Hitler's invasion, appear to have resurfaced. In 1941 Stalin refused to believe the credible reports from his intelligence agencies that Hitler's Operation Barbarossa was about to get underway; now he was inventing an incredible report and demanding that others believe it.

Nothing more is known about the details, real or imagined, of the "Plan of the Internal Blow." We know Varfolomeyev's fate, but little else. We know that he was executed not by Stalin but by those who followed him, and we know that all traces of the plot were expunged. There was no public trial. This Cold War fantasia had been utterly forgotten.

However, we also know that the Soviets took this fantasia very seriously for a time. One of the key charges against Ryumin at his trial was that "with his direct participation and at his order the confessions of the Japanese and American intelligence agent, VARFOLOMEYEV, I. I., were crudely falsified. *His case had been marked to be heard at open trial*" (emphasis added).[15] Varfolomeyev's trial, the decree of indictment continued, "would have brought harm to the world prestige of the Soviet State."[16] It might well have made the Soviet Union the laughingstock of the world as well once the shock wore off. However, the fact that it had been "marked to be heard at open trial" tells us that this was a case of exceptional importance to Stalin.

The Soviet authorities, in the wake of Stalin's death, were as concerned to seal off the Varfolomeyev case from any possibility of public disclosure as they were to abrogate all charges against the doctors. From the interrogations of Ryumin; the testimony of senior MGB investigator Pavel Grishaev, who had direct charge of Varfolomeyev; and other court proceedings, it is clear why. Although unlike the doctors, Varfolomeyev may well have been a real agent of American intelligence, the plot for blowing up the Kremlin and America's impending declaration of war against the Soviet Union were pure invention, the outcome of Stalin's and the MGB's feverish pursuit of any lead that could provide what Ignatiev needed to show: "who concretely controls the underground and how it is organized."

Varfolomeyev was the missing link that would demonstrate the active involvement of American intelligence, the American military, the American government, and American Jews in a plot to destroy not just Zhdanov and Shcherbakov, but the entire Soviet Union. As time went by, it might have been possible for Stalin to show that Hartman, Rogalsky's father-in-law, was Jewish and that he had had connections with other Jewish organizations, perhaps even those with which Mikhoels and the Jewish Antifascist Committee had dealings, and thereby a link could be made to Etinger and from Etinger to the death of Zhdanov . . . Though farfetched, it is no more farfetched than Ryumin's suggestion to Sheinin in July 1952, the same time period in which the Varfolomeyev case was being investigated, that he could confess that he became "acquainted with ETINGER maybe in 1946, perhaps when he took a vacation somewhere in the Baltics," adding "For you this has no particular significance, but to us this is important."[17]

It is not certain toward what conflagration in Stalin's mind these plots were converging, but it is clear that the entire Soviet security apparatus was engaged in developing them. After Stalin's death the enormity of what he had imagined began to dawn on those involved. "Only now have I understood in actuality," investigator Pavel Grishaev told *his* interrogators in May 1953, "what sort of provocation, into what sort of dirty whirlpool I was dragged by RYUMIN and IGNATIEV, to what grave consequences for the prestige of the Soviet government the story of the trial of the VARFOLOMEYEV case would have led had it been successfully brought about." As he put it, the Varfolomeyev case "lay like a heavy stone" on his conscience. Grishaev testified that he "in no way" wished to minimize his "sins in this shameful case," but he asked the investigation to consider that he had been given a direct order to handle the case as he did.

> Educated in a Chekist milieu, I am used to believing my comrades in work, to believe my leaders. In addition, I absolutely believed IGNATIEV and RYUMIN who I considered the conduits for the party's politics. . . . I believed that everything they ordered agreed with the authorities and consequently was the dictate of the party, the dictate of the Central Committee.[18]

Pavel Ivanovich Grishaev was no Garkusha. Born in 1918, Grishaev had graduated from college and was a Candidate of Juridical Science, the Soviet equivalent of a J.D. With the rank of lieutenant colonel, he was the deputy head of the higher school of the MVD. Grishaev was never arrested, and after Stalin's death, though released from service in the MGB, he became a respected professor of law at the All Union Correspondence University of Law. In fact, he had been chosen specifically for this important assignment because he was considered one of the most "juridically informed" investigators in the MGB, someone who would understand what was necessary to prepare a successful public trial.[19] Nevertheless, Grishaev, too, was just another cog in the vast machine of terror.

Grishaev had good reason to believe that the party and the Central Committee stood behind the plot he was now engaged in concocting. When Ryumin gave him direct instructions about Varfolomeyev, Grishaev told the investigation, he invariably "underscored that this was not his order but the order of the very highest and most authoritative people of our party."[20]

Though self-serving, Grishaev's testimony indicates the extent to which the system in which he found himself continued to compel obedience. It also suggests the extent to which the system was beginning to erode. "More than anything in the world," Grishaev lamented in 1953, "I hate lies and dishonorable people."[21] Grishaev said that there were two cases that most troubled his "party conscience"—that of Varfolomeyev and that of the Jewish Antifascist Committee.[22] It would be easy enough to dismiss Grishaev's moment of moral fervor as merely politically motivated posturing, but here, as in the case of Cheptsov, we may be witnessing another indication of Stalin's real enemy: the elementary human instinct to recognize and tell the truth. At the same time, the authority Grishaev brought to bear in his self-criticism was no abstract ideal of justice or conscience, but rather his "party conscience."

He invoked the authority of the system that was itself the cause of his "sins in this shameful case." The self-contradictory situation in which Grishaev found himself demanded that he contradict the principles his party conscience taught him to serve. Much as Vinogradov found himself in a "tragic situation," Grishaev found that all he had been taught and all he had taught others about the purpose and intent of the party was being contradicted by the party itself. Garkusha did not experience this internal conflict, but Grishaev did. The rhetoric and ideology of the party was being undermined from within by the party itself, not by the ironic verses of Mandelshtam or Akhmatova.

What did Grishaev do that lay like a stone on his conscience and would have been so harmful to the world prestige of the Soviet Union, if it had come to fruition? In February 1952, a year before Stalin's death, at the time Dr. Ryzhikov was arrested, Grishaev wrote a *spravka*, or informational memorandum, about the Varfolomeyev case that was probably directed to Ryumin and Ignatiev, although this isn't indicated on the document itself.

At the time Grishaev wrote this *spravka*, Maklyarsky and Sheinin were being interrogated concerning traitorous Jewish nationalism in the MGB; and Krongauz was being interrogated about his conversations with Dr. Etinger's son. In February 1952 Ryumin told Maklyarsky that "he intended to put the question of the expulsion of the Jews from Moscow before the government."[23] At the time Grishaev's memorandum hit his desk, Ryumin was ruminating on how to propose the expulsion of the Jews to Stalin who at that very moment was awarding the Stalin prizes in literature and scolding Soviet authorities for their anti-Semitic practices.

Grishaev's *spravka* contained the following information:

1. Varfolomeyev confessed that a Lieutenant MacPherson in the U.S. army had been sent to Tokyo soon after hostilities broke out in Korea to lead the U.S. military mission there.
2. Varfolomeyev had named Hartman as Pierre Dupont's senior secretary.
3. MacPherson and Hartman had not up to that point been fully identified and Grishaev considered it necessary to use "all the possibilities of the Military Intelligence of the Soviet army" for this purpose.
4. Grishaev considered it "expedient" to
   a. "establish and collect the necessary information concerning two or three officers or generals of the American army who occupied influential positions in the U.S. Military Ministry and are known to be particularly aggressive in relation to the Soviet Union."
   b. "to clarify with whom, among American millionaires and representatives of military circles, DUPONT maintained the closest ties."
   c. "to establish who occupied the post of American Military attaché in Moscow in 1949, and in addition to clarify who concretely in the period was accredited in the capacity of his assistants."[24]

Grishaev had been ordered to falsify the entire case. Behind Grishaev's memorandum was a conversation with Ryumin in which, according to

Grishaev, he had told Ryumin that he thought Varfolomeyev's confession had raised "strong doubts and that it would be irresponsible to take it to trial."[25] Echoing Stalin, Ryumin told him that if Varfolomeyev's testimony was not credible, it was the responsibility of the MGB to make it so. "In other words," Grishaev wrote in May 1953, "Ryumin ordered me to embellish with pure invention the confession of VARFOLOMEYEV . . . RYUMIN said that with prisoners such as VARFOLOMEYEV, it was not difficult to reach agreement on their confessions, because on more than one occasion he [Varfolomeyev] had already proposed to the investigator that he would serve 'the interests of the Soviet Union.'"[26] Grishaev believed it was absurd.

In the information they sent to the government, IGNATIEV and RYUMIN confirmed, without any basis, that ROGALSKY, who according to VARFOLOMEYEV was the source of the above detailed information about the alleged intended shooting of the Kremlin, had been living in New York. This was not in fact the case; that is, there is no indication of any kind that ROGALSKY, about whom VARFOLOMEYEV spoke, ever existed.

Despite the fact that we had exhausted all possibilities for establishing and verifying the identity of American Lieutenant MACPHERSON, who allegedly, in turn, spoke to ROGALSKY about the above mentioned "Plan of the Internal Blow," and despite the fact that it was completely clear that MACPHERSON—this mythical person—had been invented by VARFOLOMEYEV, IGNATIEV and RYUMIN in the information they sent on to the authorities stated that the verification of the existence of MACPHERSON was continuing, and thereby they created the illusion, that the MGB of the USSR still had some work to do in this direction.

All of this could not but create a false impression about the importance and truthfulness of VARFOLOMEYEV'S confession. Therefore it was not accidental, when in February 1952, information was sent to the authorities in which a short description of the case was given from the investigation over the 8 months, and a term was set for the completion of the case, that the authorities did not permit the investigation to be concluded in the case of VARFOLOMEYEV.

> Several days after the indicated information was dispatched [in February], RYUMIN called me and informed me that the, as it were, late Head of the Soviet Government proposed to the leadership of the Ministry to prepare an open or a closed trial for the VARFOLOMEYEV case, and to unmask TRUMAN for the whole world as the instigator of a new world war.[27]

As Ryumin put it later at his own trial, the purpose of the Varfolomeyev case was to unmask and "dirty" Truman and the U.S. military "before the face of general world opinion."[28] In his defense, Ryumin said Ignatiev had told him that Stalin had placed great significance on the case and thought that the Americans had sent Varfolomeyev into the rear of the USSR for the purpose of frightening the government. According to Ryumin, Ignatiev said that Stalin believed that "the Americans in actual fact might be capable of making so monstrous a provocation" as the shooting of the Kremlin, regardless of whether they were actually planning to do so.[29] The difference between planning and intention had little import to Stalin. Stalin believed the Americans were guilty as long as they were capable of imagining it.

A nation led by the idiosyncratic belief structure of the Vozhd could not but function, as Khrushchev wrote in his memoirs, in self-contradictory, if not delusory, ways: "Stalin's version of vigilance turned our world into an insane asylum in which everyone was encouraged to search for nonexistent facts about everyone else."[30] Underlying this belief structure, however, was a coherent political vision. The result for Grishaev was that his misgivings were dismissed as cowardice. Ryumin instructed him to employ the services of military intelligence to fill the gaps in Varfolomeyev's "confession," something military intelligence seems to have been reluctant to do.[31] In the meantime, Cheptsov, who at the time was skeptical about the trial of the Jewish Antifascist Committee, which got underway in March 1952, recommended against an open trial of Varfolomeyev on the grounds that the evidence appeared both dubious and skimpy and that such a proceeding "posed too great a risk."[32] Ryumin denounced Cheptsov as a "coward who was afraid of everything."[33]

Despite Cheptsov's lack of faith in the case and Grishaev's knowledge that much of the alleged hard evidence was faked, the case continued to develop over a period of months. Grishaev related that, after several discussions, Ryumin and Ignatiev agreed to a closed trial, in the manner of the trial of the American spies Samartsev and Osmanov; that is, the case would be considered in closed judicial session, and then the results of the trial would be published in the press.[34] Soviet foreign intelligence provided various names of high-ranking American military personnel, as well as prominent American industrialists with particularly violent anti-Soviet views.

Grishaev continued his own verification and independently concluded that O'Daniel, the American military attaché, had been in Moscow in 1949 but had been recalled in August of the same year. O'Daniel therefore could not have been in Moscow in 1950 when Rogalsky, according to Varfolomeyev, alleged he was. "All this," Grishaev told the investigation, "strengthened [his] opinion that the confession of VARFOLOMEYEV could not be believed."[35]

There were other troubles. From a certain Letter No. 16/712, introduced at Ryumin's trial, we learn that foreign intelligence was also not as useful as it could have been. The letter was addressed to Savchenko of the MGB, not military intelligence, as Ryumin had initially advised, and was dated February 29, 1952.[36] In it Ryumin wrote:

> The list of names you have provided for our operational needs are not appropriate because in it are included individuals known to the entire world. . . . You must find two or three people, who work directly in the apparatus of the Military Ministry of the USA, who occupy influential positions there but are not widely known beyond the limits of the USA.[37]

In other words, though the names foreign intelligence provided were appropriate, they were not acceptable because they were too well known. Therefore, the case would not possess sufficient verisimilitude. The case lacked the necessary realism! The political imperative possessed an aesthetics as well.

A more troubling problem arose. The original "Plan of the Internal Blow" as outlined by Varfolomeyev indicated that the date set for the American strike against the Kremlin was March 1952. In February 1952, the case was not yet ready to unveil, and there were no signs of an impending U.S. invasion. Ignatiev handled this by simply blackening out the March 1952 date from the memorandum he sent to the Central Committee, stating that "the Americans seem to intend to time the shooting of the Kremlin to coincide with the beginning of the war,"[38] leaving the date unfixed. Grishaev noted that he considered this "correction" by Ignatiev a particularly "grave crime, because the danger that the Kremlin would be shot at was not established, and it would continue to hang like the sword of Damocles and without doubt would have provoked continued anxiety on the part of our government."[39]

Soon the investigation had to face another significant difficulty. The Varfolomeyev plot was designed to discredit Truman and to demonstrate his imperialistic, aggressive intentions toward the Soviet Union. By summer 1952, however, Truman was no longer running for reelection. This took the MGB by surprise. The whole plot was beginning to fall apart. In light of this unforeseen situation, Grishaev proposed that he be allowed to "restore Varfolomeyev's confession to its original form; that is, to take out everything that had been embellished" with his help.[40] But Ryumin ruled this out. He told Grishaev that he had already spoken to Ignatiev who told him that Stalin—presumably some time in November 1952—had instructed him to let Varfolomeyev "sit," saying, as he had with Etinger, "he's still useful to us."[41] Conditions might change yet again. Who knew when new pieces of the vast puzzle that was coming together—and by November 1952 was reaching its penultimate form—might again yield a place for the "Plan of the Internal Blow"?

On May 28, 1953, Varfolomeyev wrote to the head of the investigative section of the newly organized Ministry of Internal Affairs that part of his confession had been "embellished" by Grishaev. He further declared that Rogalsky's story about the "Plan of the Internal Blow" had seemed fantastic to him, but that he felt it was necessary to inform the investigation about it. He agreed to make this known, he told the investigation, only after his fellow prisoner, an O. A. Shternberg, informed him that the American embassy was near the Kremlin. On September 10, 1953, Varfolomeyev gave a complete description before the Military Collegium of the Supreme Court of the Soviet Union of the process of falsifying his confession. He was shot the same day.[42]

Though the "Plan of the Internal Blow" came to nothing, Stalin made some use of it in January 1953, as we will see, when *Pravda* announced the Doctors' Plot to the Soviet public. But that was still two months away.

Ignatiev was not arrested after Stalin's death, though Grishaev's testimony in combination with Ryumin's would have justified it. In his memoirs, Khrushchev did not explain why Ignatiev was not arrested; nor did he mention that Ignatiev was removed from his post as minister of state security shortly after Stalin's death. In fact, Ignatiev was also removed, in April 1953, as a secretary of the Central Committee for having "permitted serious errors in the leadership of the former Ministry of State Security of the USSR."[43] However, after a review by the party of Ignatiev's conduct as minister of state security, he was reinstated to the Central Committee at the July 1953 plenum, one month after Beria's arrest.[44] The fact that he was not arrested but actually reinstated in the party, despite the party's knowledge of his "serious errors" demonstrates how the system replicated itself in the new regime, Grishaev's "party conscience" notwithstanding.

EIGHT

SPIES AND MURDERERS
UNDER THE MASK OF DOCTORS

NOVEMBER 1952–DECEMBER 1952

Listen, let me have [a prisoner] for one night, and I'll have
him confessing he's the king of England.

—L. P. BERIA

I know very well the value of truth.

—S. D. IGNATIEV

Despite the scruples, or mere "cowardice" as Ryumin described it, of Cheptsov, Grishaev, and others, Stalin compelled the system to serve him until the very end. His power remained undiminished.

From November 12, when Ignatiev ordered physical torture for the doctors, to March 1, 1953, when Stalin was discovered in a coma in the Blizhnyaya dacha just outside Moscow, all the different elements of the vast action that had begun like the first winds of a hurricane in August 1948 with A. A. Zhdanov's death gathered together into an immense political-social-ideological storm, largely hidden from the outside world, like the famine of the Ukraine in 1946, until the revelations published in January.

What those inside the Kremlin knew in November 1952 could not be easily communicated outside its walls even if they had wanted because no one knew all of it, and those who knew the most—Abakumov, Varfolomeyev, Kuznetsov—were either safely in prison or had been executed. By the beginning of March 1953, virtually everyone who had planned or taken part in the

Doctors' Plot or the Varfolomeyev case had been publicly silenced in one way or another.

Even after their release at the beginning of April, the doctors themselves were in no position to discuss the case. Those who had been implicated in the death of Zhdanov—Yegorov, Vinogradov, Vasilenko, Maiorov, and Karpai—had been rendered speechless by their own incomprehensible participation. Others, such as Dr. Yakov Rappoport, who had been arrested in February 1953, assumed that the plot reflected nothing more than Stalin's diseased anti-Semitic paranoia.

Beria, perhaps, knew the most; he was arrested at the end of June 1953 and executed some months later, after having done his utmost soon after Stalin's death to eradicate all vestiges of the plot from public consciousness. As a result, the Doctors' Plot has come down to us as little more than a footnote to Stalin's vicious anti-Semitism, the last crazy, paranoid conspiracy of his murderous regime. But it was much more than this, or would have been much more had he lived.

On November 13, 1952, Ryumin was dismissed; on November 14 Ignatiev had a heart attack and did not return from the hospital until January; the doctors were "beaten to a pulp"; confessions multiplied like mushrooms under birch trees; Jewish doctors and intellectuals throughout the USSR were arrested; and S. Goglidze, Beria's close associate, was installed as the head of the investigation of the Doctors' Plot.

The "confessions" of the doctors and the arrested MGB workers were predictable. On November 14, Vovsi, for instance, told the investigators the following:

> Thinking it all over, I came to the conclusion that despite the rottenness of my crimes, I must disclose the terrible truth to the investigation of my villainous work conducted with the aim of destroying the health and shortening the life of specific, leading state workers of the Soviet Union.
>
> Having become one of the executors of these rotten criminal plots, I must in the first instance talk about myself.
>
> I am an opponent of Soviet power. In the post war years my hostility and hatred to the Soviet order gained strength.
>
> From my enemy motives, from my hatred toward the leaders of

the party and the Soviet government, I turned to the profession of medicine not to improve their health but to destroy it, to shorten the lives of specific, leading Soviet and party workers.

This was accomplished using the rottenest and most villainous means. I strove to accomplish my criminal objectives by means of incorrect, depraved methods of treatment, not observing prophylactic measures during treatment or the detection of the illness, and even by means of covering up the criminal activities of other doctors that were known to me.

Knowing all my responsibility for my villainous activity, and aiming to cover up the traces of my crimes I had recourse to various kinds of tricks.[1]

He then confessed to having mistreated Andreyev with a cocaine solution while conspiring with both Yegorov and Vinogradov.

While Vovsi, Vinogradov, and other doctors testified about their medical crimes, the investigation continued to expand the scope of the conspiracy beyond the medical profession into higher government strata. On November 15 Yegorov confessed that he and Vlasik had often gotten drunk together and that he was frequently a guest in Vlasik's house. He also revealed that Vlasik routinely acquainted him with MGB memoranda providing background information on the doctors working in the Kremlin hospital system. In 1947 he read the MGB *spravki* on approximately two hundred workers in the hospital system, though he denied having access to more particular information.[2] Yegorov and Vlasik's relationship could easily have been construed as collusive and pertinent to their behavior in dealing with the Timashuk letter of August 29, 1948, as well as to Yegorov's subsequent actions.

Yegorov told the investigation that he often visited Vlasik in his MGB office and would stay up with him until five in the morning looking over materials. What they said to each other during these late-night visits about Kremlin Hospital personnel; the treatment of important personages, such as Zhdanov; or Stalin's disposition toward them is unknown. But these visits and this personal relationship may well have been one of the key factors in Yegorov's early sense of invulnerability. "Vlasik could have commissioned his subordinates to organize a verification of the Timashuk statement, but he didn't do this, and apparently he relied on me completely," he told the investigation.[3] "Vlasik's scatterbrainedness and blindness helped me for several years to hide all traces of the grave crime I had committed."[4] As late as 1951, Vlasik continued to make the MGB *spravki* available to Yegorov and to inform Yegorov of the progress of the investigation so that he could remain "on guard."[5]

Lieutenant General Vlasik was brought in for questioning on November 21, 1952. After numerous interrogations, he was accused of protecting Yegorov and the other doctors from investigation and arrested on December 16. Thus the purge of top MGB personnel would be completed. By December 1952 Abakumov, former minister of state security; Vlasik, the head of the Kremlin guards; and Yegorov, the head of the Kremlin Hospital, were all in prison. Smirnov, the minister of public health of the USSR, had been thrown out of job. Many who worked under them, such as Likhachev and Leonov in the Investigative Unit for Especially Important Cases, and Lynko, Vlasik's deputy, were also in prison. Ryumin was gone, though not yet arrested. Many others would follow. Stalin was making good his threat to launch a general, nationwide purge. Though unclear about exact dates, Sudoplatov's recollection gives a general sense of the operation.

> At first I was not aware of the scope of the MGB purge because arrests were never announced, and it took several weeks to grasp the full extent of the scourge. I sensed something was seriously wrong when I tried to contract Colonel Shubnyakov, chief of the American section in the Counterintelligence Directorate, and found he was unavailable, despite my need for information from him about an agent he was running. . . . At that point it became clear to me that we had begun to repeat the pattern of the prewar mass purges.[6]

The purge of the MGB had two nodal points. The first was Abakumov, who was exposed as a traitor, who allegedly covered up both the Etinger and the Timashuk cases; the second was Vlasik, who allegedly covered up the Timashuk case. Both were connected to the underlying, unifying Jewish plot, indirectly in the case of Vlasik through Yegorov's complicity with Jewish doctors; directly in the case of Abakumov, through his role in the Etinger investigation and his alleged relationship with the Jewish MGB officers, Shvartsman and Broverman, both of whom had been arrested in the wake of Ryumin's October 1951 denunciation letter to Stalin.

Shvartsman had worked in the NKVD under Leonid F. Raikhman and quickly, according to his testimony, came to see that Raikhman, like himself, was a "Jewish nationalist masquerading as a Soviet person."[7] Shvartsman testified that he and Raikhman "openly expressed dissatisfaction with the national politics of the party," even before World War II. But it wasn't until after the war that his Jewish nationalism allegedly became virulent.

In 1946 the Ministry of State Security was led by ABAKUMOV, who had earlier worked together with me and RAIKHMAN in the secret-political department of the NKVD USSR. Even before this, during the War I acquired new sympathizers with nationalistic convictions in the person of the former deputy head of the secretariat of ABAKUMOV for military counterintelligence—BROVERMAN; DORON, the prosecutor for special affairs; ITKIN, the head of the investigative unit of the 2nd Chief Directorate; PALKIN, the head of the department "D."

With all these individuals I maintained a close tie and exchanged nationalistic ideas.

It's particularly necessary to speak about my mutual relationship with BROVERMAN. From long ago we openly conversed on nationalist themes and expressed sympathy with Americans and Palestinian Jews.

BROVERMAN and I referred contemptuously to Russians, and in every way praised Jews, elevating their intelligence and abilities, declaring in this regard that, really, by their history Jews were chosen to rule over the world. Accordingly, we expressed the view that it was necessary for Jews, in the USSR and in other countries, to take the example of American Jews who had penetrated into all chinks of the economic and political life of the country, demonstrating influence over the foreign and domestic politics of the American government.[8]

Doron, Itkin, and Palkin were heads of important departments; Broverman was deputy to Abakumov. Together they supposedly exchanged nationalist views, spoke about Palestine and the creation of the state of Israel, discussed ways of penetrating all the "chinks" of Soviet life in order to influence policy. Why? Because "by their history the Jews were chosen to rule over the world." The "mess of pottage" for which Yegorov sold out to Kuznetsov was being stirred in a big pot. In welding the charge of anti-Soviet Jewish nationalism with Abakumov's ministry, the MGB vividly revived the prerevolutionary libel of *The Protocols of the Elders of Zion,* a text Hitler had cited to substantiate the Nazi charge of a worldwide Jewish conspiracy. Thirty-five years after the Revolution had swept Imperial Russia off the face of the earth, one of its most heinous and fraudulent inventions had come back to life.

According to the new MGB line, Shvartsman, Broverman, Raikhman, Doron, Itkin, Palkin and their likes were simply representatives of the age-old Jewish conspiracy to "rule the world." They had wormed their way into the Ministry of State Security of the Soviet Union, just as the murderers of Zhdanov, Shcherbakov, Dimitrov, and others had seized control of the Kremlin Hospital, while Kuznetsov had been a member of the Central Committee itself.

Shvartsman was interrogated by the same Pavel Grishaev who had interrogated Varfolomeyev earlier that year and prepared the February 1952 memorandum on the plot to blow up the Kremlin. Despite his retrospective self-criticism over his role in the Varfolomeyev falsifications, Grishaev subsequently professed no compunctions or second thoughts at all about having aggressively pursued a line of inquiry that tied Abakumov directly to the Jewish conspiracy to rule the world. In 1953 his party conscience was still undisturbed by blatant anti-Semitism.

Grishaev demanded to know why Shvartsman tied his "nationalist activity" to Abakumov's appointment as minister of state security in 1946. Knowing what was expected of him, Shvartsman followed Ryumin's advice to give testimony that was "pointed and politically sharp."

This is explained by many causes. First of all, from the very beginning of ABAKUMOV's leadership of the Ministry, he began to establish his own organization, beginning with the exchanging of

cadres. Under the appearance of strengthening the leadership of the apparatus of the MGB USSR with qualified workers, he took with him from military counterintelligence an entire group of people loyal and dedicated to him, placing them in leadership positions of work sections.

ABAKUMOV changed the entire leadership of the Investigative Unit for Especially Important Cases, leaving only me in place, as someone to his liking. His staff of LEONOV, LIKHACHEV and KOMAROV led the Investigative Unit. They were not distinguished by their accomplishments, but rather by their uncomplaining obedience to ABAKUMOV. They were prepared to fulfill any of his orders, little concerned with the question of whether they were correct or not. From them one might hear the words: "The Minister ordered," "The boss gave the order," and so on.

ABAKUMOV did not pass over BROVERMAN who was advanced to the position of deputy head of the secretariat of the MGB USSR. At the same time, ABAKUMOV introduced a new, or, as he said, "special" procedure for the investigation of cases in the Investigative Unit. This opened wide new possibilities for the deception of the Central Committee of the Party and the concealment of important testimony of the arrested.[9]

Broverman became famous for "cooking" the documents sent by Abakumov to the Central Committee. He was so skilled at this that his office became known affectionately as "Broverman's Kitchen." He cut and pasted dates, facts, testimony, correcting protocols to fit Abakumov's instructions, embellishing, providing more "colorful representations," sometimes, according to Shvartsman, going so far as to leave out essential information such as that the prisoner confessed to "terrorist plans and preparations."[10]

"But really," Grishaev asked him, "didn't Abakumov know of your and Broverman's nationalistic convictions?"

"Of course, Abakumov saw that both Broverman and I were Jewish nationalists." Shvartsman explained that Abakumov allowed Broverman and himself to supervise preparation of the protocols of the investigation into the Jewish Antifascist Committee.

"This means that Abakumov, knowing you to be a Jewish nationalist, entrusted the protocols of the interrogation of the Jewish nationalists to you?"

"So it happened," Shvartsman replied. In the end, he confessed that his impression was that "Abakumov, for some reason, was well-disposed to individuals of Jewish nationality. But after this event, I was finally convinced that he wanted to mask criminals in the ranks of the Jews."[11] Just as Abakumov had covered up Etinger's confessions, he was now accused of having shielded Jews throughout the Ministry of Security in their quest for influence and ultimate domination of the Soviet government. It is worth remembering that in February 1952, during his interrogation of Maklyarsky, Ryumin told him:

> In Moscow there live more than a million and a half Jews. They have seized the medical posts, the legal profession, the union of composers and the union of writers, I'm not even speaking of the trade networks. Meanwhile of these Jews only a handful are useful to the state, all the rest—are potential enemies of the state. Especially if you consider that in Moscow are to be found all the foreign embassies, foreign correspondents, etc.[12]

The groundwork for a new assault on Jews and Jewish life in the Soviet Union went well beyond the anticosmopolitan campaign to expel Jews from jobs in science, medicine, the arts, journalism, and education, on the grounds that they were not sufficiently loyal to the Soviet Union or that, like Etinger, they greatly admired the work of their peers in America or Western Europe. Now the Jewish people as a whole was being stigmatized in the Ministry of Security and throughout the Soviet government as state enemies, just as the followers of Trotsky had been twenty years before: Jews spied for the Americans and worked actively against the state. Ultimately the charge was that Jews were plotting a coup d'etat in a conspiracy involving the Kremlin doctors, at least one secretary of the Central Committee (Kuznetsov),[13] the minister of state security, and the head of the Kremlin guards (Vlasik).

Shvartsman was probably prompted by Grishaev in stating that "Abakumov was well disposed to individuals of Jewish nationality." This idea, which had no basis in reality, posed some difficulty to the investigation. Not until the actual criminal indictment of Abakumov in February could the obvious contradiction between his alleged pro-Jewish sympathies and the arrest and torture of the members of the Jewish Antifascist Committee be resolved. Whatever Abakumov's "sympathy" for Jewish nationalists may have been, it did not seem to protect any Jewish intellectuals from the anticosmopolitan campaign.

A sign that the charge of Abakumov's pro-Jewish bias was late in developing is that it was not part of Ryumin's initial denunciation of Abakumov in July 1951. In his letter to Stalin, Ryumin said only that "comrade Abakumov, as far as my observation goes, had an inclination to deceive the government organs by means of keeping quiet about serious defects in the work of the organs of the MGB." As regards Etinger, Ryumin stated only that Abakumov had protected the doctor—not because the old man was Jewish but rather because he was anti-Soviet. Ryumin did not explain why Abakumov let him die. In his interrogations Abakumov himself was not questioned about his activity on behalf of Jews, but only about his anti-Soviet ambitions.

However, by the time Grishaev wrote his draft memorandum on the Abakumov-Shvartsman case sometime in the summer or fall of 1952,[14] Abakumov was linked to the participants of "the Jewish organization of a number of youths, calling itself 'The Union for the struggle for the Revolution' (SDR) that hid terrorist plots . . . in connection with the leaders of the Party and Soviet government."[15] This was the same Jewish youth organization referred to in the July 11, 1951, decree from the Central Committee. However, in the 1951 decree, Abakumov was linked to this group not because the members were Jewish but because they were anti-Soviet. By September 1952 the linkage had changed. Abakumov was associated with "a number of Jewish nationalists, politically and morally degenerate elements, and together with them [he] conducted subversive activity against the party and government," the purpose of which was to seize "supreme power in the country."[16]

Shvartsman provided Grishaev with information about Abakumov's role in the SDR case, stating that "Abakumov gave me the order to misrepresent the 'SDR' as a local youth group from the ranks of Jews, reading Trotskyist literature, without any kind of influence for developing a path of struggle with Soviet power. . . . Abakumov stated directly to me that the

Central Committee of the party might order that the MGB would have to discover the center that led and directed the activity of the indicated organization."[17] Shvartsman understood by this that Abakumov was intent on protecting the "center" of opposition and certain individuals despite whatever order the Central Committee might issue. Shvartsman then obediently corrected the protocols of the case, deleting, he alleged, all reference to terror. He did the same, he confessed, in other instances as well, though there is no documentation to confirm this. In November 1952 the investigation of Shvartsman finally returned to one of the premises of the July 1951 secret letter stating:

> In January 1951, in Moscow participants of a Jewish anti-Soviet youth organization [SDR] were arrested. During the interrogation, several of the arrested confessed that they had terrorist designs in connection to the leadership of the party and the government. However, in the protocols of the interrogations of the participants of this organization presented to the CC VKP(b), the confessions of the participants concerning terrorist designs were deleted at the order of Abakumov.

Thus, a year and a half after they had been itemized, all the terms of the original Central Committee decree were fulfilled, proving Stalin's farsighted vision and the infallibility of his judgment.

When Grishaev asked Shvartsman whether he had had any terrorist intentions of his own, Shvartsman readily acknowledged that he had: "In 1948 the nationalists active under the protection of the Jewish Antifascist Committee were crushed, and because of this I decided to commit a terrorist act against G. M. Malenkov."

"What practical measure did you undertake for the realization of your enemy plan?" Grishaev asked.

"In August 1950, I knew that together with Leonov and Komarov, I would have to go on an official mission to Sochi for the drafting and composition of important documents with the Head of the Soviet government[18] who took His vacation in the South. Supposing that Malenkov would be on vacation at the same time as the Head of the government, I decided to use this occasion to commit the terrorist act.

"Practically speaking, I intended to realize my enemy plot, in accordance with circumstances. . . . With this aim I took with me from Moscow a personal weapon—the pistol 'Valter.'"

"You have in mind the pistol, 'Valter' No. 777602, that was confiscated from you during the arrest?"

"Yes. I wanted to use this pistol."[19]

Nothing came of this "terrorist" plot. Over the ten days in which Shvartsman was in Sochi, the right moment never seemed to present itself for shooting Malenkov. Shvartsman claimed that he tried again in 1951 to shoot Malenkov. Again nothing came of the plot. Abakumov, Shvartsman testified, was a coconspirator, and he concluded this interrogation ominously:

> As I remarked earlier, I knew ABAKUMOV since 1938. From this period, we became participants of a conspiracy directed against the interests of the Russian people.
>
> In order to fully characterize this conspiracy, to speak about its aims, and in addition to name all of the participants and to show their ties with foreign intelligence, I must begin with the leading link of the conspiracy and name names that the investigation would think I was naming with a provocational intent.[20]

The investigators did not want Shvartsman to "name names" or show them how high the threads went. Abakumov's warning that they would be "burned up" by the case must have concerned them. Shvartsman did provide Grishaev with the names of many additional Jewish accomplices that enabled the MGB to extend the "plot" throughout the length and breadth of the security organs. In 1955 the Supreme Court of the USSR sentenced Shvartsman to "VMN"—the highest form of punishment, or execution. However, there is no record that this sentence was carried out and his fate remains unclear.[21]

MGB officer Yakov Mikhailovich Broverman had a slightly different function at this time in the rapidly expanding investigation. Born in 1908, a native of Zhitomir, Broverman had been a lifelong member of the party. During the war years he, like Ryumin, had worked in SMERSH and eventually rose to become deputy head of SMERSH under Abakumov.

After the War he joined the MGB and managed the flow of information sent to the Central Committee deploying his considerable literary skills in perfecting the culinary expertise of "Broverman's Kitchen." In his November 27, 1952, interrogation Broverman spoke openly about his activities as a forger and falsifier of security documents.

> Thus in 1945 with PALKIN, the former head of department "D" of the MGB USSR, and with UTEKHIN, the former head of the 1st Directorate, and with LEONOV, the head of the Investigative Unit for Especially Important Cases, at ABAKUMOV's order, I fabricated a photo album sent to the Central Committee about the subversive work of White émigré organizations active in Manchuria.
>
> I must say that the majority of documents, photocopies of which were placed in the album, related to the 1930s and had nothing to do with 1945. However, ABAKUMOV, wanting to create an impression that the counterespionage organs of "SMERSH" were successful in allegedly completely destroying the White émigré organizations and had seized their documents about their activities in the period of the second world war, ordered us to glue dates onto the documents found in the album.
>
> Preparing for the Central Committee a summary of information about the work of the organs of the MGB concerning the search and arrest of agents of foreign intelligence and in addition the authors and distributors of anti-Soviet papers and anonymous letters, at the order of ABAKUMOV I indicated in this material only a collection of investigated spies and authors of anonymous documents and I hid the fact that for a prolonged period of time the organs of the MGB had not ascertained who they were and that several tens of thousands of these criminals indicated in the above category were still at large. This made it possible for ABAKUMOV to hide from the Central Committee the unsatisfactory situation concerning the search for spies and the authors of anonymous enemy documents.
>
> Having material about the instances of betrayal of the Motherland by Soviet citizens and military men found on the territory of Germany and Austria, ABAKUMOV, however, did not

inform the Central Committee of these enemy manifestations; but in those instances when he was forced to inform the Central Committee of these facts, ABAKUMOV with my help attempted to present the situation such that the organs of the MGB remained peripheral to the case so that the full blame would fall on other departments. Deceiving the party and the government, ABAKU-MOV and we, his associates, hid from the Central Committee the unsatisfactory situation in the MGB and the criminal collapse of the work in the Chekist organs.[22]

Broverman's testimony was collaborative, the result of what the MGB wanted and what he could invent. How much of it, if any, was true is impossible to know because it so thoroughly reflected the interests of the investigation. Broverman's last sentence, "Deceiving the party and the government, ABAKUMOV and we, his associates, hid from the Central Committee the unsatisfactory situation in the MGB and the criminal collapse of the work in the Chekist organs," summarized a key element of Ryumin's original charge against Abakumov: premeditated concealment of "the unsatisfactory situation in the MGB" from the Central Committee. It echoed Ignatiev's November 24, 1952, memorandum to Stalin on the state of affairs in the Ministry of Security and provided a confessional basis for the December 4 decree of the Central Committee stating that Abakumov "hid evidence from the Central Committee" and led it "astray" in the Zhdanov investigation.

Pavel Grishaev, one of Broverman's interrogators, was simultaneously interrogating Shvartsman and Varfolomeyev and was in an excellent position to know what was necessary for Broverman to confess. He was not satisfied with Broverman's simple account of his falsification work. "While you acknowledge your guilt in committing specific crimes," Grishaev told him, "you, nevertheless, attempt to hide the essence of your enemy activities." In other words, the testimony still lacked political point and sharpness.

"Who directed your subversive work?" Grishaev wanted to know.

"On whose behalf Abakumov acted, I have no idea," Broverman told Grishaev. "I executed the criminal orders of Abakumov; I had no other bosses."[23]

It is not clear why Broverman could not now invent a suitable falsification. Possibly Broverman's inconclusive answer provided the kind of verisimilitude Ryumin thought lacking in the list of names produced by foreign intelligence in the Varfolomeyev case. Though he could not tell who, other than Abakumov, directed his subversive work, he was able to provide the investigation with a long list of individuals who were so directed. He named fourteen members of the MGB as his close associates in this conspiracy. Most were Jewish. All were immediately arrested.

I must say that for several years SHVARTSMAN had already been known to me as the most vicious Jewish nationalist contriving criminal affairs in the organs of state security.

In conversation, SHVARTSMAN and I often poured out our hearts to each other and maliciously slandered the existing Soviet order.

We spoke with scorn about Russians and other nationalities and in every way elevated Jews for their alleged outstanding intelligence and abilities.

Enveloped by the nationalistic poison, we agreed to the blind conviction that Jews by virtue of their alleged special qualities of intelligence were called by history itself to rule the world.

As an example that demonstrated our thinking, we referred to the American Jews who dared to penetrate the sphere of management and political life of the country and directed both foreign and domestic politics of the USA.

In connection with this SHVARTSMAN declared, and I supported this, that the Jews living in the Soviet Union should take as their example the American Jews.

Our favorite theme that I discussed more than any other with SHVARTSMAN was the question of the so-called "suppression" of the Jews of the USSR.

Fully sharing these enemy views of SHVARTSMAN, in conversation with him I also slandered the national politics of the party and I said to SHVARTSMAN that the Jews in the Soviet Union were not given the chance to develop their culture, that they had closed the only Jewish theatrical studio in the USSR and the Jewish theaters.

> Reacting viciously as regards the closing of the office of the Jewish culture in the Academy of Science of the Ukraine SSR and the arrest of the leading figures of this office, I slanderously called these measures the fruit of an incorrect politics in regards to the Jews, conducted by the party, and that the struggle against cosmopolitanism qualified as "the organized persecution of the Jews."[24]

Broverman's testimony reproduced parts of Shvartsman's almost verbatim. The statement that Jews "by virtue of their alleged special qualities of intelligence were called by history itself to rule the world," shows Grishaev's guiding hand and indicates that the testimony was probably prepared for Broverman in advance, just as Ryzhikov's and Likhachev's were. But Broverman's statement, also echoing Shvartsman, that "As an example that demonstrated our thinking, we referred to the American Jews who dared to penetrate the sphere of management and political life of the country and directed both foreign and domestic politics of the USA," further supported Ryumin's allegation that Jews had "seized the medical posts, the legal profession, the union of composers and the union of writers, I'm not even speaking of the trade networks." Ryumin made this comment to Maklyarsky in February 1952; Broverman's "confession" was made November 27, 1952. Obviously the investigation was coordinating the testimony of key witnesses. This gives strong reason to believe that a public trial was being planned, as Ryumin had said there would be, at which the confessions of leading MGB officers would be made to harmonize with that of Varfolomeyev and the doctors. Though Broverman could not say who Abakumov's bosses were, this would be revealed once "The Plan of the Internal Blow" was made public and Abakumov's covert ties to Kuznetsov and the death of Zhdanov were exposed.

The case against the Jews as Jews went forward quickly now. It was no longer a matter of Jewish doctors and Jewish MGB agents bearing a grievance against Soviet policy and conspiring with Abakumov to usurp the government. The libel against the Jews took on the age-old accusation that all Jews *inevitably* felt a deep and ineradicable hatred toward the motherland. Broverman's testimony about a certain Jewish MGB agent, Belkin, illustrates how this was developing. Broverman related that in 1948 Belkin[25] was recalled from Austria for a meeting at the Central Committee. Belkin presumed that he was going to be sacked and arrested. He complained to Broverman that the sole reason behind this was the fact that he was Jewish. "Here," Broverman said, "Belkin raised the usual Jewish slander against the national politics of the Central Committee of the party and of the Soviet government, talking openly about the so-called 'persecution' of the Jews. . . . Specifically, I agreed with his concocted notion that there exists a 'directive from above' in our country concerning the delimitation of the rights of Jews and that local party and Soviet workers created a complete tyranny over the Jews, and that the CC and the government knowing about this did nothing."[26]

"Why was it forbidden," Broverman asked Grishaev, "for Jews to express their feeling of sympathy toward the representative of the Jewish state [Golda Meir], the creation of which they had dreamed of for centuries, while at the same time Russians were not prohibited from honoring the Patriarch Aleksei in Yelokhovsky Cathedral?"[27] The transition from Belkin's personal hatred toward Soviet policy ("directive from above") to the centuries-old Jewish dream of a Jewish State shows how the transition was being effectuated from a specific to an inveterate, congenital condition of opposition. Jewish hatred against the Soviet Union was age-old, undoubtedly going back to pre-Revolutionary times when the patriarch of the Russian Orthodox Church had an honored place in Russian life. This inveterate hatred would be presented as something concealed in the perverse conviction among Jews that they were being singled out by the Soviet government and discriminated against. Broverman acknowledged that such discrimination was a "concocted notion." It served as a pretext for Jewish treason and would confirm Jews, to use Ryumin's phrase, as a "spying nation." Broverman was not shot along with Abakumov. In 1954 he was sentenced to twenty-five years in the gulag but was eventually released in 1976.

Broverman's testimony would help merge the alleged MGB conspiracy with the case of Dr. Etinger, who had complained that, "if the representatives of local power knew that the government would punish them for their anti-Semitism, they wouldn't do it." The government had now gotten another Jew, Broverman, to state categorically that this "anti-Semitism" was a fraud, and the kind of complaint made by Etinger was nothing but a smoke screen for age-old "enemy intentions." Brought into the Varfolomeyev case because of his "juridical" background and skill as an interrogator, Grishaev was tying up the loose ends of this vast conspiracy for presentation to the Soviet public and world opinion.

On November 24, 1952, Goglidze, who had taken over the investigation of the doctors after Ignatiev became ill, wrote to Stalin:

> The investigation established that YEGOROV and FEDOROV [the pathologist]—were politically and morally rotten people; MAIOROV—came from a manor house circle; VINOGRADOV— in the past was affiliated with the SRs [Social Revolutionaries]; VASILENKO—since 1922 hid his expulsion from the party for deviation from party discipline, and connected with him, the Jewish nationalist KARPAI (arrested)—all of them composed the hostile group, active in the Polyclinic of the Kremlin hospital, that strove to cut short the lives of leaders of the party and government through medical treatment.

> This enemy, terrorist group of physicians worked exactly as did physicians of the people in the past—PLETNEV and LEVIN, insidiously killing V.V. KUIBISHEV, V. R. MENZHINSKY, A. M. GORKY, and his son, M.A. PESHKOV. They—YEGOROV, VINOGRADOV, VASILENKO, FEDOROV, MAIOROV, and KARPAI conducted terrorist activity by means of prescribing to the patients such treatment that ruined their health, complicated the illness, and led to their demise.

> YEGOROV, VINOGRADOV, VASILENKO, MAIOROV and FEDOROV acknowledged that they formerly were enemies of the party and Soviet state; that they, making use of the illness of com. ZHDANOV, in premeditated fashion prescribed for him a categorically contraindicated active regime, they provoked a grave heart attack and in this way killed him.

> During interrogation, VOVSI and KOGAN [B. B.] confessed that both of them, being Jewish nationalists, supported enemy ties with the leaders of the Jewish nationalist underground, operating under the cover of the Jewish Antifascist Committee.

> In this way, by means of the collected verifying documents and the
> confessions of the prisoners it has been established that in the
> Lechsanupra of the Kremlin a group of terrorist doctors was
> active—YEGOROV, VINOGRADOV, VASILENKO, MAIOROV,
> FEDOROV, LANG and Jewish nationalists, ETINGER, VOVSI,
> KOGAN and KARPAI, who strove through medical treatment to
> cut short the life of leaders of the party and the government.[28]

With the exception of Karpai, the doctors had by this point all confessed
to murder; Maiorov, who came from "manor house circles," had been a
criminal from birth because of his alien social origins; while Vinogradov
had been associated with the outlawed SRs (Social Revolutionaries), a
charge brought in at the last moment and not included in any prior interrogations; nor had this past association prevented Vinogradov from accompanying Stalin himself to the Tehran Conference in 1943 as his personal
physician or treating the Vozhd in the fall of 1951. Perhaps most importantly, Goglidze followed the line laid down in the secret letter and tied the
criminal activity of this group back to the alleged assassinations of the
1930s, performed by Drs. Pletnev and Levin.

> Among the doctors there undoubtedly exists a conspiratorial group
> of individuals, intending through medical treatment to shorten the
> life of leaders of the party and the government. It is impossible to
> forget the crimes of those well-known doctors, committed not that
> long ago, such as the crimes of doctor Pletnev and doctor Levin,
> who poisoned V. V. Kuibyshev and Maxim Gorky at the direction of
> foreign intelligence agencies. These villains confessed to their
> crimes at an open trial and Levin was shot, but Pletnev was sen-
> tenced to 25 years of prison.[29]

The Soviet people could therefore see that there was an established pattern to this deviltry. The Jews in combination with traitorous Russians had
fooled the government for many years, but at last they had been caught.
Goglidze's report went beyond the secret letter in some significant respects.
Goglidze weaved this terrorist group of "politically and morally rotten people" together with the plotting of Kuznetsov, who "in connection with his

own hostile plans was interested in the removal of comrade ZHDANOV."[30] In addition, he provided the plot with its Jewish character by identifying Etinger, Karpai, Vovsi, and Kogan as Jews and part of a "Jewish underground," something the decree had implied but left unsaid. It had taken one and a half years, but the case had finally ripened, like the fruit Stalin showed off to Smirnov in his garden in the Crimea.

Nearly everything was ready, and Stalin was poised to act.

On November 30 Goglidze prepared the penultimate document[31] in which he set out for Stalin the deplorable conditions in the MGB requiring immediate correction and pointed out that the terrorist doctors had been apprehended and made to confess (something Stalin already knew), and reconfirming that they were working primarily for "American intelligence."[32] This document complained that the Abakumov-Shvartsman case had not been fully or properly developed because of the clumsy and inept work of the MGB operatives who "work without spirit," clumsily make use of "slips of the tongue" by the prisoners, and do not "clutch like hooks at each possibility, however small, to grasp, to take the enemy into their hands—fully to unmask them." [33] This is the result, Goglidze explained, of abuses that had continued for many years and were not easily or quickly rectified. A total purge was therefore necessary.

The next day, December 1, 1952, Stalin called an emergency meeting of the Presidium of the Central Committee. The Timashuk letter had come to his attention. Stalin denounced Abakumov and Vlasik for treason and criminal negligence; the terrorist group of doctors in the Kremlin Hospital along with their Jewish nationalist conspirators were exposed; the Minister of Health of the USSR, Ye. I. Smirnov, was removed from his post. On December 4, the Central Committee issued its declaration:

> Having been apprised of the information of the MGB concerning sabotage in the doctors' plot, the Presidium of the Central Committee of the party has confirmed that in the Lechsanupra a group of criminals has been active for a long time; the former heads of the Lechsanupra Busalov and Yegorov entered this group; doctors Vinogradov, Fedorov, Vasilenko, Maiorov, Jewish nationals Kogan, Karpai, Etinger, Vovsi and others were also a part.
>
> Documented facts and the testimony of the arrested have estab-

lished that the enemy group was tied to English and American embassies, worked at the direction of American and English intelligence and had the goal of committing terrorist acts against the leadership of the Communist Party of the Soviet government.

Under the weight of the evidence, the participants of the groups have confessed that to commit sabotage they established incorrect diagnoses of illnesses, prescribed and implemented incorrect methods of treatment and by these measures brought their patients to their deaths. The criminals have confessed that by these means they were successful in killing A. A. Zhdanov and A. S. Shcherbakov. Was there the possibility expeditiously to uncover and decapitate this enemy group, active in the Lechsanupra? Yes, there were possibilities for doing this.

As far back as 1948 the MGB had at its disposal signals that manifestly spoke of the unsatisfactory situation in the Polyclinic. Doctor Timashuk turned to the MGB with her declaration in which on the basis of her electrocardiograms she confirmed that the diagnosis of com. Zhdanov, A. A., was incorrectly established and did not correspond to the facts of the inquiry, and that the prescribed treatment for the patient did him harm. If the MGB would have conscientiously investigated this exceptionally important statement, it certainly would have prevented the villainous murder of com. Zhdanov, A. A. It would have exposed and liquidated the terrorist group of doctors. This did not happen because the workers of the MGB USSR dealt criminally with the case, putting the declaration of com. Timashuk into the hands of Yegorov who was a participant in the terrorist group.

Further in 1950, the former minister of state security, Abakumov, having direct facts about medical wrecking, having received from the MGB the results of the investigation of the case of the arrested doctor Etinger of the Lechsanupra, concealed this from the Central Committee of the party and curtailed the investigation into the matter.

The former head of the Chief Directorate of the Guards, Vlasik, who should have exercised control over the work of the Lechsanupra at the commission of the MGB, fell together out of drunkenness with the now exposed leaders of the Lechsanupra and became a blind tool in their hands.

The Minister of Health of the USSR, com. Smirnov, instead of implementing control and leadership in the Lechsanupra, having entered into the system of the Lechsanupra, also because of drunkenness, fell in with the exposed head of the Lechsanupra and, despite the presence of signals of a bad situation in the Lechsanupra, he did not show vigilance or principled behavior.

After the change in leadership of the MGB USSR in July 1951, the Central Committee of the party considered it necessary to remind the new leadership of the MGB of the criminality of other well known doctors, such as Pletnev and Levin, who at the direction of foreign intelligence poisoned V.V. Kuibishev and A. M. Gorky, and who showed through this that there was a conspiratorial group of physicians striving through medical treatment to shorten the lives of the leaders of the party and government. The Central Committee then demanded that the MGB use all possible means to expose this group of enemy doctors. However, the new leadership of the MGB unsatisfactorily fulfilled these directives, demonstrated slowness, poorly organized the investigation of this important case, with the result that much time was wasted in exposing the terrorist group in the Lechsanupra.

The Central Committee declares:

1 To charge the MGB as follows:
   a. To expose to the end the terroristic activities of the group of physicians, active in the Polyclinic, and its ties to American-English intelligence;
   b. In the course of the investigation to clarify by what means and by what steps it will paralyze and correct the wrecking work in the Lechsanupra and in the treatment of patients;
2 To remove Com. SMIRNOV, Ye. I., from his post of minister of health on the grounds of unsatisfactory leadership and political negligence.

   The case of Com SMIRNOV is to be transferred for review to the Committee of Party Control.

3. To charge the bureau of the Presidium of the Central
   Committee of the party:
   a. To select and appoint a minister of health for the USSR;
   b. To work out measures for correcting the situation in the
      Lechsanupra.

This December 4 decree of the Central Committee did not identify the
plot as entirely Jewish. Even here, Stalin was careful to link the Jews with
several other interests: American intelligence, Abakumov's ambition, and
subversive Russian nationals, such as Yegorov. Only when the "plot" was
presented to the Russian public would the Jewish element burst to the fore.
The wrecking of the medical profession was associated yet again with the
"murders" of Gorky and Kuibishev. Stalin, however, had said more than
was recorded in the written decree:

> The greater our success, the greater will be the harm our enemies
> attempt to do to us. Under the influence of our great successes, our
> people forget about this; indifference, carelessness [*rotozeystvo*], con-
> ceit have come about.

> Every Jew-nationalist is an agent of American intelligence. The
> Jewish-nationalists think that the USA has saved their nation (there
> they may become rich men, bourgeois, etc.). They consider them-
> selves obligated to the Americans. Among the doctors are many
> Jewish-nationalists.[34]

Stalin told the assembled members of the Presidium of the Central
Committee[35] gathered in his office on December 1 that "every Jew-
nationalist is an agent of American intellegence," but it was quickly
emerging that every Jew was a nationalist. We shall soon see what steps he
would devise to eradicate *rotozeystvo* (carelessness or thoughtlessness) from
his society along with the other enemies of the people.

The case against the MGB hinged on its alleged criminal handling of
Timashuk's 1948 statement, which is why the question of whether
Yegorov actually saw the letter to Vlasik was so important. It was vital for
the case against the MGB to demonstrate that the MGB actually gave

Yegorov the letter, something Yegorov initially denied but eventually, and only after physical torture, admitted on February 7, 1953—two months after the assertion had been made in the December 4 decree.[36] In addition to Abakumov and Vlasik, Aleksandr Belov, Zhdanov's bodyguard and family friend since 1935, allegedly had betrayed MGB security. No one could be trusted.

The December 4 decree did not divulge what Stalin knew to be the case, that Timashuk's statement had been duly delivered to Abakumov the day Vlasik received it, and that Abakumov instantly passed it to Stalin along with copies of Zhdanov's EKGs. To achieve the necessary political aims, such facts were irrelevant. The wording of the declaration suggested that Ignatiev, who was still recovering in the hospital at the time, might well have become another victim. Though not mentioned by name, he was responsible for the "new leadership of the MGB" that

> unsatisfactorily fulfilled these directives [of the Central Committee], demonstrated slowness, poorly organized the investigation of this important case, with the result that much time was wasted in exposing the terrorist group in the Lechsanupra.

Ryumin, who had been instrumental in purging Abakumov, had himself already been eliminated. His honorable service was not mentioned. Ignatiev must have known that he could well have been next in line.

Unceasing rounds of new interrogations of the doctors and MGB operatives ensued after this December 4 decree. Vovsi, Vinogradov, and Yegorov admitted everything their persecutors put before them. The name of Solomon Mikhoels turned up in the questioning of Vovsi, who on December 8 "recalled" that "Mikhoels painted the American way of life in joyful colors." The Jewish Antifascist Committee and the Jewish underground would be made important elements in the ultimate presentation of the conspiracy to the Russian public, and eventually Vovsi confessed:

> In this way, it was then already clear to me that the name "Jewish Antifascist Committee" was only a smoke screen under which Jewish nationalists realized their anti-Soviet, nationalistic goals; and this fact that the Committee directed to the US various kinds

of information about the Soviet Union directly shows that he [Mikhoels] essentially served the interests of Zionist circles in the US.[37]

From medical wrecking in the Lechsanupra, the plot would widen to the role of Jews throughout Soviet society, reaching into "important posts in science, art, literature and education." It would include both "old positions" and those filled with "young Jews" in "various spheres." Old and young; politics, art, science, literature, and education; the scenario mapped out by Ryumin to Maklyarsky before he left his post in 1952 had now been realized. It was indeed a "spying *nation*." The Jewish *nation* had "seized the medical posts, the legal profession, the union of composers and the union of writers. I'm not even speaking of the trade networks. Meanwhile of these Jews only a handful are useful to the state, all the rest—are potential enemies of the state."

One piece of the puzzle still had to be put in place. Vlasik, Stalin's loyal bodyguard and head of the Kremlin guards, knew too much about the handling of the Timashuk letter and the situation with Dr. Yegorov. On November 26, 1952, Vlasik's deputy Lynko was interrogated and told the investigation that Yegorov and Vlasik drank together, that Vlasik allowed Yegorov to collaborate over the vetting of candidates for Kremlin Hospital positions. Maslennikov was interrogated on the same day and produced the information about "Yurina's" report, stating that it was turned over to Lynko. He blamed Vlasik and Abakumov for ineptitude and irresponsibility. Vlasik was not, however, immediately arrested. Stalin waited again. Perhaps he assumed that Vlasik's arrest could be the centerpiece of the next phase of the operation after the case became public.

Throughout the month of December the interrogations and arrests mounted. Day after day new names were added to the lists of conspirators in every walk of life. At this point Vlasik became a major focus of MGB attention. He was interrogated intensively on December 4 and 10; he was finally arrested on December 16. Much of the inquiry concerning Vlasik appears to have been directed at sealing shut the Timashuk-Yegorov-Abakumov story and demonstrating the total corruption of the security services. Moreover, it is clear that the interrogation of Vlasik was designed to distance Stalin from any connection to the plot. The story was reiterated through Vlasik's interrogation in such a way as to leave Stalin out of it and to make sure his name could not under any circumstances be invoked. Vlasik was the last man active in the Soviet government who could have made the direct connection with Stalin.

As Abakumov before him, Vlasik did not reveal—even under terrible conditions of intimidation and mental, if not physical, torture—the source from which his orders came. On December 10 Vlasik was questioned closely about his handling of Timashuk's August 29, 1948, statement and the September 6 session in the Kremlin Hospital conducted by Yegorov. The interrogators asked why he didn't organize a "Chekist" verification of her statement? They pointed out that only individuals interested in refuting Timashuk's statement took part and therefore it couldn't possibly have been conducted objectively. Vlasik remained firm: "I can only repeat myself," he told the MGB, "that my guilt consists in the fact that I believed the conclusions of the session and was convinced that the statement of TIMASHUK had been refuted, and that YEGOROV had reported this to the Central Committee. In connection with this I did not pay attention to the list of participants of the session."

His interrogators pointed out that Timashuk complained about the meeting with the doctors on September 6. Why did he still do nothing? "I don't remember this report, but if it existed, I gave LYNKO or MASLENNIKOV the order to sort it out," Vlasik replied.

"You never gave such orders. LYNKO testified that you considered the verification conducted by YEGOROV in connection with the TIMASHUK statement satisfactory and therefore no measures were taken on it."

"They never reported the TIMASHUK statement to me."

"The head of the operational department MASLENNIKOV stated that he personally reported this statement to you."

"I speak the truth. If there had been a statement, it would have been reported to me, I would have given an order regarding a verification of it."

"MASLENNIKOV testified that he put the question of conducting such a verification to you, but that you did not permit it to be made, declaring that it would arouse an undesirable response in Moscow."

"What led you to forbid the conduct of the inquiry on the TIMASHUK statement?"

"MASLENNIKOV testified falsely. I was not able to prohibit the verification of the signals received from TIMASHUK, and I did not give him this instruction."

The MGB could get nowhere with Vlasik—but this is precisely where they needed to go, even if the interrogator himself did not realize it. The situation was complex. There is no evidence that the MGB knew how the Doctors' Plot was initiated or toward what goal it was driving. The interrogators themselves were acting on orders—from Garkusha to Grishaev—and were not necessarily aware that *the threads went higher*. Had Vlasik revealed the entire story, assuming he even knew it, he probably would have been denounced as a provocateur and shot immediately.

Vlasik's cat-and-mouse game with his interrogators took another turn.

"There were other signals that obliged you to take immediate measures concerning the TIMASHUK statement, but you evaded them. Where was the autopsy of the body of comrade ZHDANOV, A. A., performed?"

"I don't know."

"Who performed it?"

"I don't know because I was busy fulfilling other assignments."

"You, as the head of the Chief Directorate of the Guard should have known where and by whom and in what circumstances the autopsy of the body of comrade ZHDANOV, A. A., was performed." Vlasik explained that there was a procedure whereby autopsies of deceased leaders of the party were performed by a government commission in combination with the head of the Lechsanupra. Vlasik was not responsible. The interrogator was incredulous. If a government commission was supposed to undertake the autopsy, why was it performed in the bathroom at Valdai by a single pathologist, Fedorov? Vlasik said only that he had to believe that it was done correctly because of the procedures in effect at the time.

"But you had at your disposal other signals about the incorrect diagnoses of patients by doctors of the Lechsanupra. This obligated you to take measures to ensure that the results of the autopsy of the body of comrade ZHDANOV, A. A., were objective." Vlasik said that there were no such signals that the autopsy in Valdai had been performed by "doubtful individuals or those who did not inspire confidence. For these reasons," he told the investigation, "I did not verify who was allowed to perform the autopsy on the body of A. A. ZHDANOV."

The argument was circular and self-contained. Despite his reputation for being "scatterbrained" and "politically blind," Vlasik proved himself to be a formidable witness. At every point where the truth could have leaked out—that Poskrebyshev had approved the autopsy and that Abakumov had indicated Stalin's reaction to Timashuk's letter—he turned the discussion toward impersonal procedures and commissions. No one could be blamed. The plot was a black hole from which no light could escape.

The interrogator was not satisfied and pressed on, pointing out that the autopsy was conducted by only one pathologist and not by a commission of authoritative specialists. "In this situation you should have been critical of the results of the autopsy," Vlasik was told.

But he was too quick for these investigators. "In connection with the fact that the official verification conducted by YEGOROV according to Lechsanupra procedure refuted the statement of TIMASHUK, I did not pay attention to the fact that the autopsy of the body of A. A. ZHDANOV was conducted by only one pathologist."

Unable to make Vlasik contradict himself, the investigators finally asked: "What explains the fact that you believed YEGOROV?"

"I can explain this," he said, "only by the fact that no one gave me any instructions concerning an inquiry into the statement of TIMASHUK after the session convened by YEGOROV." What did this mean? No instructions?

The interrogators did not even ask. Earlier, he testified that Yegorov told him that he had reported the results of the September 6 Kremlin session to Poskrebyshev, something that gave Vlasik reason to think that others in the government, Stalin in particular, knew about it. Perhaps sensing potential awkwardness, the interrogators had backed away from Vlasik's previous assertion, and now they did the same. Instead of asking what he meant by "no instructions," they wanted to find out more about his "mutual relationship" with Yegorov.

They were close, Vlasik told them, "we drank together; but nothing was ever said about official matters."

The investigators knew this was false. "This is not true. You were under YEGOROV's thumb and incorrectly, in an *unchekist* manner looked upon signals concerning the Lechsanupra, especially when they pertained to YEGOROV himself."

"I never hid the crimes of YEGOROV. Concerning material received by the Chief Directorate of the Guard about the crude relationship of YEGOROV to his assistants in the Lechsanupra, the use by him at the State's expense of one of his assistants as a nanny for his child, I pointed out to YEGOROV the impermissibility of similar facts."

This did not satisfy the investigation. Vlasik was again accused of not having undertaken the necessary verification of Timashuk's statement. In avoiding the question of "instructions" from above, the investigators nevertheless turned the inquiry back to the subject of Timashuk's statement that appeared to have been left far behind. Vlasik reminded them of this. "I have already given testimony on the questions connected with the verification of the statement of TIMASHUK." Without being able to name Stalin, Vlasik could provide nothing substantial, but the MGB did not know this. He attempted to end the interrogation with the admission that "serious signals" might have gotten past him through lack of vigilance. This was not enough for the investigators.

"MASLENNIKOV testified that the materials on YEGOROV's failure to take measures to liquidate laxness in the work of the Lechsanupra was met by you with hostility, groundlessly put under suspicion, the result of which was that necessary measures were not taken."

"MASLENNIKOV lies." In fact, Maslennikov was probably telling the truth, and the investigators knew it. "There are materials in the MGB USSR that testify, to the contrary, that your testimony does not correspond to actuality. In 1950 you were given a detailed *spravka* on YEGOROV in which there were many facts characterizing the unsatisfactory conditions in the Lechsanupra." Vlasik not only saw this *spravka*, he actually annotated it. Where it spoke of a visit by Yegorov's wife to the Syrian and Egyptian missions in Moscow, "allegedly with the aim of arranging for work there," Vlasik wrote the comment at the bottom of the page, "Pure nonsense, verify all this information, check it." Again they asked him what his relationship with Yegorov was all about.

Vlasik told them that he charged Maslennikov to check everything carefully but that he didn't check to see whether his order had been fulfilled or not. Then Vlasik revealed something that indicated that he knew far more than he would say. Vlasik claimed that, in fact, Maslennikov inexplicably delayed in verifying the incidents and, furthermore, he didn't pay attention to the results because "the plan of action for the cultivation of YEGOROVA [Yegorov's wife] was not constructed until March 1952, after which I reported it to the minister of state security of the USSR, IGNATIEV." Yegorov was not arrested until September 1952 after Ryumin's August letter to Stalin, seven months after the plan to "cultivate" his wife had been "constructed." The decision to move against Yegorov had, therefore, been planned *before* Ryumin denounced him in his report to Stalin about the "expert examination" of Zhdanov's heart.

The plan to connect Yegorov and the death of Zhdanov to Etinger and the death of Shcherbakov apparently was gaining momentum at an even earlier point than had previously been thought. To what extent Stalin's signature was engraved on this tiny golden nail in the flea's shoe we will never know, but it is doubtful that Vlasik would have undertaken such an action without approval from the very top of the Soviet government.

"I must confess," Vlasik concluded, "that I entirely believed YEGOROV. However, my close mutual relations with him did not influence the outcome of the verification of his incorrect actions."

The investigators remained unconvinced.

"Your statement that the mutual relations with YEGOROV did not influence the resolution of the questions of Chekist work in the Lechsanupra is contradicted by the facts.

"You did not demonstrate elementary Chekist vigilance. And you did not react and did not force the apparatus to react to the signals received by the Chief Directorate of the Guard."

In the end, Vlasik asked his interrogators to believe that his "mutual relations with YEGOROV did not tell on [his] reaction to signals received in the Chief Directorate of the Guard." He concluded, in standard MGB fashion, by blaming his subordinate Lynko, upon whom Vlasik relied to oversee the operational department of the guards. No one can believe anyone in Stalin's world of "*everlasting* distrust."

Although Vlasik's last act as the head of the Kremlin guards was to protect Stalin, not from physical assault, but from any taint of responsibility for the Doctors' Plot, he was not well rewarded. His fate was an important object lesson in the meaning of "party discipline."

Nikolai Sidirovich Vlasik was born in 1896 in the village of Bobynich in Belarus. He joined the party in 1918. He rose to the rank of lieutenant general. He was decorated with three Orders of Lenin and four other military orders, including the Red Banner, the Red Star, and Kutuzov First Class, and he had numerous medals for the defense of Moscow and victory over the Germans.

Vlasik was a dissolute but exceedingly loyal man, who seems to have enjoyed Stalin's goodwill over many years and was favored with both his confidence and his forbearance. He rose from his peasant origins to become the head of the Chief Directorate of the Kremlin Guards. On December 16, 1952, Goglidze wrote a terse memorandum to Stalin informing him that "the former head of the Chief Directorate of the Guard of the MGB USSR, VLASIK, N. S., was arrested today 16 December of the present year. We are commencing his interrogations."[38] Stalin's reaction was not recorded.

Molotov speculated that it was women who "got him into trouble."[39] Although Molotov undoubtedly overrated the significance of Vlasik's many affairs in comparison with the importance of the Timashuk letter, the MGB made liberal use of a vast amount of incriminating evidence to discredit Vlasik as a degenerate and a scoundrel. His corruption was relentless and astounding. He drank almost continuously and had so many women he couldn't keep track of them all. His "sexual adventures" were legendary.

Vlasik's friend Vladimir Stenberg, an artist of Swiss parentage who had studied and worked abroad in the 1930s, told the court in 1955 that it was hard for him even to estimate how many women Vlasik had. Stenberg would often arrive at his friend's dacha for an afternoon of billiards and drinking to find a different woman in every room. "What he did with them," Stenberg disingenuously added, "I have no idea."[40] However, he was able to remember the names of some: Nikolaeva, Ryzantseva, Dokukina, Loktionova, Spirina, Veshchitska, Gradusova, Averina, and Vera Gerasimovna. "I suppose that VLASIK also made love to SHCHERBAKOVA [probably not the wife of A. S. Shcherbakov], with the sisters of Gorodnicheva—Lyuda, Ada, Sonya—with Kruglikova, Sergeyeva (the wife of Kozlovsky) and her sister and with others whose names I don't know." Stenberg ultimately confessed that after they had become close friends, Vlasik proceeded to seduce Stenberg's own wife, a fact of which he cynically informed Stenberg in due course.[41]

Vlasik also didn't know or couldn't remember the names of all the women he'd seduced. But he confessed to the investigation that there had been "many." When the MGB searched his apartment, they confiscated a notebook filled with names. Most were in the masculine form.[42] The investigation became suspicious because some had two lines drawn next to them. He was questioned closely about this.

"For a long time," Vlasik confessed, "I have been morally dissolute; I had many women whom I've made love to and I noted their telephone numbers in my notebook. With the aim of keeping it from my wife, I indicated some of my lovers under the guise of male names, and then so that I would remember, I placed a conventional mark—two lines—before these names."[43]

Vlasik's alleged escapades were not limited to women, however. By his own account, Vlasik, "having lost a human conscience . . . swiped produce and wine from the kitchen of the person" he was guarding.[44] This "person" was Stalin himself; the kitchen was at the Blizhnyaya dacha outside Moscow. When Stalin learned of the matter, he reportedly reprimanded his longtime subordinate by telling him simply to cut it out. Others could have been shot the next day for such behavior.

While at Potsdam in 1945 where he was an important part of Stalin's entourage, Vlasik confessed that he was "seized by the thirst for gain" and, like Gargantua, "dragged away everything" he could. It was actually quite a bit. His assistant Fedoseyev brought him women almost daily and helped him cart off valuables that he would pack up and ship by government train back to Russia—presents for himself, his wife, and various relatives in Byelorus.

> [Fedoseyev] was a toady before me; he gave me produce and wine from the kitchen of the guarded person [Stalin], and during my stay in 1945 on an official mission in Potsdam, we got drunk together and at my request he brought me the sister of the hostess who I attempted to induce to make love to me.
>
> When I was seized by the thirst for gain, I dragged away everything that I could in Potsdam, beginning with automobiles, rugs, pictures and ending with accordions and alarm clocks, sending it all by the airplane and train of the Directorate of the Guard from

Germany to myself in Moscow. FEDOSEYEV on his own initiative selected a series of things for my wife.

Speaking honestly, the main role in carrying away state property was played not by FEDOSEYEV but by me, and my moral degeneracy during the mission to Potsdam was displayed especially sharply.

Everything that I took from there seemed little to me, and I decided to satisfy not only myself but my relatives on the state's account. With this aim, I placed my nephew on the Directorate of the Guard's airplane for Potsdam from [Belarus], and then with him on the Directorate of the Guard's train, I sent my brother, sister and nephew in Byelorus, three *korov*, pedigree bulls and two horses.[45]

Vlasik's boundless appetite, if without malice, was also without conscience of any sort. He raped everything he could get his hands on. His concupiscence was limitless and egalitarian—women, *korov* bulls, horses, automobiles, alarm clocks, china service for his wife, and accordions—all "seemed little" in his eyes. He requisitioned government airplanes for his nephew, made love to his hostess's sister, and stole food and wine under Stalin's nose.

The court asked how it was possible for him to have such an enormous amount of expensive crystal vases, goblets, and china in his house. He answered that it was all obtained in Potsdam and brought back to Moscow for Stalin's dacha. When it was found unsuitable for the dacha, Vlasik kept it for himself.[46] When the court said flatly that this was robbery, Vlasik responded, "No. It was the abuse of my position. Afterward when I was reprimanded by the Head of the Government, I cut it out."[47]

This low comedy, however, had a serious side because it was also discovered that Vlasik had taken moving pictures of government leaders, including Stalin, at various government dachas.[48] Furthermore, he had revealed to his friend Stenberg where government figures sat in the Bolshoi Theater, and had made telephone calls to the Kremlin from his dacha while Stenberg and various women were present. All of this was a serious and obvious breach of security.

But it wasn't only the physical safety of the leaders of the government that was in question. Vlasik claimed that in 1940 he and Poskrebyshev,

Stalin's secretary, often engaged in trysts with "loose women" at his dacha and drank heavily together. One of these women turned out to be a stenographer of the Special Sector of the Central Committee of the party, a Sonya whose last name Vlasik could not remember. The other, Anna Solovieva, was the typist for the Special Sector. The interrogator wanted to know whether these women spoke of their work in the Special Sector of the Central Committee. Vlasik said they hadn't. They only drank and "had fun." It eventually emerged that while at his dacha this Sonya actually lost one of the notebooks with the stenogram of a session of the Central Committee in it. Vlasik denied any knowledge of what the notebook contained. He claimed to have heard later that it had somehow been recovered.[49]

Unlike Abakumov, Ryumin, Leonov, and Likhachev, Vlasik was not shot in the aftermath of the Doctors' Plot. He was, however, placed in solitary confinement after his arrest and stayed there throughout the course of his trial, after which he went into exile in Siberia. In a letter to Kliment Voroshilov, the president of the Supreme Soviet Presidium, dated April 6, 1955, Vlasik complained of what he believed was his incomprehensibly inhumane treatment. Much of the supposed "testimony," he wrote, had been extorted from him or simply falsified. The MGB, he told Voroshilov, wanted to arrest Stenberg so that they could compromise him. He said that he was thrown in manacles "that bit to the bone" and was forced to sign horribly compromising protocols that were "embellished 90% with lies." He suffered a nervous breakdown.

> Of course, because of my age and health I could not endure it. I had a nervous breakdown, complete shattering and I lost absolutely all of my self-control and healthy thinking. And then an infarct followed because with these frightening experiences there came an intensification of my illnesses—headaches, sheer hallucinations, and nightmares. For months I was without sleep. In this condition, they fabricated the previously prepared protocols.[50]

"You will die like an animal in prison," Beria had told him in 1953, but in the end it was Vlasik who survived and Beria who died like a dog. What astounded Vlasik was that by April 1955 all the doctors had been rehabilitated, Stenberg had been released from prison, Ryumin had been shot, Beria and "all his scum" had been exposed, but only he, Vlasik, remained in unbearable incarceration. For what purpose? he wanted to know. He told Voroshilov that his hallucinations and nightmares caused him to "go out of [his] mind"; his head "swims with irrational thoughts." Why, he asked, did the new regime act just like the previous one and not believe him?

Even in the nearly incoherent prose of this letter, Vlasik recognized the inevitable conclusion: "It was important for them to dirty me; this is what they did and they achieved it."[51] Though he didn't, in the end, die like an animal in prison, he was completely neutralized both physically and morally. Deprived of his rank, and his military and government awards, and expelled from the party, Vlasik was exposed as utterly corrupt, his word counted for nothing. No one could possibly believe him again.

"In this condition not much life remains to me although by the decision of the court there are still two years and nine months in which I must remain in exile. This means that I will die far from my family." He begged for Voroshilov's intervention to allow him to return to Moscow. Vlasik was not shot and probably at Marshal Zhukov's request was released and returned to Moscow in 1955. He was by now a very sick man and disappeared from public view. On May 15, 1956, he was pardoned; in 2000 the verdict against him was quashed and the case canceled because of the absence of any crime, thirty-three years after his death in 1967.

Vlasik was not spared because of his dedicated service to the party or his longstanding relations with Voroshilov, Zhukov, or others in the Central Committee. Vlasik possessed information useful against Beria, and though he was accused of having aided Yegorov in 1952, in 1955 this now stood to his credit. He had protected Stalin to the end, but he also gave those who followed something useful in exposing the "cult of personality" as corrupt: Vlasik's own gargantuan corruption, if only by association, suggested corruption in the Great Leader himself. His women, his drinking, his thievery and abuse of power were all larger than life, a comic subtext to the greater abuses with which Khrushchev would shock the world ten months after Vlasik's letter to Voroshilov.

NINE

THE GREAT STORM

JANUARY—FEBRUARY 1953

Condemn not the king, no not in thought;
Curse not the rich in thy bedchamber;
For a bird of the air will carry the voice;
And that which hath wings will tell the matter.

—ECCLESIASTES

The Court is impervious to proof.

—KAFKA, *THE TRIAL*

The private hallucinations and nightmares of Lieutenant General Vlasik would soon be experienced by Soviet society at large. His sleepless anxiety would soon be shared not only by those who lay in manacles in refrigerated prison cells but by millions of people around the world who went to sleep every night in their own beds. Soviet society was on the verge of a massive nervous upheaval.

December was snowy and cold. There were no thaws and the ice on the streets of Moscow grew thicker and thicker. Even before January 13 when *Pravda* announced the vast nature of the Jewish conspiracy against the Soviet Union, rumors abounded—some continue to this day—of a certain "X" day when millions of Jews would be "voluntarily" deported through-out the length and breadth of the Soviet Union.[1]

Each day brought new reports of terrible threats, expulsions, plots. Checks were being conducted by spontaneously formed people's "com-missions" that would investigate ordinary citizens from apartment building to apartment building to find out who might have had a baptized grandmother or grandfather, who was a true Russian and who was not. Lists of Jews were being drawn up, it was said, in Leningrad, Moscow, and other major cities.

A Moscow journalist who lived through that time, Zinovii Sheinis, has related the story of a pure Russian editor in a foreign publishing house who was hounded by a "commission" seeking to verify whether he was Jewish. After prolonged and repeated inquiries about one thing or another, he lost patience and shouted at his tormentors, "Here have a look for yourselves . . . I'm not circumcised!" Unwilling to accept this answer, the "commission" sent an urgent inquiry to the remote peasant village from which the editor said he had come. It was far from Moscow. Hardly anyone there could even read. Three months passed before the answer arrived. The people in the village claimed they had never heard of Jews. "What is a Jew?" they asked. "We don't know. Perhaps it's a new kind of cow. We have practically no cows left. They all died from lack of fodder."[2] By then Stalin was dead and no one cared any longer whether the editor was baptized or not.

To some it seemed as though the action, whatever it was going to be, had already commenced. In his memoirs, Yakov Rappoport recalled this dark period:

> Events were mounting to a climax. Horrible news was passed by word of mouth. The MGB had disclosed a Jewish conspiracy at the Moscow automobile plant. Mass arrests had been made, wreaking havoc on the leading engineering and technical personnel. . . . More Jewish plots were unearthed—in the Moscow Metro and elsewhere. Sinister rumors crept about Moscow, which it was difficult (and dangerous) to try and check. It was said that Vovsi, Kogan, Vinogradov, and Feldman had been arrested. One dared not try and check—even uttering their names aloud was dangerous. But gradually confirmation of the rumors did arrive. . . . The medical world was not simply deflated, it was crushed . . . and everyone in [the top medical group] who was still at large expected arrest each night.[3]

The past October at the time of the 19th Party Congress, Stalin denounced Voroshilov as an English spy. At the December meeting of the Presidium of the Central Committee at which the Doctors' Plot was revealed, he had likewise denounced Mikoyan and Molotov. It was rumored that a thick file was being prepared on Beria in connection with the so-called Mingrelian Affair. Stalin's purpose according to Sudoplatov was simply "to get rid of Beria."[4] There is much to attest to this. The situation at the time has been described as follows:

> By 1953 the security organs had collected a dossier on [Beria] with such a collection of compromising documents that it would have been more than enough to have him arrested and physically eliminated. Stalin had openly called Voroshilov, one of the old members of the Politburo, an "English spy." In this way, not a single man among those in the circle closest to the leader, who together with Stalin had gone down the long path of the 20s and 30s, could be considered by the Central Committee, by colleagues in the Presidium of the CC as the leader of the party or someone who could occupy the post of General Secretary. Thus, the situation of the mid-twenties was repeated after thirty years.[5]

Inside the Kremlin, Stalin worked tirelessly to ensure that every detail of the plot and its exposure would achieve maximum impact while possessing total credibility. Sheinis entitled his book on that period, *Provocation of the Century* (1992, Moscow). There is much to support this claim. Stalin knew that the eyes of the world would be watching in a way far different from the way the world received news of the murder of 20,000 Polish officers at Katyn, the deportation of the Crimean Tatars or Chechens, or the devastation of the peasantry in the 1920s. The Holocaust had been universally condemned and the State of Israel now existed. But Stalin had little regard for world opinion.

In January 1933, Stalin wrote to Molotov that he had read a newspaper account of one of Molotov's speeches, which he praised for its "contemptuous tone with respect to the 'great' powers, the belief in our own strength." He particularly enjoyed the "delicate but plain spitting in the pot of the swaggering 'great powers'—very good. Let them eat it."[6] There is no reason to think that Stalin would not spit in the pot of the great powers yet again.

But if the planned deportation truly was a *provocation*, rather than simply an act of retribution or insanity, what was it a provocation to? War with the United States? These questions are difficult to answer. We know that the Varfolomeyev plot was a provocation designed to "dirty Truman" and stun world opinion, turning it against America. Who or what would the Doctors' Plot dirty?

Though it appears to have been the work of madness, the plot was slowly and meticulously constructed and, though it moved awkwardly like Frankenstein's monster, it developed a life of its own. Unlike the massacre in Katyn, the Doctors' Plot was not designed simply to eliminate the army of a hostile state. These enemies had been *invented* by the state of which they were citizens. Every indication points to the validity of the view that "the situation of the middle 20s was repeated after 30 years." The country that took pride in having realized more fully than any other Marx's vision of history was now poised to undergo a bloody repetition.

Stalin wanted his name kept out of the investigation until the very last moment, but this moment was fast approaching. On January 6, Vinogradov, who had been in custody since the preceding November, was intensively questioned about his treatment of Stalin's son, Vasily Iosifovich.

"As is known, you treated VASILY IOSIFOVICH and by your criminal actions you damaged his health. Can you deny this?"

"It is true that I had something to do with the treatment of VASILY IOSIFOVICH, beginning in 1930 and until quite recently. However, I did not harm his health."

The investigators were not satisfied. "Now you will tell us who, together with you, is guilty in the subversion of VASILY IOSIFOVICH's health."

"I don't know a single fact of wrecking or criminal conduct by a physician in the treatment of VASILY IOSIFOVICH."

They tried again: "What did you undertake in the Crimea [in 1952] to harm the health of VASILY IOSIFOVICH?"

"Nothing . . ."[7]

In November Yegorov had been questioned about the treatment of Svetlana Alliluyeva. Clearly, the investigators were hoping to be able to make the case publicly that the entire Stalin household was under attack by these medical murderers. Both cases, however, were quietly dropped and never mentioned in any public statement. Even at this next to last moment, with all the pieces in place, Stalin again chose to distance himself and his family from the conspiracy, though Malenkov had been identified by Shvartsman as a target and other government leaders had been publicly named.

On January 9, a meeting was called of the Central Committee to discuss a draft of the article on the doctors' plot for TASS. Those attending included Dimitry Shepilov, now the editor-in-chief of *Pravda*; Beria; Bulganin; Voroshilov; Kaganovich; Malenkov; M. G. Pervukhin; M. Z. Saburov; and Khrushchev. Goglidze and Ogoltsov from the MGB were invited to the meeting. Ignatiev was not present. Presumably he was still recovering from his heart attack. Stalin's name was listed, but had been crossed off.[8] Kostyrchenko believes plausibly that Stalin did not attend this meeting because by not attending he further removed himself from direct complicity in the plot. Alternatively, he may have been there but subsequently removed his name for the same reason.

In any event, we know from Shepilov's recently recovered, handwritten notes, that the article slated for *Pravda* was extensively edited and revised by Stalin personally.

"To Comrade STALIN," Shepilov wrote,

I am presenting the draft article, "Spies and Murderers Under the Mask of Doctors," D. SHEPILOV
January 10, 1953[9]

This draft consisted of the typeset text of a news article, the boldface headline of which read: "BASE SPIES AND MURDERERS UNDER THE MASK OF PROFESSOR-DOCTORS." Stalin returned the draft to Shepilov with numerous marginal corrections and insertions in his own hand. The next day, Shepilov sent Stalin the revised text that would appear in *Pravda* on January 13. In this text, rather than in the TASS news release, the question of what sort of provocation Stalin had in mind becomes clearer.

Though he may have been absent from the January 9 meeting of the Central Committee, Stalin controlled the shape and extent of the information given to the public. The difference between the TASS release and the *Pravda* article leaves no room to doubt this. Shepilov was present when the

TASS release was planned in the Central Committee and he would never have altered that plan for the *Pravda* article by himself and then sent this altered account of the plot to Stalin for corrections without some higher authorization. Stalin must have orchestrated both the TASS release and the *Pravda* article.

Many details suggest that the TASS release and the *Pravda* article were coordinated in order to suggest that the government—as reflected in TASS—remained calm and in control, while the dirty truth of the matter—as reflected in *Pravda*—was indeed incendiary. In the TASS release, there are few rhetorical flourishes, adjectives or adverbs to describe the doctors or their plot. They were called "subversive," but this only accorded with the charges brought against them; they were said to have "heinously undermined the health" of their patients; at one point they were called "culprits" and their activities "insidious." TASS became grandiloquent only in the last paragraph:

> It has been established that all these killer doctors, these monsters who trod underfoot the holy banner of science and defiled the honor of men of science, were in the pay of foreign intelligence services.[10]

But this rhetoric is nothing in comparison with the *Pravda* article.

> The unmasking of the band of doctor-poisoners dealt a shattering blow to the American-English instigators of war. . . . The whole world can now see once again the true face of the slave master-cannibals from the USA and England.[11]

These doctors Vovsi, Kogan, Feldman, Grinstein, Etinger and unnamed others were "the paid agents of American intelligence." They constituted an alleged fifth column of bourgeois secret enemies of the people. In the case of Vovsi, the *Pravda* article wrote that he obtained his directives for "the extermination of the leaders of the cadres of the USSR" from the "spying terrorist organization the 'Joint,'[12] Dr. Shimeliovich and the well-known Jewish bourgeois nationalist, Mikhoels." The transition from "killer doctors" to "a band of doctor-poisoners" also betokened that the doctors were engaged actively and aggressively in murdering the Kremlin leaders, as opposed to the more passive crime of negligence. They were dangerous to the general Soviet population.

Nothing is explicitly said in the TASS release of a fifth column bent on overturning the Soviet Union. But in the *Pravda* article Stalin's overall intent is clear.

> The bosses of the USA and their English "junior partners" know that success in ruling another country cannot be achieved by peaceful means. Feverishly preparing for a new world war, *they urgently sent their spies into the rear of the USSR* and into the countries of People's Democracy; they attempted to implement what had been destroyed among the Hitlerites—to create in the USSR their own subversive "fifth column." [Emphasis added.]

These doctors who were really beasts masquerading as human beings, cloaking their hatred of mankind in the guise of helping the sick, were not simply intent on harming the leaders of the nation. In what foreshadowed the as yet undisclosed Varfolomeyev case, the article notes that, "they urgently sent their spies into the *rear* of the USSR,"[13] and proclaimed that at the bidding of their American and English masters, these saboteurs and murderers intended to overturn the government as a whole. Jewish nationalism was now no longer a matter of potentially invidious ethnic identification or pride, but outright treason, and constituted an act of war.

In the TASS release, these enemies of the people included:

> Professor M. S. Vovsi, a therapist; Professor V. N. Vinogradov, a therapist; Professor M. B. Kogan, a therapist; Professor B. B. Kogan, a therapist; Professor P. I. Yegorov, a therapist; Professor A. I. Feldman, an otolaryngologist; Professor Y. G. Etinger, a therapist; Professor A. M. Grinstein, a neuropathologist; G. I. Maiorov, a therapist.[14]

Vovsi, who was not arrested until mid-November and who had not taken part in the treatments of either Zhdanov or Shcherbakov, was listed first, followed by Vinogradov. Although Yegorov and Maiorov are here, Vasilenko is not.

A notable absence is Ryzhikov, Shcherbakov's doctor, who was among the first to be arrested and whose forced confession about the murder of Shcherbakov was a key element in driving the case forward. Grinstein and Feldman had no connection whatever with either of the principal cases. Obviously, Stalin wanted to exploit as many opportunities as possible for future development of the case by including these individuals, but he also had to show that it was a mass action undertaken not simply by Vovsi, Etinger or Karpai.

The TASS list of names demonstrated that the case included both Jewish and non-Jewish doctors. Things were quite different in *Pravda*. Here the list of doctors was shorter but, as Ryumin put it, more "pointed and politically sharp." The only names prominently mentioned now were: "Vovsi, B. Kogan, Feldman, Grinstein, Etinger and others . . ." The names of Vinogradov and Yegorov appear later in the column buried in parentheses as simply "other participants of the terrorist group" connected to the Zionist, bourgeois, espionage organization, the "Joint."

The TASS release ended simply by saying, "The investigation will be concluded soon." The Pravda article sounded a much more ominous note. Though implicit in the TASS release, the role of Jews became explicit in *Pravda*, where the Jews were identified as enemies along with American and English imperialism—the slave master-cannibals of America and England. "All of this is true," the article concluded, but there is another enemy as well.

> It is also true that, besides these enemies, we have still one more enemy—the thoughtlessness (*rotozyestvo*) of our people. Do not doubt but that where there is thoughtlessness (*rotozyestvo*) among us—there will be wrecking. Consequently: in order to liquidate wrecking it is necessary to finish off thoughtlessness in our own ranks.

The enemy was within as well as without. It was the Jewish doctor-murderers, of course, and their bourgeois bosses in America and England. But the enemy was also to be found in the MGB, the Central Committee, and in the Soviet people themselves—the enemy was in their very souls, as Stalin had said in his 1937 toast. To finish off thoughtlessness is to finish off thoughtless people—this was one of the crimes Vlasik had been accused of. The MGB had shown continued, mulish thoughtlessness in the face of obvious "signals" coming from the Kremlin hospital. MGB officers did not work with "true Chekist spirit." Stalin had warned Ignatiev that they could "see nothing beyond their own noses . . . they are degenerating into ordinary nincompoops, and . . . they don't want to fulfill the directives of the CC." He told Goglidze

> that the investigators worked without spirit that they clumsily used contradictions and slips of the tongue of the prisoners for exposing them; they clumsily posed questions, not seizing like hooks on every, even a tiny, possibility, in order to seize, in order to grasp the prisoners in their hands, etc., etc.[15]

At the December 1 meeting of the Presidium of the Central Committee, Stalin had denounced Vlasik for conspiring with Abakumov over the Timashuk letter, but he also berated all those present saying, "Here, look at you—blind men, kittens, you don't see the enemy; what will you do without me?—the country will perish because you are not able to recognize the enemy." Those who do not see the enemy must be finished off. The time was now at hand for another complete accounting.

Stalin's reckoning began to assume definite shape in the days immediately following the January 13 revelations. On January 15, there appeared a page 1 article in *Izvestia*, entitled, "To raise political vigilance." Much of the article was taken word for word from the previous article on the doctors' plot, but the Jewish doctors—and by now, they were all Jewish doctors—play only a small role in what the article described as the much larger problem of American-English espionage and subterfuge in the Soviet Union, all of which was aimed at destabilizing and ultimately bringing down the Soviet government.

> The spies, diversionists sent by imperialist intelligence agents into our midst or recruited from within our country from among the secret enemies of the people, the unannihilated remainder of anti-Soviet riff raff, do not come out frankly and boldly. They work "on the sly," they masquerade in the guise of Soviet people in order to penetrate our institutions and our organizations, worming their way in to develop their traitorous work.[16]

But the article did not confine itself to enmeshing the Jewish doctors in the much wider plot of imperialist Western aggression against the Soviet Union. The last two thirds of it bear on the subject of the vigilance the Soviet population must take in dealing with foreign spies and provocateurs as well as with the thoughtless, irresponsible gossipers in their own ranks.

> Thoughtlessness (*rotozyestvo*) and chatter—these are the entryways for enemies. We still have among us complacent and negligent people who safeguard the most important documents with *criminal carelessness*. Wishing to boast of how well informed they are, they gossip from time to time about information containing state secrets. In this way, these people become aides to the enemy.
>
> Party, Soviet, social organizations must conduct an uncompromising, relentless struggle with thoughtlessness and complacency, to educate Soviet people in the spirit of the most stringent maintenance of state secrets in the spirit of the highest revolutionary vigilance. Wherever a Soviet man works—in a state institution, in an economic organization, in a business, in a kolkhoz, in transportation, in a scientific institution—everywhere and in every way he must remember and unshakably fulfill the demands of the party and the Soviet state—to be vigilant.
>
> Comrade Stalin mercilessly thrashing the idiotic sickness of thoughtlessness (*rotozyestvo*), spoke at the February-March Plenum of the Central Committee of the Party *in 1937*:
>
> *"And when we are finished with this idiotic sickness, we will be able to say with complete confidence that we are afraid of no enemy, neither internal nor external; their attacks do not frighten us, because we will destroy them in the future as we destroy them now, as we destroyed them in the past."*
> [Emphasis added.]

In 1937 approximately 1 million Soviet State and Party workers were arrested; many were shot outright. Citing the year 1937 would alone have terrified virtually all the readers of this issue of *Izvestia*. Soviet society was at the threshold of another great tragedy.

On January 20, one week after the first article on the Doctors' Plot appeared in *Pravda*, Lydia Timashuk was called to the Kremlin by

Malenkov. Her vigilance had at last been rewarded. She alone, with her EKG, saw into the heart of darkness in Valdai in August 1948 and now she would receive the state's highest honor, the Order of Lenin.

Yakov Rapoport recalled that, "the Soviet press went into raptures about the perspicacity and courage of this paragon of virtue." Poems honored her. This "great daughter of the Russian people" was compared to Joan of Arc.[17] Rapoport also remembered the profound resentment shown to all physicians at the time.

> Every physician was regarded as a potential murderer. I shall never forget the face of my laboratory assistant, distorted with fury and hatred, as she hissed through clenched teeth: "Damn intellectuals, they all deserve to be cudgeled." . . . Meetings were held at all factories and offices, some organized, some spontaneous, and almost all openly anti-Semitic. Speakers would vehemently demand that the criminals should be put to a terrible death. Many went so far as to offer their services in carrying out the actual executions.[18]

A large article appeared in *Pravda* on January 31 with the headline: *Thoughtless People—Accomplices of the Enemy.* It contained yet another warning that this "idiotic sickness" was tantamount to moral degeneracy and treason, specifically that it played into the hands of the "terrorist group of doctor-saboteurs, working in the service of foreign intelligence" recently been uncovered by the MGB.[19]

The possibility of spontaneous, righteous pogroms against Jews grew by the day. One Jewish physician in Kiev refused to attend the ailing child of a local Russian politician, stating that he was no longer allowed to treat Russian children. The mother insisted. Eventually he went to their house and examined the child. Afterward, he was asked to stay a moment longer to speak privately with the father. He did so, not knowing what to expect. "Without explaining anything, [the father] asked the professor if he could see a way to leave Kiev at once and settle in a some out-of-the-way town. It was a friendly warning."[20] There were many such warnings and stories of mothers who refused to take their babies to clinics even though they were suffering from pneumonia. There were stories of beatings, dismissals, arrests.

On February 7, Ignatiev signed an order for the arrest of Dr. Maria Weitsman, sister of Chaim Weizman, the first President of Israel. The arrest order noted that over a period of years the MGB, by bugging her apartment, had learned of "the Zionist agitation she conducted," the "enemy positions" she had taken critical of "Soviet reality" and the "rotten slander" she had uttered against the leaders of the party and the Soviet government, including "the most extreme animosity shown toward the Head of the Soviet government." Her husband had been arrested in 1949 by the MGB for anti-Soviet agitation. Her famous brother was invoked as a British agent who had met Mikhoels and Fefer in America during their 1943 tour. Ignatiev considered it urgent to arrest her now because she had expressed the wish to go to the Israeli mission in Moscow to seek diplomatic protection.

On February 12, Eleanor Roosevelt appealed publicly to President Eisenhower to protest the treatment of Jews in the Soviet Union. Stalin broke off relations with the State of Israel the same day.

An even more threatening development had recently occurred that was not reported in the newspapers. More may have been discussed at the December 1 meeting of the Central Committee Presidium than simply the exposure of a Jewish medical conspiracy. On January 30, 1953, a memorandum was sent to Malenkov from the MGB outlining the circumstances and number of "German, Austrian and other foreign criminals" who had been arrested in the zones of Soviet occupation in Germany and Austria.[21] Apparently, these criminals were taken back to the Soviet Union and held in Soviet prisons. There were, by the MGB's count, 5,337 individuals in all. Of these:

Spies, terrorists, diversionists	3,141
Germans guilty of war crimes against Soviet citizens	1,047
Anti-Soviet agitation	633
Banditry, theft, speculation, and other crimes	516
	5,337

This was followed on February 17 with another memorandum to Malenkov suggesting a plan of action.[22] This document was signed by Ignatiev, Suslov, and Serov, demonstrating that it had been discussed within the administration—in the MGB, MVD (Serov), and in the Agit-prop sector of the Central Committee (Suslov). Accordingly, condemned Germans, Austrians and foreign nationals arrested in Soviet zones of occupation would be separated from other criminals in the gulag system, namely Russians. In order to accomplish this, it was recommended that four new camps be built especially for this purpose: "to organize a system of MVD special camps: in Komi ASSR – 2 camps; in Kazakhstan – 1 camp; in the Irkutsk region – 1 camp." It was decided to transfer all the prisoners to these camps over a six-month period. This recommendation was sent on to the Council of Ministers, which accepted the proposal from the MGB.

What was the sudden need for these special camps? Their specific purpose was never discussed in any of the memoranda, nor was the reason why at this particular moment it was deemed necessary to take special steps to isolate so far from European Russia these especially dangerous criminals, who could all have fit into one large Moscow apartment complex. Kazakhstan, Irkutsk, Komi—all were Central Asian or Far East locations. Surely the camps in Kolyma or Archangelsk would have been frigid and terrible enough and could easily have been redesigned in some measure to separate these prisoners from others.

Little more is known about these camps except that construction for them got underway immediately and rumors spread among those directing the construction that they were for the Jews.[23] Such a scenario makes sense. The deportation of the Jews and the conditions in which they would be held had to be kept top secret from the Russian population for fear that the sheer, gross inhumanity of it did not destabilize the population.

No other documentation exists on these camps whose construction was initiated three weeks before Stalin's death. It is not known whether construction continued after his death. What is known is that Beria, who took over the Security Service after March 5, 1953, did not concern himself with these exceptionally dangerous criminals, but rather recommended a general amnesty, that envisioned freeing some 1,000,000 individuals from the gulag.[24]

By mid February, a storm of *"everlasting* distrust," as Bukharin had imagined, engulfed the country, and at the same time the interrogations of the doctors continued. Clearly Stalin worried that the case might not yet be completely sewn together. The assertion in *Pravda* for instance that Vovsi had confessed to having worked for the "Joint" had not yet been extracted from him when it was made on January 13. Vinogradov finally confessed to Zhdanov's murder only on February 12, but on February 18 at a face-to-face encounter between Karpai and Vinogradov, Karpai continued to refuse to admit any guilt. They could not go to an open trial with the doctors until each told the same story. By now, Vovsi was clearly cooperating, as had Yegorov, Vinogradov, Vasilenko and Maiorov. Of the doctors originally arrested only Sophia Karpai refused. Therefore, only one of the Jewish doctors originally arrested and directly involved with the Zhdanov and Shcherbakov cases remained alive and she did not cooperate.

The purpose of the February 18 face-to-face was clearly to pressure Karpai in a way that Yegorov earlier had hoped to accomplish with Timashuk. It did not succeed. Vinogradov began by explaining that his relationship with Sophia Karpai was precisely what it had been with Yegorov, Vasilenko, and Maiorov—they were his "associates in the wrecking treatment of A. A. ZHDANOV."[25] Vinogradov attempted to force Karpai to admit that the EKGs she took in July and August showed "a recent myocardial infarct." He said that, "this was confirmed by the clinical picture of the illness."[26] He pointed out that Karpai altered her conclusions in her desire to join the medical conspiracy. Karpai denied there was any such data from the EKG or that she ever changed any of her conclusions. There was no conspiracy, she told the investigators. "No. I don't confirm it."[27] Later she stated: "I deny my participation in the criminal treatment of A. A. ZHDANOV. According to the electrocardiogram, I did not find an infarct. I do not deny the clinical facts; but I want to say that I observed attacks of cardiac asthma in the patient without the formation of a recent myocardial infarct."[28]

This, she claimed, was the conclusion she delivered to the secretariat of the Kremlin Hospital on August 8, 1948. Vinogradov disputed this, but in the end nothing he or the interrogators said could compel her confession. Like the other doctors, she had no doubt been brutally beaten and intimidated in Lefortovo. Her unrecorded heroism in the face of this is astounding, because she could have assumed only one thing: if Vinogradov had confessed and was pressuring her to confess, then she truly was all alone. None of her fellow doctors held out so long; none of the MGB officers, like Broverman or Shvartsman, did either. The only other person who would not yield to the MGB's torture was Abakumov himself, who to the end continued to deny that Dr. Etinger had said anything whatever on which a case could be built. The delays caused by Sophia Karpai's heroism, Abakumov's refusal and other small bureaucratic missteps, like grains of sand in the gears of the huge machine that had been set in motion, prevented another catastrophe in Soviet society and politics generally, and saved the lives of thousands, if not millions, of innocent people.

The legend of day "X" as told by Zinovii Sheinis arose at this time and traveled from Jewish household to Jewish household like cholera. It has never been verified. No document has ever come to light to substantiate it. It is worth repeating largely because rumors of day "X" and the expected deportations have played a large, almost mythological role, in post-Stalinist Jewish life. The question of the possible deportations requires further study and search for additional documents. However, the argument now current in some circles that the absence of an order authorizing deportations proves none was planned is not persuasive. The order authorizing the deportation of the Chechens in 1944 to Kazakhstan was signed a week *after* the deportations had occurred.

According to the scenario outlined by Sheinis, Dmitri Chesnokov, a professor of philosophy and the editor of the journal, *Questions of Philosophy*, had been enlisted by Stalin at the time of the 19th Party Congress in October 1952 to provide a theoretical basis for the deportations grounded in Marxist-Leninist thought. Chesnokov's assignment was to show why the deportations were inevitable.[29] No evidence of any sort exists connecting Dmitri Chesnokov to this conspiracy. Chesnokov himself had a Jewish wife who bore him two children who would have been victims of any such action.

Sheinis and others maintain that by the beginning of February, Chesnokov had completed his work. The pamphlets allegedly were printed and hidden in one of the security vaults in the basements of the MGB in Moscow. At the order of Stalin on day "X" they would be brought out and distributed throughout the entire country. At the same time, the national press would produce a corresponding document for world consumption. This pamphlet has never surfaced. It is not even known how this rumor got started.

Arkady Vaksberg has written that the trial of the doctors was planned for March 1953.

> The trial was planned for March, but by February thousands of bar-
> racks unsuitable even for cattle had been hammered together in
> Birobidzhan . . . reserve tracks around Moscow were filled with
> freight cars, militia headquarters in large cities were writing lists of
> citizens subject to deportation—those 100 percent and 50 percent
> Jewish blood . . .[30]

Like Sheinis, Vaksberg has produced no evidence to support the date of the
trial, the reported barracks in Birobidzhan, or the alleged reallocation of
railroad facilities around Moscow. Any sort of change or movement gave
rise instantly to such ideas. Rumor substantiated rumor and beliefs were
taken as facts. The Jews might well have said about Stalin what he had said
about the Americans in connection with Varfolomeyev—It makes no difference whether or not Stalin *actually* planned what they most feared,
because he was capable of doing it.[31] Vaksberg's description of the "reserve
tracks around Moscow . . . filled with freight cars," however, is undoubtedly another unnecessary embellishment. There was no need for such
reserve tracks or freight cars. Once the deportations began, the railway system could have been redeployed within twenty-four hours. Much debate
still exists over the question of whether such deportations were planned at
all. The testimony of Maklyarsky concerning Ryumin's threats gives ample
reason to believe that, as in so many other cases, a policy was being developed without explicit written directives, that it had emanated from the
Central Committee and had penetrated the investigative units of the security services.[32] Ultimately, this policy gave rise to the written directive to
build the camps. That order was written because the camps would be visible and needed to be prepared beforehand. People would wonder about
them, and an explanation was necessary. Stalin undoubtedly hoped that the
deportations would be "forced upon him" by the will of the people, a step
he would take only at the last possible moment, much as the charade about
the "discovery" of Timashuk's letter in December 1952 provoked the
Central Committee to issue its decree.

The story emerged that the Jews would be "voluntarily" deported,
because they recognized the righteous indignation of the Russian people
and feared reprisals. Furthermore, they would ask to be deported in an
open letter to Stalin. There are several accounts of this fantastic story. Ilya

Ehrenburg, who received the Stalin Peace Prize in another cynical gesture on January 27, 1953, claimed that he was approached to sign a letter requesting that Jews be saved from the "wrath of the people," by being deported to Birobidjan.[33] Somewhat more evidence exists for a version of this story, than for Chesnokov's "X" day.

Sheinis has described a meeting between himself and Ehrenburg in July 1953 at Ehrenburg's Moscow dacha. Ehrenburg was depressed and appeared agitated.[34]

"They came to see me at my dacha," Ehrenburg began.

"Who?" Sheinis wanted to know.

"Academician Mintz; the former General Director of TASS, Khavinson, and one other. This was back in January or February 1953."

Ehrenburg related that these two respected members of the Soviet-Jewish intelligentsia came to see him with the purpose of getting him to sign a letter they had prepared, apparently under pressure from above, that would request protection from "the great and wise leader, Comrade Stalin." The essence of the letter, Ehrenburg explained consisted in the humble supplication made to Comrade Stalin on behalf of the Jewish community of the Soviet Union: The doctor-murderers had been unmasked; the anger of the Russian people was justifiable. Perhaps Comrade Stalin could consider it possible to show graciousness and "'save the Jews from the righteous anger of the Russian people'; that is, to deport them."[35] The authors of the letter, Ehrenburg told Sheinis, agreed to the deportation of the people in the hope that they personally would not be executed. Ehrenburg recounted that he had not been the only one approached to sign this letter. A certain distinguished professor Yerusalimsky had been asked, but had refused.

Ehrenburg claimed that he also refused to sign.

However, he quickly wrote a letter to Stalin. In order to get it to Stalin as quickly as possible, Ehrenburg went to the editor of *Pravda*, Dimitry Shepilov, and asked his advice. Shepilov agreed to help and told Ehrenburg to write the letter in his office. This Ehrenburg did. It was a short letter, the essence of which was that Ehrenburg urged Stalin to consider the very complicated situation with the Jews internationally at the present moment. There would be international consequences. We, Ehrenburg said, would lose the friendship of the entire world if such an action were undertaken.[36]

According to Sheinis, Ehrenburg told him that the letter was delivered to Stalin through Poskrebyshev. This detail seems unlikely and may cast some doubt on the credibility of the entire story because according to various sources, Poskrebyshev had by that time already been dismissed from his position as secretary to Stalin and was in some danger of being arrested.[37] It is doubtful that Shepilov did not know this and therefore it would have been improbable for him to send Ehrenburg's letter through such a discredited source.

Nevertheless, according to the story, Ehrenburg received no reply for some days. Then Malenkov called him. He went to see Malenkov in his Kremlin office. Kaganovich was there. Kaganovich accused Ehrenburg of wanting to spread a rumor that there was anti-Semitism in Russia. The conversation otherwise was desultory, lacking focus. Nothing happened. The letter was never published. Later, Ehrenburg wrote, "I thought at the time that I dissuaded Stalin with my letter; now it seems to me the whole business was delayed and Stalin did not succeed in doing what he wanted to do."[38]

Did any of this happen? Every day, Sheinis wrote, the Jews of the Soviet Union waited for the trial of the murderer-doctors to begin. Nothing happened. Jewish children were kept out of school, where they would be taunted and verbally abused. Often they would be beaten. People stayed indoors. They waited. One document has come to light that puts Sheinis's story about Ehrenburg in some perspective. Published in the Russian journal, *Istochnik*, in 1997, was an undated letter to *Pravda* by I. I. Mints and Ya. S. Khavinson, the two Jewish intellectuals mentioned by Ehrenburg. The proof text lists the signatures of some fifty-eight leading Jewish intellectuals—including Ehrenburg. It is not known whether these individuals actually signed the letter. It is not known whether this was a revision of an earlier draft. We reproduce the letter in full. It was set in type for publication but never appeared.

LETTER TO THE EDITORIAL BOARD OF *PRAVDA* (no date)

In this letter we consider it our duty to express the feelings and thoughts that make us anxious in connection with the complicated international circumstances. We would like to summon toiling

Jewish people in various countries of the world to reflect upon several questions touching upon the present interests of Jews.

There are people, who, presenting themselves as "friends" and even as representatives of all the Jewish nation, declare that there exists unity and general interests among all Jews, that all Jews are united among themselves by a common goal. These people— Zionists—are the helpers of the Jewish rich and are villainous enemies of the Jewish workers.

Every toiling person understands that there is a difference between one Jew and another Jew, that there is not and there cannot possibly be anything in common between people who get their bread by their own labor and financial bosses.

Consequently, there are two camps among the Jews—the camp of the workers and the camp of the exploiters—those who suppress the workers.

An unbridgeable gulf separates one from the other. Together with all toiling masses and with all progressive forces, Jewish-toilers are vitally interested in strengthening peace, freedom and democracy. We know that in the camp of the fighters against the instigators of war an active role is played by the representatives of Jewish workers.

As for the Jewish industrialists and bank magnates, they go a different way. This way of international adventurers and provocateurs, of espionage and diversion, is the way of unleashing a new world war. A war is necessary to Jewish millionaires and billionaires, as it is to rich people of other nationalities, because it is for them a source of great profit. The politics conducted by the Jewish rich are deeply hostile to the vital interests of Jewish workers. It is fraught with ruinous consequences for Jewish workers.

Where is there here a common path, where is there "unity" and the common interests of all Jews which the sham "friends" of the Jews—the Zionists—endlessly repeat?

Deceiving with hypocritical words about the "common way," "common interests" of the Jews, the leaders of the state of Israel allow themselves to claim that it is they who express the interest of all Jews. But let us look into who in actuality the leaders of the state of Israel are and whom they serve. Is it not a fact that in Israel only a small bunch of rich people make use of the goods of life at the same time that the suppressed majority of Jews and Arabs suffer from huge need and deprivation, and drags out a semi-indigent existence? Is it not a fact that leaders of Israel have tied Israeli workers to two oppressors—Jewish and American capitalism?

It turns out the state of Israel, *just like any bourgeois state in any part of the world*, is a kingdom of exploiters of the common mass of people, a kingdom of profit for a small bunch of rich people. It turns out that the ruling clique of Israel represents not the Jewish people, consisting of the majority of working people, but Jewish millionaires connected with the monopolists of the USA.

This has defined all the politics of the *present* Israeli leaders. They have perverted the state of Israel into a weapon for the development of a new war, into one of the vanguard outposts of war provocateurs. The state of Israel is a bridgehead for US aggression against the Soviet Union and all peace-loving peoples.

Not so long ago all the honorable people of the world were shaken by the news of the explosion of a bomb on the territory of the mission of the USSR in Tel-Aviv. In fact those who organized and inspired this explosion are now the present rulers of Israel. Playing with fire, they increase the tension in the world situation, created by American-English warmongers.

Further, isn't it so that the international Zionist organization "Joint" that defends the interests of Jews is affiliated with American intelligence? As is known, not long ago in the USSR the espionage group of doctor-murderers was uncovered in the USSR. The criminals, among whom the majority consisted of Jewish bourgeois nationalists, were recruited by the "Joint"— M. Vovsi, M. Kogan, B. Kogan, A. Feldman, Ya. Etinger, A. Grinshtein. They set as their aim to sabotage the treatment and to cut short the life of leaders of the Soviet Union, to disable the leading cadres of the Soviet Army and moreover to undermine the defense of the country. Only people without honor and conscience, having sold their souls and bodies to imperialism would commit such monstrous crimes.

It is entirely clear that the leaders of the state of Israel, the leaders of the "Joint" and other Zionist organizations carry out the will of highhanded Jewish imperialists and of those who are their authentic masters. It is no secret to anybody who this master is— American and English billionaires and millionaires, thirsty for the blood of the people in the name of new profits.

We, the undersigned, reject the laughable claims of the Ben-Gurionists, charadists and simple warmongers to represent the interests of the Jewish people. We are deeply convinced that even those Jewish workers who up to now have believed in the sham unity of all Jews, upon reflection, will join with us in our evaluation of the traitorous essence of the politics of the Jewish rich and their supporters.

Turning the state of Israel into the homeland of America, the leaders of Zionism imagine imperialist America to be the "friend" of Jews and they lead a campaign of slander and hatred against the Soviet Union—the champion of social equality among all peoples. Let us look into this question.

Who doesn't know that in actuality in the USA there is a category for Jewish workers, oppressed by the most horrible machine of capitalist exploitation? Who does not know that precisely in this country flowers the most unbridled racism and in this also anti-Semitism? Who, finally, does not know that, in addition, anti-Semitism is the distinguishing feature of all fascist gangs that universally are supported by the imperialists of the USA?

At the same time, it is known to the entire world that the people of the Soviet Union and primarily the great Russian nation with its selfless, heroic struggle saved humanity from the yoke of Hitlerism, and the Jews—from complete destruction and annihilation. In our days the Soviet nation was in the first rank of fighters for peace, staunchly defending the affairs of the world in the interests of all humanity.

In the Soviet Union the authentic brotherhood of peoples, great and small, has been realized. For the first time in history, working Jews together with all workers of the Soviet Union have acquired a free, joyful life.

Is it not clear that the legend that imperialist America is "the friend" of the Jews is a conscious falsification of the facts? Is it not clear in addition that only undoubted slanderers might deny the solidity and indestructibility of the friendship among the peoples of the USSR?

The enemies of freedom of nationalities and the friendship of peoples, something confirmed in the Soviet Union which aspires to give Jews a consciousness of the high social duty of Soviet citizens, want to turn Jews in Russia toward espionage and toward becoming enemies of the Russian nation, and, moreover, to create the grounds for the revival of anti-Semitism, this horrible relic of the past. But the Russian nation understands that the great majority of the Jewish population of the USSR is the friend of the Russian nation. The enemy using any kind of trick will never be able to undermine the trust of the Jewish people in the Russian people, never succeed to set us at odds with the great Russian nation.

Working Jews of the whole world have only one enemy. This is the imperialist oppressors, in the service of which are to be found the reactionary bosses of Israel, and in addition the spies and diversionists—all of the Vovsis, Kogans, Feldmans, etc. Working Jews of the entire world have only one goal—together with all peace-loving peoples to protect and strengthen the affairs of peace and freedom of people. It is impossible to defend the vital rights of Jewish workers in capitalistic countries, it is impossible to be an authentic fighter for the affairs of peace and freedom of nations without conducting a struggle against Jewish billionaires and millionaires and their Zionist agents.

The vital interest of working Jews consists in strengthening the friendship with working people of all nationalities. The stronger the unity of workers of all nationalities, the more solid are the affairs of peace and democracy.

Let all working Jews, who want to work for peace and democracy, unite their strengths and present a single, broad united front against the adventurist politics of Jewish billionaires and millionaires, the leaders of Israel and international Zionism.

Considering the importance of the consolidation of all the progressive powers of the Jewish people, and in addition with the aim of veracious information about the situation of working Jews in various countries, about the struggle of peoples for the strengthening of peace, we consider that publication [of this letter] in the newspapers of the Soviet Union would be expeditious, intended for a wide stratum of the Jewish population in the USSR and abroad.

We are certain that our initiative will meet with the strong support of all working Jews in the Soviet Union and the whole world.

Volfkovich, S. I.	Faier, Yu. F.	Rubenshtein, M. I.
Dragunsky, D. A.	Grossman, V. S.	Roginsky, S. Z.
Ehrenburg, I. G.	Gurevich, M. I.	Kassil, L. A.
Kreizer, Ya. G.	Kremer, S. D.	Khavinson, Ya. S.
Kharitonsky, D. L.	Aliger, M. I.	Leider, A. G.
Kaganovich, L. M.	Trakhtenberg, I. A.	Chizhikov, D. M.
Reizen, M. O.	Nosovsky, N. E.	Veits, V. I.
Vannikov, B. L.	Oistrakh, D. F.	Fikhtengolts, M. I.
Landau, L. D.	Kaganovich, Maria	Koltunov, I. B.
Marshak, S. Ya.	Lipshits, M. Ya.	Yerusalimsky, A. S.
Romm, M. I.	Bul, B. M.	Gelfond, A. O.
Mints, I. I.	Livshits, S. B.	Messerer, S. M.
Raizer, D. Ya	Prudkin, M. I.	Shapiro, B. S.
Lavochkin, S. A.	Smit-Falkner, M. N.	Zolotar, K. I.
Tsyrlin, A. D.	Lantsman, N. M.	Bruk, S. I.
Churlinonskaya, O. A.	Gilels, E. G.	Smirin, M. M.
Dunayevsky, I. I. [sic]	Rozental, M. M.	Lokshin, E. Yu.
Briskman, M. N.	Blanter, M. I.	Shafran, A. M.[39]
Raikhin, D. Ya.	Talmud, D. L.	
Landsberg, G. S.	Yampolsky, A. I.	

As reproduced in *Istochnik*, the *Pravda* letter was accompanied by a letter from Ehrenburg to Stalin dated February 3, 1953. Ehrenburg assured Stalin that he understood that the only solution to the Jewish question was complete assimilation in Russian society, which was "urgently necessary," he wrote, "in the struggle against American and Zionist propaganda that attempts to isolate people of Jewish nationality." Tactically, however, he argued that publishing the *Pravda* letter might provoke a counterproductive international reaction. He concluded:

You understand, dear Iosif Vissarionovich, that I myself cannot decide these questions and therefore I dare to write to You. There is talk of an important political act, and I decided to request that you charge a leading government figure to inform me—whether the publication of this document is desirable and whether it is desirable to have my signature on it. I perfectly understand that if it could be useful for the defense of our Motherland and the movement for world peace, I will immediately sign the "Letter to the editorial board." With deep respect

I. Ehrenburg[40]

Ehrenburg's behavior is complicated and it is beyond the scope of this book to judge the various interpretations of his role in the Stalinist regime or the Doctors' Plot. Nevertheless, he like many others seems to have been ready to play the age-old, hopeless role of court Jew, a willing servitor with the illusion or hope of exerting a moderating influence. As early as September 1948 he publicly demonstrated his support for Stalin's view of the state of Israel, in his *Pravda* article, "The Union of the Snub-nosed" [*Soyuz kurnosykh*]. Ehrenburg argued that the charge of anti-Semitism, discrimination, and the suppression of the rights of Jews in the Soviet Union was nothing more than malicious fabrications by enemies of the Soviet order. He argued that Israel was nothing but a bourgeois state, incompetent to decide the Jewish question; nor could it unite Jews around the world. This position was not vastly different from the one expressed in the printed draft of the letter attributed to Khavinson and Mints.

There is still no satisfactory answer to the question of whether a first letter existed which Ehrenburg refused to sign. Did anyone actually sign anything? Or were the names printed at the end of the long anti-Zionist screed merely set there with the expectation that signatures would be forthcoming, just as *Pravda* had announced confessions by several doctors before they actually existed? We will never know the answer to this question. We do, however, know that such a letter was envisioned and that such signatures, as Politburo member Lazar Kaganovich, world famous violinist David Oistrakh, poet Samuel Marshak, and Ehrenburg were intended for it. Yerusalimsky, the historian, who Sheinis said had thrown Khavinson and Mints out of his house when he was first approached, was also there.

If Stalin thought publishing the letter was a good idea, it is doubtful he would have abandoned it. Rather, he acted characteristically—as he had with Etinger and Varfolomeyev, for instance—to delay tactically until he thought the moment was right. He did not live to see this moment. Ehrenburg was no doubt correct that something other than his cautionary words caused Stalin to hesitate. A more likely reason for delaying publication of the letter was that the trial itself had been delayed because all the pieces were not yet in place.

While the Jews of Moscow waited, and Khavinson and Mints were recruiting signatures of prominent Jews for their letter, the doctors were still being beaten in Lefortovo prison, Sophia Karpai, whom Vinogradov had described as nothing more than "a typical person of the street with the morals of the petty bourgeoisie,"[41] was kept in a refrigerated cell without sleep to compel a confession. She did not confess. It satisfies the imagination to think that the fate of the Jews of Russia might have depended on this latter day, unknown Esther. Though Sophia Karpai's courage was great, she was in the end not significant enough by herself to hold up Stalin's plans indefinitely. The decree on the termination of the case issued on March 31, 1953, suggests a different reason for Stalin's delay besides the resistance of Sophia Karpai.

This document was prepared by five senior members of the newly reorganized Ministry of Internal Affairs (MVD); it was approved by Beria and distributed to the members of the Politburo. In it, Beria recommended that the case against the doctors be immediately terminated and that they be released from prison. The document shows that Stalin was hoping to develop the Doctors' Plot further before it came to trial, and it suggests why nothing else happened between mid-February and the dictator's death.

Vovsi had already confessed to ties with Solomon Mikhoels, the Jewish Antifascist Committee, and the Joint. This was an important confession because it demonstrated that the Jews, through the JAC, were connected with what had been alleged to be a definite American intelligence organization, the Joint. At his trial, Isaac Fefer, one of the members of the JAC, stated that, "The Joint is a bourgeois Zionist organization."[42] If the Jewish terrorist organization in the USSR could be shown to be directed by the Joint, the question of control could be answered. Vovsi had cooperated, but one Jewish doctor did not make a conspiracy. Etinger had died before such a confession could be extracted from him. Stalin needed more confessions. Therefore, Kogan, Feldman, Grinshtein and many other Jewish doctors were arrested soon after the January announcement in *Pravda*. Despite the assurance given in that article that Kogan, Feldman and Grinshtein were connected to "the international Jewish, bourgeois-nationalist organization, the 'Joint,'" they had not yet made such confessions. Beria's March 31 decree makes this point:

> Meanwhile KOGAN, B. B., FELDMAN, A. I., GRINSHTEIN, A. M. and other doctors, arrested in the present case, did not provide confessions of any sort of ties with the Jewish bourgeois-nationalist organization the "Joint," especially concerning their terrorist aims and espionage activity.[43]

The Decree noted that only Vovsi had made such confessions and it points out that "after the publication of the [article] in Pravda, the investigators tried to obtain these confessions from the prisoners . . . about their ties to the organization the 'Joint,' but the prisoners, despite physical pressure, did not give these confessions." The Decree concluded that, "In this manner, the information published in the newspaper Pravda did not correspond to actual facts."[44] Sophia Karpai was not alone.

In his haste, Stalin uncharacteristically had moved too quickly and now had to delay until the necessary confessions had caught up with the allegations. Had he miscalculated, despite all the painstaking care? This explains why, as Kostyrchenko, Deriabin and others have pointed out, there was a lull in press coverage of the Doctors' Plot during mid-February. More and more Jewish doctors would be arrested, tortured and questioned in order to establish as quickly as possible the answer to Ignatiev's question, "who concretely controls the underground and how it is organized."

Up to the end, Stalin did not lose his grasp on supreme power; he never deviated from the objective he had set for himself at the outset of the Doctors' Plot in 1951. But he also found it difficult to get exactly what he wanted. The Vozhd too was running out of time. Questions remained. Up to the end, Stalin was inventing answers. Stunning proof of this is provided by the draft of the Criminal Indictment of Abakumov presented to Stalin by Ignatiev sometime after February 17. We know that it was sent after the seventeenth because there is another document addressed to Stalin from Ignatiev (with a copy to Malenkov), dated February 17, 1953, containing an outline of this draft. This document definitively answers two important questions: A trial was being planned that would merge the Abakumov-Shvartsman case with the case against the doctors; this trial was being planned for the very near future, probably by end of March 1953.

The February 17 document begins, "To Comrade STALIN, I present to You the draft of the criminal indictment of the enemy group ABAKUMOV-SHVARTSMAN."[45] The fact that a copy of the draft went to Malenkov shows that the entire Central Committee was by now fully informed. "In addition to ABAKUMOV and SHVARTSMAN," the draft continued, eight additional MGB officers were indicted: Raikhman, Leonov, Likhachev, Komarov, Chernov, Broverman, Sverdlov, Palkin. Belkin, Komarov, Leonov, Palkin, and Raikhman had been denounced by Broverman. Sverdlov's name was underlined, and Stalin placed a large X next to it. In the margin, Stalin wrote in a strong hand, *Ne malo?* "Isn't it few?" suggesting that more names should be added. Colonel Andrei Sverdlov was the son of the first Soviet president, Yakov Sverdlov, a Jew. On page two, Ignatiev informed Stalin that "A proposal for the judicial review of the investigative cases of the remaining prisoners" would soon be prepared. Stalin underlined the words "remaining prisoners" and in the margin wrote: *Kakiye ostalniye?* "How many remain?" It is significant that Stalin also wrote in the margin: *Nado skazat' zakrytii sud* "It's necessary to say 'closed trial.'" Apparently, the MGB officers would be tried separately from the doctors. Their trial would be closed. The doctors' trial from all available evidence was intended to be an open "show" trial. Abakumov and his adherents were all sentenced in advance to *vysshaya mera nakazahiya—rasstrelu,* "the highest measure of punishment—shooting."

The actual draft indictment dated February "___," 1953, recapitulated the preliminary report to Stalin of the seventeenth. However, it went into much greater detail about the charges and how they would be presented in court; it was heavily annotated throughout by Stalin. Occasionally, he would simply correct grammar or improve the rhetorical effect of the language; at other times, Stalin would substantively amend the document. For example, Stalin wanted the ten defendants to be referred to not simply as "traitors to the Motherland," but as "former workers in the MGB who conducted wrecking, subversive work…" Kuznetsov was referred to in the document as part of a "traitorous group working in the party and Soviet apparatus." Stalin changed this to read "traitorous group working in the party and Soviet apparatus *in Leningrad*." He repeatedly crossed out or queried the word *smazyvat'*—"to smooth out" or "slur over"—in favor of finding more pointed language. Stalin was no impotent, old man waiting for the end, cut off from his power base, stripped of his security and isolated by his followers, as he has sometimes been depicted.[46] He was guiding the most important state action of the postwar period in his characteristic fashion.

The content of the draft indictment is equally revealing. After the recitation of the charges against Abakumov, Shvartsman and Vlasik (who was not indicted here) concerning the Zhdanov and Shcherbakov cases, the document explains two of the abiding difficulties of the investigation: how Abakumov could have condemned the members of the Jewish Antifascist Committee to death while at the same time being a Jewish sympathizer, and how he could have done the same to Kuznetsov and the other members of the Leningrad party.

Acting like a subversive, ABAKUMOV and his associates LEONOV and KOMAROV [Stalin crossed out his associates] ignored the orders of the [Central Committee] concerning the investigation of the ties with foreign intelligence of the enemy of the people KUZNETSOV and the participants of his traitorous group, operating in the party and Soviet apparatus [Stalin inserted: "in Leningrad"]. With criminal intent, they [Stalin crossed out "they" and inserted "he"] oriented the investigators to think of the case of KUZNETSOV and his followers *as a local, isolated group, without foreign ties*.

The accused KOMAROV testified as regards this:

". . . He [ABAKUMOV] directly characterized the case of KUZNETSOV and his enemy group as something local and [insisted] that among the arrested there were not and couldn't be any spies with foreign ties . . ."

The result of the enemy activity of ABAKUMOV, LEONOV,

KOMAROV was that the espionage activity of the participants of the KUZNETSOV group were not investigated and in this manner the investigative case was slurred over [Stalin crossed this word out and wrote in "concealed"].

The accused ABAKUMOV together with those drawn into the present case, LEONOV, SHVARTSMAN, LIKHACHEV, KOMAROV, BROVERMAN sabotaged the investigation of the criminal activity of the arrested American spies and Jewish nationalists, acting under the cover of the Jewish Antifascist Committee. After superficial interrogation of the arrested, in the course of which their espionage activity was not clarified in full and the question of terror in general was not investigated, the indicated case was closed by the investigation and was ignored for a long time.[47]

Abakumov's crime was that he protected Kuznetsov and the Jews by not revealing the extent of their crimes, by conducting a "superficial" investigation that would not disclose the full dimensions of the threat to the USSR. The Jews were protected because the few members of the Jewish Antifascist Committee were nothing but scapegoats for an immense Jewish plot against the government. By eliminating individual scapegoats, Abakumov allowed the underground to continue to flourish.

Not until the last sentence of the indictment did Stalin finally reveal the card he had withheld so long in playing.

Set before the CC KPSS, were the exceptional confessions by criminals of terrorist plots against the leader [*Vozhd*] of the Soviet people in 1950 at the direct order of ABAKUMOV and LEONOV, as `obtained from the protocols of interrogations of the arrested agents of American intelligence GAVRILOV, LAVRENT'EV and others.

"Terrorist plots against the leader of the Soviet people" would be unmasked at the trial. The doctors, Shvartsman, Abakumov, Kuznetsov, the entire Jewish nation in league with the Americans had one paramount aim: to murder Stalin. Only now would Stalin allow himself to be named.

TEN

THE END

———

At Stalin's grave the words were pronounced: "We are the true servants
of the people, and the people wants peace and hates war.
Yes, all of us are dedicated to the desire of the people not to allow the
spilling of blood of millions and to safeguard the peaceful structure of a
happy life." These were the thoughts of comrade Stalin, his care,
his will and his comrade the head of the Soviet government
said these words. These words touch every simple person
and together with us they say: "Stalin lives."

—ILYA EHRENBURG, PRAVDA, MARCH 11, 1953

Comrade Khrushchev, have we enough courage to tell the truth?

—ARISTOV TO KHRUSHCHEV AT FEB. 1, 1956, SESSION OF THE
PRESIDIUM OF THE CENTRAL COMMITTEE TO DISCUSS
RAISING THE QUESTION OF THE CULT OF PERSONALITY
AT THE TWENTIETH PARTY CONGRESS

In the end it wasn't the Americans, the Jews, Kuznetsov, Abakumov, or
their "adherents" Stalin had most to fear. His real enemies were old age
and those closest to him in the Politburo. He did not live to see Abakumov
shot or the traitors put on trial. He did not live to see his enemies exposed.
He could not outmaneuver or terrorize death.

It is a moot question whether Stalin's mounting anxiety over another
potentially botched attempt at creating the conditions for mass terror and the
reassertion of his total mastery over the Soviet Union contributed to his own
collapse ten days after Sophia Karpai's last encounter with Vinogradov, while
the criminal indictments were being prepared. With Stalin's death, at the age

of seventy-three, the Doctors' Plot ended much as it began, in mystery. Though the last extant protocol of interrogation was the February 18 face-to-face confrontation between Sophia Karpai and Vinogradov, we know that doctors were being interrogated right up to the end. Stalin died at 9:50 P.M. on March 5.[1] Dr. Ryzhikov was interrogated twice that day, once from 1:10 P.M.–4:45 P.M. and again from 9:20 P.M.–1:45 A.M. We do not know the content of these interrogations.[2] There is more that we don't know about Stalin's death than that we do. No one knows exactly how Stalin died. The newspapers reported a cerebral hemorrhage at about the same time that Stalin's son, Vasily, was said to have come running into the dying man's room, shouting, "They've killed my father, the bastards!" Each published version of his death significantly contradicts every other.

In this vacuum of information and consistency rumors and myths have abounded for the last fifty years. Dimitri Volkogonov, who had access to the most secret and select documents among any researchers Russian or foreign, has written that the moment of Stalin's death "came at 9:50 A.M. on March 5, 1953." Yet, the universally accepted time of his death was 9:50 P.M. The oral testimony provided to Volkogonov by A. I. Rybin, a guard at the Bolshoi Theater, flatly contradicts many of the points of the testimony Rybin had published several years earlier.[3] Svetlana Alliluyeva's various accounts, though broadly corroborating Khrushchev's memoirs, contain many unsubstantiated assertions. Khrushchev's narration of the days preceding Stalin's death is a masterpiece of obfuscation and self-serving hyperbole.[4] Recently, Amy Knight has provided a balanced summary of the circumstances surrounding Stalin's death that may help explain some of these discrepancies.

> Members of the leadership may have deliberately delayed medical treatment for Stalin—probably for at least ten or twelve hours—when they knew he was seriously ill. They then covered up this delay. (Whether Stalin would have died anyway is of course a matter of conjecture.) Though Alliluyeva seems to blame Beria above all for this, her account does not absolve the other members of the leadership who were also present. She does not say who these men were, but implies that at least three others—Malenkov, Khrushchev, and Bulganin—were there, which Khrushchev attests to. . . .

> It is not difficult to come up with motives that Stalin's subordinates
> might have had in denying him medical treatment for so long. They
> might have been uncertain as to what to do with him for fear they
> would be held responsible for any mistakes in his treatment. . . .
> Then, of course, they may have wanted him to die.[5]

One persistent rumor, Amy Knight's cautions notwithstanding, is that
he was murdered, possibly by Beria, but no hard empirical evidence supporting this has been unearthed to date.[6] The argument that he died naturally of a brain hemorrhage, as *Pravda* reported, is the most credible explanation, yet many details of his death remain obscure. The mystery of
Stalin's death is, in some ways, only deepened by the discovery of a new
document providing a history of his illness from the time the doctors
arrived at his dacha on March 2 until, according to the medical report, he
breathed his last at 9:50 P.M. three days later.

The ten doctors who cared for Stalin during his illness began composing this twenty-page document in the days after his death and completed it
only in July 1953. Who supervised the final draft of this report is not
known, but it was probably Beria or his people. After going through at least
two drafts, which vary from each other in significant respects, this document was stamped *sovershenno sekretno* ("top secret") and submitted to the
Central Committee, where both drafts have been preserved, unpublished
and apparently unread, for nearly fifty years. Entitled "The History of the
Illness of J. V. Stalin, from March 2 to 5, 1953," this document tells a story
significantly different from any that has yet come down to us and provides
a fitting conclusion to the story of the doctors.

According to all first-hand accounts, Stalin became ill in Blizhnyaya, his
so-called nearby dacha in Kuntsevo, following a dinner with Beria,
Malenkov, Khrushchev, and Bulganin that had begun the night of Friday,
February 28, and ended in the early morning hours of Saturday, March 1.
Rybin stated that the dinner was over by 4 A.M.; but Khrushchev recalled
that dinner had lasted until "five or six in the morning" of March 1.
Khrushchev made the further claim that Stalin was "pretty drunk and in
very high spirits. He didn't show the slightest sign that anything was wrong
with him physically."[7] Rybin, however, wrote that Stalin drank nothing
but fruit juice that night, a more likely scenario because, as it has been

widely reported, Stalin drank little hard liquor, preferring to see his guests' tongues loosened.[8]

Khrushchev wrote that "after this particular session we all went home happy because nothing had gone wrong at dinner. Dinners at Stalin's didn't always end on such a pleasant note." Yet Volkogonov, citing Rybin's oral testimony, described an angry, paranoid Stalin, obsessed with the progress of the Doctors' Plot, stating that if Ignatiev did not procure the desired confessions, "we'll reduce his height by a head."[9] Stalin, Volkogonov wrote, "was visibly irritated with the company. Only Bulganin escaped his angry reproach. But this angry outburst was to be his last. Breaking off in midsentence, he suddenly got up and went to his room. The others silently dispersed and went home. Malenkov and Beria traveling in the same car."[10] Volkogonov may well have picked up this last detail from Khrushchev's memoir, but it is the only point on which they agree.

Matters become still cloudier during the period after the dinner party dispersed and the time Beria, Khrushchev, Bulganin, and Malenkov reconvened at Blizhnyaya to witness Stalin's death agony which lasted from March 2 to March 5. Rybin wrote that at midday on March 1 there was no movement in Stalin's quarters into which servants or other personnel were not allowed except at appointed times or with special approval. Though the staff was becoming worried, no attempt was made to ascertain the state of Stalin's health. At 6:30 P.M. a light came on, indicating that Stalin was awake and working.

Rybin noted that Stalin was found by a guard, a certain Lozgachev, at around 10:30 P.M., sprawled on the carpet beside his writing table. Elsewhere, he said that another of Stalin's guards, Starostin, found Stalin lying in pajama bottoms and undershirt on the floor with a copy of *Pravda* beside him. He stated to Volkogonov that Stalin "must have been lying there a long time, as the light had not been turned on." Yet previously he had noted that a light had come on in Stalin's room at 6:30, which allayed the fears of the household staff.

Svetlana Alliluyeva, by contrast, wrote that Stalin's body had been found at 3 A.M. on the morning of March 2,[11] not at 10:30 the night before. Khrushchev wrote that Stalin's body was found late in the evening of March 1 by Stalin's faithful maid, Matryona Petrovna. It was at this time that the Chekist bodyguards lifted Stalin up and put him on the divan.

Alliluyeva subsequently amended her account to say that it was Stalin's maid Valentina Istomina who found Stalin on the floor in the evening of March 1 and requested that a doctor be found. She wrote that the doctors did not arrive until 10 A.M. on March 2.

Rybin, a Bolshoi Theater guard, never actually attended Stalin. He repeated only what he claimed had been told to him, highlighting the importance of the MGB—*they* found the body; *they* took the initiative to call Malenkov; *they* sat with Stalin during the long night waiting for the doctors to arrive. Khrushchev and Alliluyeva told the story to downplay the role of the bodyguards and to damn Beria. In their accounts the MGB guards did not appear except to lift Stalin to the divan where the doctors found him. Ignatiev, head of state security, was not referred to at all.

Khrushchev suggested that no immediate action was taken to call the doctors because Stalin might have had a hang over: "We decided that it wouldn't be suitable for us to make our presence known while Stalin was in such an unpresentable state."[12] But in what way was Stalin "unpresentable"? Khrushchev doesn't say.

Given the charged political atmosphere of Kremlin politics in the early 1950s, the degree of passion aroused in Stalin by the Doctors' Plot, and what we know of his hectoring and abusive manner, Rybin's basic account, despite his self-contradictions and self-protective instincts, appears more persuasive than Khrushchev's. Furthermore, it is difficult to guess Rybin's reason to note that Stalin had had nothing to drink that night except the "young Georgian wine" Stalin liked to drink, described as fruit juice.

Khrushchev wrote that he had been called to the *Blizhnyaya* dacha twice, once on the night of March 1 and then in the early morning hours of March 2. According to Rybin's written account, Khrushchev appeared only once, at 7:30 A.M. on March 2. According to Rybin's oral testimony, however, this appearance did not take place until an hour and a half later, about 9 A.M. Rybin wrote that Beria telephoned the dacha at around 11 P.M. on March 1 and warned the guards to say nothing to anyone about Stalin's illness. According to his oral testimony, from 11 P.M. until around 3 A.M. on March 2, Beria couldn't be located and therefore the doctors could not be called. Even then Beria showed no sign of wanting to obtain medical help.

The same events, narrated by Khrushchev, unfold much differently. Late on the evening of March 1, he claimed to have been called by Malenkov, who reported that Stalin appeared ill. Khrushchev went to the dacha and observed that "apparently Stalin had gotten out of bed and had fallen" and that in his present condition it would be unseemly for Khrushchev, Beria, Malenkov, and Bulganin to be found on the premises. They left. Later that night Malenkov called again more urgently. Khrushchev made a second trip to the dacha and this time they called the doctors.

Paradoxically, Rybin's account, though favoring the MGB, presents Beria as even more plotting than does Khrushchev. This may seem odd because Khrushchev had much to gain from blackening Beria and took advantage of every opportunity he could find to do so. Could it be that Khrushchev did not wish to blacken Beria too much for fear of turning suspicion on the rest of the quartet of guests at Stalin's last dinner? In one account, medical assistance was delayed by at least four or five hours while Beria was being tracked down. In the other, medical aid was procured as soon as it became clear it was needed. In one account, the security services were entirely to blame for the endlessly protracted process of caring for Stalin; in the other, the system worked. The veracity of Khrushchev's account is further undermined by the fact that a standing order existed in the Kremlin Guard that if any Kremlin official showed signs of illness, doctors were to be called immediately *by the guards themselves.* There was no need to go through Malenkov, Beria, or Khrushchev to obtain medical assistance for Stalin. Either the guards had been instructed to deviate from this standing order by members of the Politburo, or their call for help was countermanded. In either case, complicity at the highest level of Soviet government appears to have ensured that Stalin would die.

The tears Khrushchev and others shed at Stalin's dacha on the evening of March 5 were probably as much from fright at what they had collectively accomplished as sorrow at the passing of their great leader. We will never know for sure. However, from "The History of the Illness of J. V. Stalin," we can now begin to establish a rough set of temporal, geographical, and medical coordinates corresponding to objective events surrounding Stalin's death.

On the night of March 2, 1953, comrade Stalin experienced a sudden loss of consciousness and paralysis of the right hand and leg developed.

> The first examination of the patient was performed on March 2 at
> 7 o'clock in the morning by Professors Lukomsky, P. E.; Glazunov,
> I. S.; Tkachev, R. A.; and docent Ivanov-Neznamov, V. I., in the pres-
> ence of the Head of the Kremlin Polyclinic, Comrade Kuperin, I. I.

When the doctors came upon the scene at 7:00 A.M. on March 2—not
10:00, as Alliluyeva has stated; nor much earlier in the morning, as
Khrushchev has led readers to suppose[13] —Stalin "lay in a state of uncon-
sciousness on the divan, dressed in his clothes. His clothes were soaked in
urine, which indicated that involuntary urination had occurred." Perhaps
this was the "unpresentable state" Khrushchev meant. Both his right arm
and leg were paralyzed; the lines around his nose on the right side were
smoothed out. His pupils were dilated; tendon reflexes were negligible on
the right side, while on the left they were normal; the right side showed
Babinski's reflex.[14] Stalin's pulse was 78 beats per minute; his blood pressure was 190/110. The doctors diagnosed general arteriosclerosis "with the
primary lesion of the blood vessels of the brain. Right side hemiplegia, cardiosclerosis and sclerosis of the kidneys." Stalin was in a "deep unconscious
state" from which he never recovered. "Absolute quiet" was ordered for
the patient, who was left on the divan where they found him. Eight leeches
were applied behind his ears ("leeches were preferred for bloodletting,
because the sharp fluctuations in blood pressure that would have resulted
from bloodletting [in the normal manner] were considered undesirable").
They gave him cold compresses and a hypertonic microenema consisting
of a sulfate of magnesium. The old man's false teeth were removed, and a
round-the-clock watch was established with a team of nerve pathologists,
therapists, and nurses. They sponged his body down with aromatic vinegar
and ventilated the hall in which Stalin lay by opening a window in an adja-
cent room. The enemas of magnesium sulfate, Vaseline or glucose in addition to the leeches continued to be administered. Cheyne-Stokes respiration[15] appeared at 2:10 on the afternoon of March 2 and Stalin's blood
pressure reached 210/120.

Over the course of the next two days, Stalin's vital signs continued to
deteriorate. There were sharp fluctuations in his blood pressure and his
heart had a "hollow tone." They gave him lemon juice mixed with glucose
to drink, and injected him with camphor to stimulate his breathing, which

had become more and more irregular. His left side would from time to time break out in agitated movement, while there were no reflexes at all on the right side. In the early evening of March 3, Stalin managed a brief recovery. After an enema, he experienced diarrhea "with meager, watery feces," and he was taken off the oxygen, which the doctors had been giving him. "During this period, short moments of glimpses of consciousness could be observed. The patient reacted with open eyes to the speech of his comrades who surrounded him." This did not last. He lost consciousness completely later that evening and never regained it. The enemas were discontinued in order not to "subject the patient to superfluous disturbance."

They gave him a combination of carbon dioxide and oxygen to counteract the interruption of his breathing. As Stalin's condition grew worse, he was given injections of caffeine, camphor, and cardiozol. His breathing became ever more confused, and in the early morning of March 4, Stalin developed cyanosis. He began to hiccup uncontrollably and sweat profusely. By 8:20 A.M., the cyanosis had become acute. He experienced "agitated movement." According to the document, this is when the vomiting of blood began which the doctors attributed to a stomach hemorrhage "dependent on vascular disruption."

By 7:15 in the evening of March 5, Stalin's pulse was fluctuating sharply and reached 118 to 120 beats per minute; his blood pressure had gone down to 150/100; by 8:10 P.M., his pulse was 140–150 beats per minute and the great leader was completely comatose. At 9:20, the doctors gave him a 5 percent solution of glucose under the skin; by 9:30, the sweating was even more profuse and a pulse was undetectable. Cyanosis became even sharper. They administered oxygen. At 9:40, Stalin was given camphor injections and adrenaline along with carbon dioxide and oxygen.

At 9:50, Stalin was dead.

The stated time of the doctors' arrival at Stalin's bedside, in "The History of the Illness of J. V. Stalin," provides the first documentary confirmation that while Rybin's account may be inaccurate, Khrushchev's account is false. But the first sentence of the report proves there was a cover-up. The report states unequivocally: "On the night of 2 March 1953, Comrade Stalin experienced a sudden loss of consciousness, and paralysis of the right hand and leg developed." Yet the two eyewitness accounts, Khrushchev's and Rybin's, relate that Stalin became ill on the evening of March 1, not March

2. If it was true that the guards found Stalin unconscious at 10:30 P.M., as Rybin related, eight and a half hours passed before medical treatment arrived. The assertion that Stalin suffered his stroke "on the night of March 2" would clear Beria, Khrushchev, and the others of having delayed medical help. The long hours from the evening of March 1 to 7 A.M. March 2, when the doctors appeared on the scene, and the anxiety of the security service no longer required explanation. "On the night of March 2, 1953..." marks this as yet another document designed to be used in future political struggles. Khrushchev subsequently intended to show that the top leadership acted responsibly and responsively, caring for the life of their leader, when in fact something else had occurred. Rybin's accounts place the blame for delayed medical treatment on Beria. Could Khrushchev have been covering for Beria? If so, Khrushchev himself might have been complicit. But complicit in what—neglect of the dying Stalin, or actual murder?

Was Stalin's death a coup d'état? Molotov claimed that Beria told him on May 1, 1953, that he had been responsible for Stalin's death: "'I did him in!'" Beria boasted. "'I saved all of you!'"[16] Malenkov, Molotov supposed, "knows more, much more, much more." But Malenkov left no memoir. Even a hint of malfeasance on Khrushchev's part would have brought terrible retribution during his lifetime. It is significant in this respect that in Khrushchev's condemnation of Beria, he did not even raise the question of Stalin's death or Beria's possible role in it. Though he suspected Beria of much, he never publicly accused him of this crime. Yet little would have prevented Khrushchev from framing his narrative of Stalin's death so as to incriminate Beria fully—either in the wake of Beria's arrest in June 1953 or nearly twenty years after the event when he composed his memoirs.

Khrushchev's account is made more questionable by the fact that the doctors did not find a high alcohol level in any of their numerous analyses made of Stalin's urine or blood. Nothing supports Khrushchev's statement that Stalin was "pretty drunk" by the time their dinner broke up. Nothing confirms Khrushchev's story that before he died, Stalin pointed to a picture of a "little girl" feeding a lamb from a horn in order to show appreciation to his colleagues who were caring for him. "Then he began to shake hands with us one by one. I gave him my hand, and he shook it with his left hand because his right wouldn't move. By these handshakes he conveyed his feelings."[17] Touching but not true.

According to the doctors' record, Stalin began to vomit blood; there was blood in his urine as well. The doctors also noted that Stalin suffered stomach hemorrhaging.[18] In the first draft, the doctors concluded their report with the paragraph:

> **Arising on March 5 in connection with the basic illness—hypertension and the disruption of circulation in the brain—a stomach hemorrhage facilitated the recurrent collapse [of the patient], which ended with death.[19]**

In the final draft of the report submitted to the Central Committee, this paragraph was cut. All mention of the stomach hemorrhaging was either deleted or vastly subordinated to other information throughout in the final report. It was never reported in any public statement about his death and obviously was something the authorities wanted suppressed, possibly because it might have raised the question of whether the stomach hemorrhaging and bleeding was stress induced or had some other cause, perhaps poison. Whether it had already been minimized in the draft report remains unknown.

This important discrepancy between the drafts demonstrates the care with which "The History of the Illness of J. V. Stalin" was prepared in order to serve new political realities. Beria was arrested on June 26, 1953, and the report was submitted to the Central Committee with the date "July 1953." Its ambiguities may have been intended both to underscore Beria's claim that he had saved the party and to provide deniability. The first draft was preserved, possibly because the new leadership might have toyed with the idea of adding the charge of poisoning to the long list of Beria's crimes, though in the end, they may have thought this was too risky. The autopsy report may have been withheld for the same reasons that these contradictions were never made public. Timashuk's Order of Lenin had been revoked; another cover-up had begun.

Stalin was seventy-three and suffered from hardening of the arteries, as did virtually every other Kremlin leader of the time. Shcherbakov, Zhdanov, Abakumov, Ignatiev, Yegorov, Vinogradov, and Etinger had all either died from heart attacks or suffered from them. A year before, Vinogradov had advised Stalin to retire out of concern for his general physical condition. He was ready to die. The question is whether he was ready to die just then, two weeks before the Jewish doctors were allegedly to be put on trial. The fact that no two accounts of his death correspond and that much of the final medical report was never disclosed may be the result of faulty memories and the inveterate conspiratorial nature of the Soviet government. However, these circumstantial facts also raise again the question, not simply whether he helped to die, as Amy Knight suggests, but whether he was murdered.

Rybin reported that Stalin drank only fruit juice—young Georgian wine—the night before. One conceivable scenario is that Beria, with Khrushchev's knowledge, slipped some poison, such as transparent crystals of warfarin, into the wine. Used as rat poison, warfarin is a tasteless and colorless blood thinner, administered to patients with heart disease. It was patented in 1950 and marketed aggressively internationally shortly thereafter. The right dosage over a period of five to ten days could have induced hemorrhaging as well as a stroke in a patient suffering from acute arteriosclerosis. However, without further hard medical evidence, such as a tissue sample, this remains only speculation.[20]

Whatever happened, Khrushchev would have known about it, could have participated in it, and then could have sought to erase all trace of it—both by arresting and executing Beria (along with many other members of Beria's entourage) and by falsifying the details of Stalin's last illness in his memoir. The autopsy report, if found and published, may provide additional clues. How ironic, yet fitting, that the man who subjected his people to the horror of his great purges and was preparing them for yet more should himself, at the end, have been purged continually by the magnesium sulfate enemas administered by doctors seeking to prolong his life, while his heart had a "hollow tone."

Immediately following Stalin's death on March 5, all the furniture was removed from the Blizhnyaya dacha and all the personnel who had cared for Stalin during his last days were discharged. Most were sent out of Moscow. The deputy head of the guards was sent to the Tula region and lives there to this day. He has refused to tell his story for the last fifty years.

The importance of the doctors' case for the new leaders of the Soviet Union can be judged by the extraordinary haste with which measures were taken to dispose of it in the aftermath of Stalin's death. Within a week of the funeral, Beria initiated a "complete" review. One of his first official acts as the new minister of the newly created MVD [Ministry of Internal Affairs] was to issue a top secret order to "hasten the review of investigative cases" then underway in the ministry. The first case on the list of three important cases was the Doctors' Plot. To expedite the process, Beria appointed special commissions. In the case of the Doctors' Plot, the commission consisted of K. A. Sokolov, A. V. Levshin, and V. V. Ivanov.[21] It was headed by L. E. Vlodzimirsky.[22] Much like Yegorov's special commission in the Kremlin hospital that "investigated" Timashuk's allegations, three of Beria's appointees had taken part in the case themselves. Sokolov and Levshin had both interrogated the doctors. Yegorov wrote to Beria that Levshin was "a sick man—a sadist capable of extreme evil."[23] Of Sokolov, Yegorov wrote:

> Once Lieutenant Sokolov declared to me that, "we will beat you everyday, we will tear out your arms and legs, but we will all the same learn everything there is to know about the life of A. A. Zhdanov and all the truth."[24]

Ivanov was the investigator Konyakhin referred to in his testimony who had discovered the falsifications of inspector Garkusha and brought the matter to inspector Konyakhin's attention, who then brought it to Ryumin's. The commission, therefore, knew the case from the inside, and was in a special position to create a particular view of it. These men could be counted on to provide as much information about Ryumin and the torture of the doctors as possible, and as little about the larger affairs of the ministry or its relations to the Central Committee. It also meant that information about the case was not shared with anyone not already familiar with it and could be more easily controlled. The next case for which Beria created a similar commission was that of the members of the MGB who had been arrested in so-called Abakumov-Shvartsman case. The system endlessly replicated itself.

On March 13, Beria investigated the charges against Yegorov and interrogated one of the nurses on duty at Valdai. On the fourteenth, he interrogated a second nurse. On the seventeenth, Ryumin was arrested, accused of falsifications and perversions in the doctors' case and other matters of state security.

On the eighteenth, Ryumin offered his first confession in which he claimed that Yegorov worked at the order of Kuznetsov, who was a paid American agent. He claimed that he never beat any of the prisoners. On March 24, Ryumin stated:

> In conclusion I say that if by some monstrousness I must fall into the hands of the Abakumovshchina and they will punish me at the stake, then my last words would be—In 1951 I went to the Central Committee in truth.
>
> When I must die, independent of who or whatever circumstances may cause it, my last words will be—I am loyal to the party and its Central Committee.
>
> At the present time I believe in the wisdom of L. P. BERIA and the present leadership of the MVD of the USSR and I hope that my case will have a just outcome.[25]

On March 24 Beria received Yegorov's statement on his relations with Vlasik; on March 26 he heard from Ryzhikov; and on March 27 he received the first of two statements from Vinogradov. On March 28 Beria received a statement from Sheinin on Ryumin and on March 29 he received another statement from Yegorov. The interrogation of Ryumin continued.

On March 31 a "Decree on the termination of criminal prosecution and the freeing of the prisoners in the doctors' plot" was issued by the MGB under Beria's direction. It was signed by investigator Sokolov, among others, who had previously taken part in the interrogation of Yegorov and had threatened him with physical torture. The decree concluded:

> On the foregoing grounds and taking into account that all the arrested doctors in the present case were illegally imprisoned, and that the basis of the charges against them of anti-Soviet terroristic and espionage activity are lacking:
>
> WE DECREE:

In view of the absence of crimes, being guided by statute 4, p. 5 URK, to free from custody and with full rehabilitation those imprisoned in this case: [all the names of the 37 defendants are then listed].[26]

On April 6, 1953, *Pravda* ran an article putting the whole matter publicly to rest: "Soviet Socialist Law Is Inviolable":

> The result of a review of the Doctors' Plot showed that the doctors had been arrested by the former Ministry of State Security (MGB) incorrectly, without any legal basis. As they say in the reports of the Ministry of State Security, the examination showed that accusations brought forward against the accused were false and the documentary base on which the security agents relied was insufficient.[27]

Pravda blamed the MGB for being "remote from the people, from the party. They forget that they are the servants of the people and duty bound to guard Soviet law." It noted that Ryumin was a "secret enemy of the government" who became a criminal adventurist, and it attacked Ignatiev on the same grounds. Furthermore, the article exonerated Solomon Mikhoels and by extension all those associated with him—the Jewish Antifascist Committee and its Jewish nationalist bourgeois agents of Western imperialism. It concluded like a Shakespeare play with the words:

> Each citizen of the great Soviet State may be assured that his rights are guaranteed by the Constitution of the Soviet Union, that they will be sacredly guarded and preserved by the Soviet government.

It was over. Macbeth was dead. Life could begin again.

Behind the *Pravda* article was a top-secret note Beria had written to Malenkov, dated April 1, that provided both an official explanation for the complete rehabilitation of the doctors, and a publicly acceptable view of the plot. The difference between what Beria obviously knew to be the case and this official account can be taken as a guide to the difference between what Khrushchev knew to be the case in the matter of Stalin's death and the official version. The premise of Beria's note to Malenkov was that the "so-called" Doctors' Plot began in the MGB in 1952. "Sensational importance was attached to this case," he wrote, and he reminded Malenkov that it received a TASS news story as well as editorial columns in *Pravda, Izvestia*, as well as other "central newspapers."

"In light of the special importance of this case," Beria noted, the ministry undertook a careful verification that revealed that from beginning to end it was an intentional provocation dreamed up by Ryumin. This, Beria wrote, was the beginning of the Doctors' Plot. To supplement his case and give it greater credibility, Ryumin connected the unrecorded testimony of Etinger to the letter of Timashuk written in 1948 concerning the treatment of A. A. Zhdanov, which, Beria noted, was given to Stalin and then sent "into the archive" of the Central Committee. Beria did not explain why Stalin did nothing about Timashuk's letter. Nor did he explain how Ryumin dug it out of the Central Committee archive and made use of it in 1952.

As Beria put it, setting out on the course of deception and falsification, purposely misleading the Central Committee, Ryumin then took every step necessary to procure the confessions of the doctors, including the most horrendous physical torture: "the leadership of the MGB practiced in its investigative work various forms of torture, horrible beatings; the application of manacles that produced tormenting pain; and the prolonged deprivation of sleep from the prisoners."[28]

"It is necessary to remark," Beria wrote, "that in the Ministry of State Security [Ryumin] found propitious circumstances for all this. All the attention of the Minister and the leading workers of the Ministry was swallowed up by 'the doctors' plot.'" The entire ministry was "swallowed up" by the Doctors' Plot because of the machinations of Ryumin and the irresponsible leadership of Ignatiev. Stalin was deceived, so was the Central Committee. Ryumin acted alone. Beria concluded this letter with the thought that the Ministry of Internal Affairs would take measures to exclude the possibility that a similar "perversion" of Soviet law could ever happen again in the organs of state security.[29]

The process of perversion, however, continued in the very letter in which Beria decreed it would end. We know from all that has been told in this book that the Doctors' Plot did not begin in 1952 and that Ryumin did not act alone, as Beria knew full well. As Ignatiev's March 27, 1953, declaration to Beria testified, it was Stalin who acted alone and was able, ultimately, to pull everyone else along with him. The perversions would not cease until the system that made it possible for Stalin to act alone ceased. It was a system that not only allowed but required absolute power; a system continually fuelled with enemies both from within and without; a system that those in charge understood was inherently illegal.

The drive to legitimate the government and its new leaders was clearly behind Beria's letter to Malenkov and a subsequent letter, on April 3, to the Central Committee concerning the falsifications of the Doctors' Plot. This letter had seven points:

1. To fully rehabilitate the doctors and arrest all those formerly in the MGB responsible for the plot.

2. To confirm the text of the document that would be issued from the Central Committee on the plot.

3. To demand that Ignatiev, the now former Minister of Security, give a complete accounting of how the plot developed and how the Ministry tolerated such total dereliction.

4. To take into consideration measures undertaken by the new Ministry to ensure that such dereliction does not happen again.

5. To abrogate the January 20, 1953, award of the Order of Lenin to L. F. Timashuk on the grounds that it was "incorrect, in connection with actual circumstances that have arisen at the present time."

6. To expel Ignatiev from his post in the Central Committee on the grounds of gross incompetence.

7. To distribute this decree along with the letter of Beria and the conclusions of the investigating commission to all members of the Central Committee of the party, first secretaries of the Central Committees of the Communist parties of the united republics and to other responsible member of the party.

On April 2 Beria wrote a letter to the Presidium of the Central Committee fully exonerating Solomon Mikhoels, his cousin, Dr. M. S. Vovsi, and all other members of the Jewish Antifascist Committee, though the names of those executed on August 12, 1952, were not mentioned. The fact that the doctors were so visibly and immediately rehabilitated while the thirteen members of the Jewish Antifascist Committee were only amnestied but not formally rehabilitated until the late 1980s demonstrates that Beria's reorganization and search for truth could go only so far. Rehabilitation was a political weapon, among other weapons, to consolidate power.

The article that broke in *Pravda* on August 6, 1953, announcing the end of the Doctors' Plot, bore the headline: SOCIALIST LAW IS INVIOLABLE. The cynical irony of this headline was probably not apparent to those who wrote it. Stalin had hoped to capitalize on a terrified nation to reassert total control over the government, expunge his enemies, and perhaps lead his nation into another international conflict, possibly even nuclear war. The final irony of the conspiracy against the Jewish doctors is that the new Soviet leadership made use of it to accomplish much the same end of consolidating power and punishing enemies. As Churchill recognized, it "cut very deeply into communist discipline and structure." Dismantling it signaled that, for the time being, the new leaders were intent on averting open conflict with America.

Stalin's funeral took place on March 9, beginning at noon. The photograph published in *Pravda* shows Stalin's coffin being borne on the shoulders of his loyal comrades whose somber faces convey the depth of their grief. On March 10, the funeral orations of Molotov, Beria, and Malenkov were published in *Pravda*. Molotov bid farewell to his beloved Vozhd with the prophetic words:

> The deathless name of Stalin will always live in our hearts, in the hearts of the Soviet people and of all progressive humanity. The fame of his great deeds in the service and happiness of our people and the workers of the whole world will live forever!
>
> Long live the great, all-conquering teachings of Marx-Engels-Lenin-Stalin!
>
> Long live our powerful, socialistic Motherland, our heroic Soviet people!
>
> Long live the great Communist Party of the Soviet Union!

Just days before he spoke these words, Molotov's life literally had hung by a thread and his wife, Zhemchuzhina, still "sat" in Stalin's gulag. It was not mere hypocrisy that moved Molotov to cherish Stalin's "deathless name" in his heart, or for Ehrenburg to proclaim two days after his funeral: "Stalin lives." Like so many before him, Molotov would surely have gone to his death with Stalin's "name on [his] lips," if Stalin had required it. To search for the causes of this transcendent loyalty would take us truly into the depths of the heart of darkness. In investigating the great conspiracy of Stalin's Doctors' Plot we have descended only to the threshold of those depths.

CONCLUSION:
THE CONSPIRATORIAL MIND

Stalin will be rehabilitated, needless to say.
—V. S. MOLOTOV

Much remains unseen and unknown in Stalin's terrible world. The mechanism of his power cannot be understood apart from the system he and his comrades created in the early years of the Revolution, based on conspiracy and the exercise of brute force. Power was never conferred; it was seized. Nothing legitimated the government except this act of seizure—not the church, not elections, not the "will of the people," not an aristocratic oligarchy or the interests of an economic elite. The system required enemies who would destabilize social and political conditions so that power could be seized and held. Political stability depended upon crises. Without crises, stagnation, not stability, resulted.

Though cruel, inefficient and self-defeating, the Stalinist system, inherited from Lenin, managed to work for nearly seventy-five years and inspired intense loyalty both in the Soviet Union and around the world. This was not simply an aberration in human history. Though in Stalin's time, such loyalty was directed overwhelmingly toward Stalin himself, its origins went much deeper than simple adoration of a great man or slavish devotion to a cruel dictator. Over one billion people around the globe, from Cuba to the People's Republic of China, still wake up each morning to find themselves in this inverted world.

It worked partly by virtue of the kind of perverse upward mobility that made Ryumin Deputy Minister of State Security, Yegorov the head of the Kremlin Hospital system, and Vlasik the Head of the Kremlin Guards. Others benefited as well. Jews among them. The Jewish poet Itzik Fefer, executed in 1952 with his colleagues from the Jewish Antifascist Committee, said before his death, "It seemed to me that only Stalin could correct the historical injustice committed by the Roman kings. . . . I had nothing against the Soviet system. I am the son of a poor schoolteacher. Soviet power made a human being out of me and a fairly well known poet as well." Many Jews felt this way, which helps explain why so many could be found in the Soviet security organs and throughout the governmental bureaucracy. Jews had advanced with extraordinary speed from second-class citizens in Tsarist Russia to the plenipotentiaries of a great world power: Trotsky, Litvinov, Kamenev, Zinoviev, Yagoda, Kaganovich, and Solomon Lozovsky (executed as a member of the Jewish Antifascist Committee in 1952) were only a few of the Jews who rose through the system to the very top and exercised more real power in the Soviet Union than Jews had for nearly two millennia anywhere else in the world. Many others can be found in this book: Shvartsman, Broverman, Palkin, Raikhman, Sverdlov, Sheinin, Maklyarsky, Ehrenburg, Zhemchuzhina. It is a sad story but well worth telling. At his trial, the poet Peretz Markish, another member of the Jewish Antifascist Committee, told the court: ". . . Jews here have closer ties to Russian culture [than to Yiddish]. They want to be part of a great culture; they don't have any nationalistic enthusiasm. They don't speak Yiddish . . . Ten years from now their children aren't going to be speaking Yiddish."[1] Tragically, Markish wasn't entirely wrong. The Revolution gave Jews a chance to be part of a great culture, with nearly fatal consequences. Many, like the writer Isaac Babel, understood the horror but could not dissociate themselves from it.

Loyalty to the system, Grishaev's "party conscience," replaced any other guiding moral perspective. Although Ignatiev claimed to Beria, "I know very well the value of truth," this knowledge did not prevent him from sanctioning the most horrific lies, reviving the ancient blood libel against the Jewish people and spreading the preposterous indecencies of *The Protocols of the Elders of Zion*. Such things did not contradict his "party conscience." Crimes of this sort were vindicated by the power of the Soviet state that had created "equality" throughout the society and whose military had crushed Hitler's Germany. As the unpublished Khavinson-Mints letter to *Pravda* put it,

. . . it is known to the entire world that the people of the Soviet Union and primarily the great Russian nation with its selfless, heroic struggle saved humanity from the yoke of Hitlerism, and the Jews—from complete destruction and annihilation. In our days the Soviet nation was in the first rank of fighters for peace, staunchly defending the affairs of the world in the interests of all humanity.

Stalin's system created the simulacrum of a social contract without the reality of one. Combating the enormity of its lies with reason is almost impossible because its greatest lies were driven not by a misrepresentation of facts but by a false vision of the world. As Ryumin said, "the question of whether you are guilty is decided by the fact of your arrest." Kafka could not have put it more concisely. Facts, logic, reason, truth play no role in the matter. It was a closed system, developed by intellectuals in defiance of all the dictates of rationality, except the entirely rational premise that power must be retained at all costs. A sad paradox of the modern period. Only intellectuals, like Dostoyevsky's Raskolnikov, could claim that the very things they held most dear—truth, justice, reason—were subordinate to this "higher truth" that justified any kind of lie, criminality, and illogic. The "higher truth" was cloaked, in the words of the Khavinson-Mints letter, as "the authentic brotherhood of peoples, great and small . . ." How could anyone reasonably object to this? "For the first time in history, working Jews together with all workers of the Soviet Union have acquired a free, joyful life." On March 11, 1953, six days after Stalin's death, Ilya Ehrenburg wrote an article published in *Pravda*, "The Great Defender of Peace," in which he praised Stalin's vision of peaceful coexistence with the United States of America. He concluded almost poetically: "In these difficult days, as little children play . . . Simple people live, and in them lives Stalin."[2]

Stalin's vision of the "free, joyful life" offered what Milan Kundera has called the totalitarian idyll that lulled so great a part of western intellectual life to sleep for over a century and continues to shape the rhetoric of militant Arabs around the globe against Israel, America and the west, while its mental inclinations, to use the locution of the historian John Lukacs, still dominate a large portion of the western intelligentsia in Italy, France, Germany, England and America. Breaking through this idyll is not easy. As Lysenko's stunning victory at the 1948 conference proved, facts and reason alone will not prevail against deeper political motivation and "party con-science." Nevertheless, facts and reason are all we have. That is the prem-ise of this book.

Stalin's Doctors' Plot was not the haphazard product of a vindictive, paranoid mind. It had a shape forged in a conspiratorial system by a highly calculating, masterful intellect. The Plot was roughly diamond-shaped: Stalin at the apex; the conspiracy against the doctors on one side; the conspiracy against the MGB on the other; America, the ultimate enemy, diametrically opposite Stalin at the nadir. There was never a straight line from Stalin to his enemy. Stalin rarely, if ever, acted *directly* against those he destroyed. He invariably used proxies. Thus, he would not directly accuse America; rather, he worked through the doctors and the MGB to form the accusation. The line along which he would strike was oblique. Tracing it back to him was therefore always difficult. Stalin instigated the plot against the doctors and the plot against Abakumov in order eventually to move against America. Churchill intuited this but could not quite grasp its meaning. In attacking Molotov, first Stalin arrested Zhemchuzhina and then he had Poskrebyshev denounce him. In attacking the doctors, the same diamond-like configura-tion exists. Stalin at the apex; Zhdanov (Timashuk) to one side; Shcherbakov (Etinger) to the other; the doctors at the bottom. Attacking the Jews, Stalin made use of the same technique: Stalin at the apex; the doctors' plot and the MGB at the meridian; the Jews at the antipode.

Stalin was interested in art—simple, classical, direct, tunes the people could hum, plots whose meaning the common man could easily grasp. His own plots and tunes had a different aesthetics. The Jews were culpable because the doctors were killers and the MGB had planned a coup. The doctors were culpable because both Zhdanov and Shcherbakov had allegedly been murdered. Even Ryumin's half-literate letter denouncing Abakumov had this bifocal character: Abakumov was guilty of mishandling the Etinger case *and* the case of Salimanov. To get Abakumov, Stalin employed Ryumin, on the one hand, and Shvartsman on the other. Stalin's conspiratorial plots were like military campaigns with the difference that the goal was always disguised. The attack on America would emerge as necessary *because* it was discovered that America was behind the Jewish doctors and the sinister plot of Abakumov. Whereas, the Varfolomeyev case in combination with a wealth of other evidence demonstrates that America was the target from the beginning and, as this book has shown, the connection with the Jews and Abakumov (through Kuznetsov) was established over time. Even more tragically for the Soviet people, the attack against America would have been united with an attack against, as Stalin put it in 1937, "anyone who, by his deeds or his thoughts—yes, his thoughts—threatens the unity of the socialist state." No one would be exempt.

Plot fit inside of plot, down to the tiniest detail of Ryumin's denunciation of Abakumov and Stalin's order to arrest Yegorov, which occurred only after both Timashuk's denunciation *and* the "expert examination" of Zhdanov's heart. Stalin's signature on the head of each nail is to be found in the replicating, formal elegance of his conspiracies. They were not the product of blind hatred or thuggish rage. Their characteristic shape points to an intellect capable of a very high order of abstraction and geometrical ratiocination. The shape was teleological and always had political intent. This, as much as his cruelty, was fundamental to Stalin's makeup. He could see his plots unfold in time and space. He conceptualized the mutual relationships among the various elements of each plot. For this reason, it was *necessary* that the doctors were guilty and that the Jews were in league with American intelligence and that Kuznetsov inspired Yegorov to murder Zhdanov. Empirical reality meant nothing because the form created its own necessity.

In this sense, Stalin was a seer and a prophet. He could see the enemy. The "blind men, kittens"—Molotov, Mikoyan, Malenkov, Beria and Khrushchev—could not. He was a mighty adversary because of his ability to conceptualize the larger framework of his actions. But he was a false prophet. His Nietzschean will could not obliterate the tiny facts of day-to-day existence: the famine in the Ukraine; the falsity of Lysenko's oblong red potato-tomatoes; the fact that Lammot Dupont had died in 1884 some 65 years before he was to have participated in the "Plan of the Internal Blow"!; the fact that Timashuk's EKG did *not* prove Zhdanov's heart attack; the fact that Garkusha had falsified Ryzhikov's testimony; the fact that Sheinin had never met Etinger or that Etinger had never confessed to having murdered Shcherbakov; the fact that Karpai would not confess.

"A fact is the most stubborn thing in the world," Bulgakov wrote. These tiny facts built up over time and in the end overwhelmed the great lie of the system. Perhaps the most stubborn fact Stalin could not destroy, however many people he killed, was the spontaneous human need to see them. His greatest enemy was human nature itself and the millions of years of evolution that had produced the capacity for rational thought. Even Lysenko could not overcome this.

Stalin was a powerful adversary but he overestimated his strength against these stubborn facts. He miscalculated when the January 13 *Pravda* article appeared and the MGB had not yet obtained all the necessary confessions. He may have made another, even graver miscalculation in thinking that he could control Beria, Khrushchev, and Malenkov indefinitely. He well knew how he and others had isolated Lenin at the end of Lenin's life, deprived Krupskaya of any role in government, and kept their beloved leader in a state of near imprisonment waiting for death. Stalin did not want, to adapt Ryumin's classic formulation, his loyal followers to do to him what he had in his time done to Lenin. Had he waited too long to eliminate Beria? Circumstantial evidence suggests that he did. But we have no definitive answer to this question.

Molotov remarked that, "Stalin will be rehabilitated, needless to say." True. But only if the system remained intact by which Stalin rose above all the institutions and structures of the state, like his monumental image floating over Red Square—the system whereby Josef Vissarionovich Djugashvili became Stalin. Stalin was more than a man; he represented a form of power deriving from the heroic, revolutionary ambition, so prominent in the modern period, to sweep all previous human history aside in the interests of an ideology. Italian Futurism vowed to destroy all the museums, to eliminate the past. It did not succeed. The Russian Revolution went far beyond Marinetti's aesthetic proclamations and almost did succeed. But not quite. The past is a difficult thing to be rid of.

Stalin is a perpetual possibility. To eliminate Stalin means to eliminate the form of power he represented. Up to Gorbachev's reforms, and possibly not even then, Soviet leaders were unwilling or unable to do this. But they were also notably unwilling or unable to exercise fully the power Stalin had put into their hands. Long years of stagnation resulted in which Stalin was neither alive nor dead.

Josef Vissarionovich Djugashvili died on March 5, 1953, but Stalin did not disappear from human affairs. The Stalinist system survived him in a way Hitler's did not. The published accounts of his death present a man dying of a stroke. But Stalin cannot be reduced to his physical reality any more than the Doctors' Plot can be reduced to Ryumin. Stalin remains perpetually in the background of history much as he would sit at the October Hall during Bukharin's show trial, watching from an unlit recess, smoking his pipe, considering his opportunities.

GLOSSARY OF NAMES AND ORGANIZATIONS

V. S. Abakumov (1908–1954)—1943–1946, head of SMERSH (Death to Spies); 1946–1951, minister of state security USSR; July 1951, arrested; December 1954, shot

L. P. Beria (1899–1953)—1946–1953, member of Politburo (Presidium of the Central Committee); 1938–1945, minister of internal affairs (state security); March–June 1953, minister of internal affairs; 1945–1953, head of the atom bomb project; July 26,1953, arrested; December 23,1953, shot.

Ya. M. Broverman (1908–?)—1943–1946, deputy head of the Secretariat of Directorate for Counterintelligence, SMERSH; 1946–1951, deputy head of Secretariat of MGB;1951, arrested; 1954, sentenced to twenty-five years in prison; 1976, released.

A. A. Cheptsov (1902–1980)—1948–1957, chairman of the Military Collegium of the Supreme Court, USSR.

I. G. Ehrenburg (1891–1967)—Soviet novelist, journalist.

Ya. G. Etinger (1887–1951)—Leading Soviet Jewish physician; November 1950, arrested; March 2, 1951, died in Lefortovo prison.

Ya. Ya. Etinger (1929–)—adopted son of Ya. G. Etinger

I. S. Fefer (1900–1952)—Yiddish poet, Secretary of Jewish AntifascistCommittee; 1948, arrested; August 1952, shot.

S. A. Goglidze (1901–1953)—1951–1952, minister of state security, Uzbekistan; 1952, head of the Third Chief Directorate and deputy minister of state security, USSR; 1952–1953, first deputy minister of state security, USSR; March–June 1953, head of the Third Directorate of the MVD USSR; June 1953, arrested; December 23, 1953. shot with Beria.

S. D. Ignatiev (1904–1983); 1952–1953, member of the Politburo (Presidium of the Central Committee); 1950–1952, chairman of the Department of Party, Profsoyuz and Komsomol Organs of the CC while from 1951 to 1953 minister of state security; March–April 1953, secretary of the CC; in 1953, 1957–1960, the first secretary of the Tatar Regional Committee of the Party; 1960, retired.

L. M. Kaganovich (1893–1991)—1930–1957, member of Politburo (Presidium) of the Central Committee; 1961, retired and expelled from Party.

S. Karpai—head of cardiographic unit of Kremlin Hospital. Student of Ya. G. Etinger; formerly the personal physician of Zhemchuzhina; July 1951, arrested; died, according to Kostyrchenko, some years after her release from prison in 1953.

N. S. Khrushchev (1894–1971)—1939–1964, member of Politburo (Presidium of the Central Committee); 1953–1964, first secretary of the CC; 1964, retired.

V. I. Komarov (1916–1954)—1946–1951, deputy of Investigative Unit for Especially Important Cases; terrible anti-Semite; 1951, arrested; 1954, shot with Abakumov.

A. A. Kuznetsov (1905–1950)—1945–1946, first secretary of the Leningrad Regional Committee; 1946–1949, secretary of the Central Committee and head of the administration of cadres of the CC; August 1949, arrested; October 1950, shot as part of Leningrad affair.

A. G. Leonov (1905–1954)—1943–1946, head of the investigative department of SMERSH unit of counterintelligence under Abakumov; 1946–1951, head of the Investigative Unit for Especially Important Cases of the MGB; 1951, arrested; December 1954, shot with Abakumov.

M. T. Likhachev (1913–1954)—1946–1951, deputy head of Investigative Unit for Especially Important Cases of the MGB; 1951, arrested; December 1954, shot.

V. S. Lynko (1911–?)—deputy head of Chief Directorate of the Guards, MGB, under Vlasik; October 4, 1952, convicted and sentenced to ten years; subsequent fate unknown.

A. Lozovsky (Dridzo, S. A.) (1878–1952)—1921–1937, general secretary of Profintern; 1941–1945, deputy head of Ministry of Foreign Affairs; 1945–1948, head of Sovinform; January 1949, arrested in Jewish AntifascistCommittee case; August 1952, shot.

G. I. Maiorov (1897–?)—physician; attended Zhdanov in 1948; 1951, arrested; 1953, released.

G. M. Malenkov (1902–1988)—1946–1957, member of the Politburo (Presidium of the Central Committee); 1939–1946, 1948–1953, secretary of the CC; 1953–1955, chairman of the Council of ministers; 1961, expelled from party and retired.

S. M. Mikhoels (Vovsi) (1890–1948)—actor, director, teacher, People's Artist of the USSR; from 1929, the artistic director of the Moscow Yiddish Theater; January 1948, assassinated in Minsk.

A. I. Mikoyan (1895–1978)—1936–1966, member of the Politburo (Presidium of the Central Committee); 1955–1964, first deputy chairman of the Council of Ministers of the USSR; 1964–1965, chairman of the Presidium of the Supreme Soviet USSR; 1965–1974, member of the Presidium of the Supreme Soviet; 1975, retired.

V. M. Molotov (Scriabin) (1890–1986)—1926–1957, member of the Politburo (Presidium of the Central Committee); 1939–1949, minister of foreign affairs; 1953–1957, first deputy chairman of the Council of ministers; 1953–1956, minister of foreign affairs; 1957–1960, ambassador to Mongolia; 1962, expelled from party and retired; 1984, reinstated in party.

Ye. P. Pitovranov (1915–)—1946–1950, head of the Second Chief Directorate MGB; 1951, deputy minister of state security; October 1951–November 1952, imprisoned; March–June 1953, deputy head of the Second, then the First, Chief Directorate MVD.

N. Poskrebyshev (1891–1965)—1924–1929, assistant secretary of Central Committee; 1929–1934, deputy head of secret sector of CC; 1934–1952, head of special sector of CC; 1952–1953, secretary of Presidium and Bureau of the Presidium of the CC; 1929–1952, Secretary of Stalin. In 1953 Poskrebyshev was about to be put in jail but was allowed to go to Nizhny Novgorod and retire; he stayed in the MGB hotel, was driven by MGB car to the small village where he was born and was prohibited from coming out of his house by day—only allowed out by night.

A. V. Putinstev (1917–)—1946–1948, senior investigator in the Investigative Unit for Especially Important Cases MGB; 1948–1954, assistant head of the Investigative Unit for Especially Important Cases; 1954, transferred to reserves.

L. F. Raikhman (1908–1990)—1941, deputy head of the Counterintelligence Department of the NKGB; 1946–1951, deputy head of the Second Chief Directorate MGB; 1951–1953, imprisoned; May–June 1953, head of the control board of the MVD; 1956, arrested and convicted; 1957, amnestied.

R. I. Ryzhikov (1899– ?)—deputy director of the medical sector of the Barvikha sanatorium of the Kremlin Hospital system; February 1952, arrested; March 1953, released.

M. D. Ryumin (1913–1954)—1941–1945, investigator, head of investigative department of the special department of the NKGB—the department of counterintelligence, SMERSH of Archangelsk, then Belomorsky region; 1945–1947, senior investigator, deputy head of a division of the department of SMERSH (the Third Chief Directorate); 1947–1951, senior investigator of the Investigative Unit for Especially Important Cases, MGB; 1951–1952, deputy minister MGB and head of Investigative Unit for Especially Important Cases; March 1953, arrested; June 1954, shot.

A. S. Shcherbakov (1901–1945)—1941–1945, secretary of the Central Committee; first secretary of Moscow Region and Moscow Party; 1942–1945, head of the Chief Political Directorate of the Red Army; May 10,1945, died of heart attack.

L. P. Sheinin (1906–1967)—Writer, dramatist; in second half of the 1940s, investigator in the Investigative Unit for Especially Important Cases; October 1951, arrested; November 1953, released.

D. T. Shepilov (1905–1995)—1956–1957, candidate member of the Politburo (Presidium of the Central Committee); 1946–1947, editor of *Pravda*; 1947–1952, worked in the apparatus of the CC: first deputy head of the Directorate of Propaganda and Agitation (agit–prop); 1952–1956, editor–in–chief, *Pravda*; 1957, expelled from CC; 1962, expelled from party; 1976, reinstated in party; 1982, retired.

F. G. Shubnyakov (1916–)—1943–1947, head of the department of the NKGB–MGB; 1949–1950, deputy head, then head of Second Chief Directorate of MGB; 1951, arrested with Abakumov; 1953–1954, deputy head of the First Chief Directorate MVD, then deputy head of the Second Chief Directorate KGB.

Ye. I. Smirnov (1904–1989)—February 1947–1952, minister of health security, USSR; December 1952, arrested; April 1953, released from prison; returned to work in the health ministry.

L. L. Shvartsman (1907– ?)—Beginning in 1925, literary assistant, then head of editorial departments of *Kievsky proletarii* and *Moskovsky Komsomolets* and other journals; 1941–1951, deputy head of Investigative Unit for Especially Important Cases, MGB; March 3, 1955, convicted by Supreme Court and sentenced to execution; there is no record that the sentence was carried out.

D. M. Sukhanov (1904–)—To March 1953, assistant secretary of the Central Committee for G. M. Malenkov.

L. F. Timashuk (1898–1983)—1926–1964, physician; head of EKG department in Kremlin Hospital until September 1948.

V. Kh. Vasilenko (1897–1987)—physician; 1948–1952, head of the department of therapy of the First Moscow Medical Institute; 1950–1952, chief therapist of the Kremlin Hospital.

V. N. Vinogradov (1882–1964)—Therapist, professor of medicine; 1952, arrested; March 1953, released and returned to practice in the Kremlin Hospital.

N. S. Vlasik (1896–1967)—1935–1936, head of the personal guards of the operation department of the OGPU–NKVD; 1936–1938, head of the operational group and head of a division of the First Department of the First Chief Directorate of the NKVD; beginning in 1943, head of the Sixth Directorate and head of the First Department of the NKGB; 1951–1952 head of the Chief Directorate of the Guards, MGB; May–November 1952, deputy head of the directorate of the gulag camp in Asbest; December 1952, arrested; 1955, exiled for five years in Krasnoyarsk, deprived of military rank and medals; 1956, released from further prosecution.

L. Ye. Vlodzimirsky (1903–1953)—1947–1953, head of the Directorate of the cadres; 1953, head of the Investigative Unit for Especially Important Cases; July 1953, arrested; December 1953, shot.

M. S. Vovsi (1897–1960)—1941–1950, chief of therapy for the Soviet Army, consultant–therapist for the Kremlin Hospital; 1952, arrested; March 1953, released.

N. A. Voznesenksy (1903–1950)—1942–1949, chairman of Gosplan; 1947–1949, member of Politburo (Presidium of the Central Committee); 1949, arrested in Leningrad Affair; October 1950, shot.

G. G. Yagoda (Iyeguda, Ye. G.) (1891–1938)—1934–1936, commissar for internal affairs NKVD; April 1937, arrested; March 1938, shot.

P. I. Yegorov (1899–?)—Professor of medicine, therapist; 1950–1952, head of the Kremlin Hospital; September 1952, arrested; March 1953, released.

N. I. Yezhov (1895–1940)—1935–1939, secretary of the Central Committee; 1936–1938, commissar of internal affairs NKVD; June 1939, arrested; February 1940, shot.

A. A. Zhdanov (1896–1948)—1939–1948, member of Politburo (Presidium of the Central Committee); 1934–1944, first secretary of the Leningrad party; 1946–1947, chairman, Supreme Soviet.

Yu. A. Zhdanov (1919–)—1947–1950, chairman of the department of science in the Directorate of Propaganda and Agitation, Central Committee; 1957, rector of Rostov University.

P. S. Zhemchuzhina (Karpovich) (1897–1970)—Wife of V. M. Molotov; 1939–1948, head of the Chief Directorate of Textile Industry and minister of light industry; 1949, arrested and sent into exile; March 1953, rehabilitated and released.

CC Central Committee of the Communist Party; in the 1930s all party matters were decided by the Politburo, the Orgburo, and the Secretariat, consisting of thirty to forty members; this changed after the Nineteenth Party Congress when the CC and Politburo were expanded and the Presidium was established.

Jewish Antifascist Committee Among five antifascist committees establish by USSR in 1942, to win support of the West.

Orgburo Organizational Bureau of CC; Zhdanov headed the powerful Orgburo after the war; decisions taken by the Orgburo would then be discussed by the Politburo under which it worked.

Politburo Political Bureau of the CC; highest organ of the Party; met rarely after WWII.

Presidium Supreme governing body of the CC after Nineteenth Party Congress, October 1952.

Profintern Red International Trade Unions.

Secretariat Third highest organ of the party with Politburo and Orgburo.

SMERSH Death to Spies (Soviet military counterintelligence, 1943–1946).

Sovinformburo Soviet Bureau of Information.

State Security Founded in 1918, as Extraordinary Commission for Combating Counterrevolution and Sabotage (Cheka); took many other forms as GPU, OGPU, NKVD, MGB, and KGB.

ENDNOTES

The Soviet documents used in this book come from the following archives:
    Archive of the President of the Russian Federation
    Russian State Archive of Social and Political History (RGASPI)
    Central Archive of the Federal Security Service (FSB)
    Russian State Archive of Contemporary History (RGANI)

INTRODUCTION: THE INVERTED WORLD

1. Churchill to Eisenhower, April 11, 1953.

2. When the criminal indictment against Abakumov was drafted.

3. For a recent account, see Sergo Beria, *Beria, My Father: Inside Stalin's Kremlin* (London: Duckworth, 2001), 243–44. See also Yakov Rapoport, *The Doctors' Plot of 1953: A Survivor's Memoir of Stalin's Last Act of Terror Against Jews and Science* (Cambridge, Mass.: Harvard University Press, 1991), 82–85

4. In M. Gorky, MCP "Interkontakt" *Nesvoyevremennye mysli* (Moscow, 1990), 9; quoted from Chaliapin, F. I., *Maska i dusha* (Moscow, 1989), 239

5. Excerpts from the decree on the termination of the criminal prosecution and the release of the prisoners from custody, confirmed by the Minister of Internal Affairs of the USSR, L. Beria, March 31, 1953, 22

6. "Wrecking" or sabotage is the translation of the Russian word *vreditelstvo* and was commonly used to denounce all manner of spies, saboteurs, and other such enemies of the people.

CHAPTER ONE: THE UNTIMELY DEATH OF COMRADE ZHDANOV

1. *Poems of A. A. Akhmatova,* selected, translated, and introduced by Stanley Kunitz with Max Hayward (Boston: Little, Brown, 1973), 23.

2. John Lewis Gaddis, *We Now Know: Rethinking Cold War History* (New York: Oxford University Press, 1998, pbk. edition), 46

3. See Svetlana Alliluyeva, *Twenty Letters to a Friend* (New York: Harper and Row, 1967), 192–93.

4. Timashuk to Vlasik, August 29, 1948, 2.

5. Rapoport, *The Doctors' Plot of 1953,* 77.

6. Timashuk to Vlasik, August 29, 1948.

7. Protocol of interrogation of TIMASHUK, October 17, 1952, 3.

8. Ibid., 3.

9. Ibid., 5.

10. Two dates are given for this letter; the first seems to be a mistake by Timashuk on the letter itself in which she writes "7/IX–48" (September 7, 1948). But this cannot be correct because the letter relates information that she herself says occurred on September 8 (her firing). In her subsequent oral testimony to the MGB of October 17, 1952 (p. 5), she says she wrote the letter on September 15, which makes more sense.

11. Timashuk was to write two more letters in 1948 in a vain attempt to attract the attention of Kremlin leaders. The second and third letters were written to A. A. Kuznetsov, secretary of the party, who was himself arrested a year later, tried for treason and shot. See Viktor Malkin, "*Syem' pisem Lidii Timashuk,*" in *Novoye Vremya,* 1993, No. 28, 38–41.

12. *Out of the Red Shadows: Anti-Semitism in Stalin's Russia,* by Gennadi Kostyrchenko (Amherst, N.Y.: Prometheus Books, 1995, originally published in Moscow as *V plenu u krasnogo faraona* [*In the Captivity of the Red Pharaoh*]), 263.

13. "Stenogram of the Session in the *Lechsanupra* of the Kremlin Concerning the diagnosis of the illness of A. A. Zhdanov, in connection with the conclusions of Doctor Timashuk," September 6, 1948, 2.

14. Timashuk to Vlasik, August 29, 1948, 1.

15. Ibid.

16. Protocol of interrogation of Timashuk, October 17, 1952, 2-3.

17. Timashuk to Vlasik, August 29, 1948, 1.

18. Protocol of interrogation of Timashuk, August 11, 1952, 8-9.

19. Handwritten testimony of Belov, October 18, 1952, 4.

20. Malkin, "*Syem' pisem Lidii Timashuk,*" 41.

21. Ibid., 2.

22. "Stenogram," September 6, 1948, 3.

23. Ibid., 4.

24. Ibid.

25. Ibid., 10.

26. Ibid., 5.

27. Ibid.

28. Ibid., 3-5

29. Ibid., 5

30. Ibid., 5-6

31. October 11, 1952, Interrogation of Timashuk, 11.

32. October 17, 1952, Interrogation of Timashuk, 2

33. In his testimony before an MGB interrogator on October 6, 1951, Belov had difficulty recalling the details of the events in August 1948, citing a poor memory. If anyone had asked him after Zhdanov's death what circumstances existed in Valdai at the time and how events unfolded, he could have told much, he informed them. And he says that "when the details and events were fresh in my memory and might have been reconstructed . . . no one from the leadership staff of the Chief Directorate of Security ever discussed the matter with me"—something he found not a little surprising. Handwritten testimony of Belov, October 18, 1952, 6.

34. Protocol of interrogation of Belov, October 6, 1951, 12.

35. Ibid., 13.

36. Protocol of interrogation of Yegorov, October 20, 1952, 7.

37. Protocol of interrogation of Yegorov, February 7, 1953, 2-3.

38. Yegorov described his relationship to Vlasik in detail at his November 15, 1952, interrogation. See *Protocol of Interrogation of Yegorov,* November 15, 1952, 1-3.

39. N. S. Khrushchev, *Khrushchev Remembers,* with an Introduction, commentary and notes, by Edward Crankshaw, Translated and Edited by Strobe Talbott (Boston: Little, Brown, 1970), 283-84

40. From November 7, 1937, entry in *The Diary of Georgi Dimitrov,* edited by Ivo Banac (New Haven: Yale University Press, forthcoming). A different version of this toast is discussed by Robert C. Tucker in *Stalin in Power: The Revolution from Above, 1928-1941* (New York: W. W. Norton, 1990), 483-84.

41. Protocol of interrogation of Timashuk, August 11, 1952, 13-14.

42. Protocol of interrogation of Yegorov, February 9, 1953, 1. "I had an exclusively official, mutual relationship with ABAKUMOV. Several times he called me on the telephone, having to do with the condition of the health of several leaders of the Party and the government, and workers of foreign companies."

43. Timashuk to Vlasik, August 29, 1948, 1-2.

44. September 6, 1948, 10.

45. Ibid., 11

46. Ibid., 12

47. "Stenogram of Council in Lechsanupra of the Kremlin," August 31, 1948, 7, 8, 9,12.

48. Ibid., 2

49. In fact, Vinogradov at one point mentioned an EKG but without specifying which one it was. Ibid., 6.

50. Protocol of interrogation of Vinogradov, November 18, 1952, 6-7.

51. Ibid., 12–13.

52. Stenogram, August 31, 1948, 6.

53. Stenogram, September 6, 1948, 13–14.

54. Ibid., 15.

55. Ibid., 17.

56. Ibid.

57. Ibid., 17–18.

58. Timashuk to Vlasik, August 31, 1948, 2.

59. Interview with Lawrence S. Cohen, M. D., Ebenezer K. Hunt Professor of Medicine, Special Advisor to the Dean, Yale School of Medicine, May 18, 2001.

60. Stenogram, September 6, 1948, 12.

61. Ibid.

62. Ibid., 12–13.

63. Ibid., 13.

64. Ibid., 14.

65. Ibid.

66. Ibid.

67. Ibid., 14–15.

68. Ibid., 16–17.

69. Ibid., 16.

70. Ibid., 17.

71. Protocol of interrogation of Timashuk, October 17, 1952, 5.

72. W. A. Mozart, *The Magic Flute*, Act II, Scene 30.

73. Protocol of interrogation of Timashuk, October 17, 1952, 5

74. See Malkin, "*Syem pisem Lidii Timashuk*," 40.

75. Ibid.

76. Only Vlasik and Poskrebyshev remained alive after 1954. Vlasik remained in exile for years far from Moscow and returned psychologically broken, physically sick and morally discredited. Poskrebyshev disappeared from public view after 1952 and though not arrested or shot, never resurfaced.

77. "Concerning the history of the illness of A. A. ZHDANOV," March 27, 1953, 4.

78. Protocol of Interrogation of Timashuk, October 17, 1952, 5.

79. Protocol of Interrogation of Vlasik, January 24, 1953, 2.

CHAPTER TWO: STALIN'S SILENCE

1. V. M. Molotov, *Molotov Remembers: Inside Kremlin Politics*, conversations with Felix Chuer, edited with an introduction and notes by Alber Resis (Chicago: Ivan Oee, 1993), 222.

2. Ibid., 221.

3. Kostyrchenko, *Out of the Red Shadows,* 248.

4. Louis Rappoport, *Stalin's War Against the Jews: The Doctors' Plot and the Soviet Solution* (New York: Free Press, 1990), 132

5. Rapoport, *The Doctors' Plot of 1953,* 217. Unfortunately, Rapoport did not treat Zhdanov or any of the other Politburo leaders alleged to have been murdered. He was not among the inner circle of Kremlin doctors and was arrested only in February 1953. Therefore, as valuable as his memoir is, it sheds little light on the thinking of Yegorov, Vasilenko, Vinogradov, and Maiorov in Valdai in 1948 or on the intricate relations among the Lechsanupra, the security services, and Stalin that play so important a role in the story.

6. "Nevertheless it is necessary to acknowledge that the autopsy of A. A. ZHDANOV, performed on 31 August, disclosed a fresh myocardial infarct." See March 27, 1953 statement of Vinogradov to Beria, 4.

7. Declaration from Yegorov, to Beria, March 13, 1953, 1–2.

8. Ibid., 2.

9. Interview with Lawrence S. Cohen, M.D.

10. Excerpts from the history of the sickness of comrade Zhdanov from August 18, 1948, to August 30, 1948.

11. Protocol of interrogation of Timashuk, August 11, 1952, 6.

12. Excerpts from the history of the sickness of comrade Zhdanov, August 30, 1948, 2.

13. Protocol of interrogation of Timashuk, August 11, 1952, 13.

14. "Stenogram of the Council Consisting of Deputy Head of the Lechsanupra of the Kremlin, Dr. Markov, A. M.; Academics, Vinogradov, V. N., Zelenin, V. F.; Professors Nezlin, V. E., Etinger, Ya. G.," August 31, 1948, 1.

15. Statement of Vinogradov to Beria, March 27, 1953, 2.

16. Protocol of a face-to-face confrontation between Vinogradov and Karpai, February 18, 1953, 3–4. Unfortunately, the transcript of the July EKGs taken by Karpai have not been found.

17. Ibid., 7.

18. Ibid., 9.

19. Interrogation of Vinogradov, November 18, 1952, 6.

20. Stenogram, August 31, 1948, 10.

21. Ibid.

22. Protocol of Interrogation of Yegorov, October 20, 1952, 6.

23. Protocol of Interrogation of Yegorov, February 7, 1953, 5–6; see also protocol of interrogation of Turkina, January 30, 1953, 6; and protocol of interrogation of Panina, March 13, 1953, 6.

24. Excerpts from the protocol of interrogation of Panina, March 13, 1953, 5–6.

25. See protocol of interrogation of Yegorov, February 7, 1952, 6; and statement of Yegorov to Beria, March 13, 1953, 12.

26. Protocol of interrogation of Yegorov, February 7, 1953, 6. "When I arrived in Valdai, apart from VASILENKO and MAIOROV, VOZNESENSKY was there. He had arrived some hours earlier with an unknown objective."

27. Protocol of interrogation of Yegorov, February 7, 1953, 2.

28. *Khrushchev Remembers*, with an introduction, commentary, and notes by Edward Crankshaw, translated and edited by Strobe Talbott (Boston: Little, Brown, 1970), 284.

29. "Stalin's old collaborators, people who could best tell how he had slowed down, had in any case outlived their usefulness. The Mikoyans and Molotovs had been around too long. The same was true for his immediate entourage. . . . Could one trust them? Of course not." Adam Ulam, *Stalin: The Man and His Era* (Boston: Beacon Press, 1989), 704.

30. Khrushchev relates the following scene in his memoirs: "Then cannibalism started. I received a report that a human head and the soles of feet had been found under a little bridge near Vasilkovo, a town outside of Kiev. Apparently the corpse had been eaten. There were similar cases. . . . 'I found a scene of horror' [one witness reported to Khrushchev]. The woman had the corpse of her own child on the table and was cutting it up. She was chattering away as she worked, "We've already eaten Manechka. Now we'll salt down Vanechka. This will keep us for some time." Can you imagine? This woman had gone crazy with hunger and butchered her own children!'" See *Khrushchev Remembers*, 234–35.

31. In a radio address on February 9, 1946, Stalin signaled that there would be no liberalization of the Soviet system. He called for unrelenting hard work, austerity, and discipline and further sacrifices by the Soviet people. In 1947 the currency was devalued by 1000 percent.

32. This and other information concerning Stalin's early Cold War thinking is to be found in the communication he carried on, largely by telegram, with Politburo leaders from his Crimean headquarters in those years. These telegrams have recently been discovered as part of what is called the *Lichnii fond Stalina*, Stalin's personal archive, formerly housed in the presidential archive. They will appear in both Russian and English editions as part of Yale University Press's Annals of Communism series. See an account of this meeting in Ulam, 660–61. Malenkov accompanied Zhdanov on this trip to Poland. Stalin was alarmed and offended by the public discussion of his health after the war. An article published in the *Chicago Tribune* from 1945 speculated that Zhukov and Molotov would lead the country after Stalin's imminent retirement.

33. Naumov, 26–27.

34. Ibid., 29–30.

35. Though it was a position of great importance and prestige, Stalin may have worried about Beria's grip on security.

36. Unpublished telegrams between Stalin and the Politburo in the Stalin archive, Moscow.

37. Molotov was officially relieved of his duties as foreign minister in March 1949 when Vishinsky assumed that position. Ulam, 705.

38. See *Tak bylo,* by Anastas Mikoyan (Moscow:Vagrius, 1999), 535 for a substantially different account of this incident. However, Mikoyan's own working notes are probably more reliable. See his diary entries located in Mikoyan's published memoirs assembled by his son. RTsKhIDNI, op. 3, d. 190, f. 84.

39. See *Molotov Remembers,* 319–20, for Molotov's somewhat sketchy account of this.

40. Voroshilov, Andreev, Rykov, Kirov, Kalinin, and Dimitrov all had Jewish wives.

41. Zhemchuzhina means "pearl" in Russian. She was never in a concentration camp. Her sister lived in Bridgeport, Connecticut.

42. See *Molotov Remembers,* 324–25; See also Kostyrchenko, *Out of the Red Shadows,* 119–24.

43. *Molotov Remembers,* 325. ". . . when she was arrested in 1949 they accused her of plotting an attempt on Stalin's life."

44. "Kuznetsov" is A. A. Kuznetsov, to whom Timashuk wrote her letters.

45. Cf. *Khrushchev Remembers,* 228–44.

46. Dimitri Volkogonov, *Stalin: Triumph and Tragedy* (New York: Grove Press, 1991), 528.

47. Zhores A. Medvedev, *The Rise and Fall of T. D. Lysenko,* translated by I. Michael Lerner, with the editorial assistance of Lucy G. Lawrence (Garden City, N.Y.: Anchor Books, 1971; originally published by Columbia University Press, 1969), 103.

48. Ulam, 651.

49. *O polozhenii v biologicheskoy nauke* [On the Situation in Biological Science]: *stenograficheskii otchyot sessii vsesoyuznoy akademii cel'skokhozyaictvehhikh nauk 31 iyulya—7 avgusta 1948.* (Moscow: OGIZ - SELKHOZGIZ, 1948), 5.

50. *O polozhenii v biologicheskoy nauke,* 529–531.

51. After neo-Darwinist August Weismann.

52. After the geneticist T. H. Morgan.

53. *O polozhenii v biologicheskoy nauke,* 512–13.

54. David Joravsky, *The Lysenko Affair* (Chicago: University of Chicago Press, 1970; pbk ed., 1986), p. 317 ff.

55. Protocol of interrogation of Krongauz, February 7, 1952, 1–3.

56. *I primknuvshii, k nim Shepilov* (Moscow: Zvonnitsa-MG, 1998), 269 ff. Shepilov's account accords with that of V. A. Malishev, who took notes of the meeting. See *Istochnik*, May 1997, no. 30., 135.

57. *I primknuvshii*, 277.

58. Ibid., 280.

59. Ibid., 275–81.

60. Ibid., 9–10.

61. After Zhdanov's death, Shepilov returned as an inspector in the Central Committee. In 1952 he became the editor of *Pravda*.

62. "My father had been very fond of Andrei Zhdanov. He respected the son and had always hoped the two families might one day be linked in marriage. It happened in the spring of 1949 as a matter of hard common sense but without any special love or affection. I thought I would gain at least a little freedom by moving to his house and that it would give me the access to people that I lacked. . . . My father knew everyone in the Zhdanov household and couldn't stand either the widow or the sisters." See Alliluyeva, *Twenty Letters to a Friend*, 192–93.

63. Ibid., 521.

64. Protocol of interrogation of Belov, October 6, 1951. At this interrogation, Belov produced his notes from July 1948. It is from Belov that we know it was Shepilov who called.

65. Stenogram, August 31, 1948, 1.

66. *Izvestia*, September 3, 1948, 1.

67. "*To Comrade I. V. Stalin*," *Pravda*, August 7, 1948, no. 220, 5.

68. P. M. Zhukovsky in his statement repudiating his former "incorrect" ideas cites the appearance of Yuri Zhdanov's letter that morning, but points out that his sincere transformation was the result of a "great moral, principled and political" awakening. See *O polozhenii v biologicheskoy nauke*, 523.

69. Protocol of interrogation of Vinogradov, February 12, 1953, 7.

70. Ibid., 7–8.

71. *Eugene Onegin* is a narrative poem written by Pushkin. *Reabilitasiya: Kak Eto Bylo*, General editor, A.N. Yakovlev (Moscow: Mezhdnnayodnii and "democratiya," 2000) 129–130.

72. Ibid., 3–4.

73. Protocol of face-to-face confrontation between Vinogradov and Karpai, February 18, 1953, 3.

74. Ibid., 3–4.

75. Protocol of Interrogation of Timashuk, August 11, 1952, 11.

76. Protocol of face-to-face confrontation between Vinogradov and Karpai, February 18, 1953, 3–4.

77. Interrogation of Vinogradov, February 12, 1953, 7.

78. Protocol of face-to-face confrontation between Vinogradov and Karpai, February 18, 1953, 3.

79. A. Malenkov, *O moyom ottse, Georgii Malenkov* (Moscow, 1992), 55

CHAPTER THREE: THE PYGMY AND THE TERRORIST

1. "To the secretary of the Central Committee (VKPb), Malenkov, G. M. (personal)," from B. Shimeliovich, July 11, 1952.

2. See Rapoport, *The Doctors' Plot of 1953,* 33.

3. For an account of this massive anti-Semitic campaign, see Kostyrchenko, *Out of the Red Shadows,* and Rubenstein and Vladimir P. Naumov, eds., *Stalin's Secret Pogrom: The Postwar Inquisition of the Jewish Anti-Fascist Committee* (New Haven, Conn.: Yale University Press, 2001). See also *Special Tasks: The Memoirs of an Unwanted Witness—A Soviet Spymaster,* by Pavel Sudoplatov and Anatoli Sudoplatov, with Jerrold L. and Leona p Schecter (Boston: Little, Brown, 1994), 285 & ff. The campaign ceased only with Stalin's death in 1953.

4. See Sudoplatov's account in *Special Tasks,* 296–97.

5. Rapoport, *The Doctors' Plot of 1953,* 36.

6. Emelia Teumin was not technically a member of the committee, but she had worked for Lozovsky in the Sovinformburo and had helped prepare some materials for the Committee. See Rubenstein and Naumov, *Stalin's Secret Pogrom,* 111 & ff.

7. For a full account of the liquidation of the Jewish Antifascist Committee see Rubenstein and Naumov, *Stalin's Secret Pogrom.*

8. Rapoport, *The Doctors' Plot of 1953,* 49.

9. Vladimir P. Naumov, et al., eds., *Nepravednii sud* (Moscow, 1994).

10. Protocol of interrogation of Etinger, December 11, 1950, 2.

11. Protocol of interrogation of Etinger, December 6, 1950, 2.

12. Excerpt from the protocol of interrogation of Fefer, April 22, 1949, 1–2.

13. Fefer confessed to this at a closed session of the JAC trial on June 6, 1952. His code name was Zorin. Rubenstein and Naumov, *Stalin's Secret Pogrom,* 324. Mikhoels and Fefer toured America, Canada, Mexico, and England from June to December 1943.

14. Rubenstein and Naumov, *Stalin's Secret Pogrom,* 93.

15. See the excellent account of this in Rubenstein and Naumov, *Stalin's Secret Pogrom.*

16. Excerpt from the protocol of interrogation of Fefer, April 22, 1949, 3–4.

17. "*Moim naparnikom po kamere okazalsya Lev Sheinin*" ["My Cellmate in Prison Was Lev Sheinin"] Interview with Yevgeny Pitovranov in *Rossiiskoye Obozreniye,* Globus, 1993, 1.

18. During the entire period between the end of the war and his 1951–1952 imprisonment, Yevgeny Pitovranov headed the Second Chief Directorate (counterintelligence) of the MGB. As of January 3, 1951, he was a deputy minister in the MGB and a member of the Collegium of the MGB. In this position, he directed all internal security operations of the MGB. His immediate superior from 1946 to July 1951 was Viktor Abakumov. Pitovranov's arrest may have been connected to his having been a member of the special commission dealing with the Doctors' Plot investigation, although in subsequent interviews with him concerning his discussions with Stalin in the summer of 1951, he asserted that Stalin's main concern was the effectiveness of the Second Chief Directorate. According to the publication of the Federal security service, *Liubianka 2: Iz Istorii Otechestvennoi Konterrazvedk* (Moscow, 1999), 258–62, this directorate was very effective indeed, particularly in work against the American embassy. Pitovranov survived Beria and later served in Beijing.

19. Ibid., 1.

20. Vladimir P. Naumov interview with Yakov Y. Etinger. Minsk was liberated in 1944.

21. She has been memorialized in Yad Va Shem, in Israel.

22. Rubenstein and Naumov, *Stalin's Secret Pogrom,* 93.

23. Protocol of interrogation of Etinger, December 4, 1950, 1–2.

24. Rubenstein and Naumov, *Stalin's Secret Pogrom,* 109 ff.

25. Ibid., 111.

26. No document with such an order has been found, but the order has been attested by several different sources—the interrogations of Abakumov, Leonov, and M. D. Ryumin.

27. "Approximately 28–29 January 1951, Leonov called me to see him at the order of Abakumov. He ordered me to cut short the work with prisoner Etinger and said that the case was to be put on the shelf, as Leonov expressed himself." From the confession of M. D. Ryumin, no date. A. G. Leonov was the head of the so-called Investigative Unit for Especially Important Cases. He could have received this order only from Abakumov.

28. Protocols of interrogation were not simple transcripts of interrogation sessions but usually were composed after the fact by the interrogator, often with some additional assistance, in which only salient facts, names, or confessions were written down. Generally speaking, these protocols reflected the direction in which the investigation *wished* to proceed and they were often doctored to include materials and admissions not provided by the prisoner. In the case of the Jewish Antifascist Committee trial and the Doctors' Plot, Stalin reviewed these protocols daily and provided guidance to the MGB on how to interrogate the prisoners. The protocol of review is the only official document noting Etinger's association with the Jewish Antifascist Committee. This document was prepared not by the MGB but by the Ministry of Justice.

29. Sudoplatov, *Special Tasks,* 301.

30. *Khrushchev Remembers*, 257–58.

31. Dimitri Volkogonov, *Stalin: Triumph and Tragedy* (Rocklin, Calif.: Prima Publishing, 1992), 520.

32. Sudoplatov makes this point very strongly, stating that "the Doctors' Plot was . . . part of a struggle to settle old scores in the leadership [of the Party]." Sudoplatov, *Special Tasks,* 298. In doing so, however, Sudoplatov gives the impression that the Doctors' Plot was nothing more than a smoke screen to purge the MGB, which is also mistaken.

33. Volkogonov, *Stalin: Triumph and Tragedy,* 520–24; Walter Laqueur puts the number at 2,000 in accord with a 1988 article in the Russian journal *Argumenty I fakti.* See *Stalin: The Glasnost Revelation* (New York: Charles Scribner's Sons, 1990), 116. The party leaders were shot and the rest imprisoned.

34. See Molotov's account in *Molotov Remembers,* 291 & ff. See also *Khrushchev Remembers,* 246–57, and Volkogonov, *Stalin,* 520 & ff.

35. See the Diary of D. T. Shepilov (unpublished manuscript), 199.

36. Volkogonov, *Stalin,* 520.

37. *Molotov Remembers,* 292.

38. Adam B. Ulam, *Stalin: The Man and his Era* (Boston: Beacon Press, 1989), 653.

39. *Khrushchev Remembers,* 370.

40. *Reabilitatsiya,* 115–16.

41. See discussion of Ryumin's role in the interrogations of the JAC defendants in Rubenstein and Naumov, *Stalin's Secret Pogrom,* 54.

42. Ibid., 54–56.

43. See Naumov's discussion in ibid., xiii–xv.

44. Ibid., 63.

45. Protocol of interrogation of Sheinin, December 15, 1953, 1–2.

46. Koba was Stalin's party name before the 1917 revolution. Afterward it was used affectionately only by close comrades.

47. In J. Arch Getty and Oleg V. Naumov, *The Road to Terror: Stalin and the Self-destruction of the Bolsheviks, 1932–1939* (New Haven, Conn.: Yale University Press, 1999), 557.

48. *Reabilitatsiya,* 371.

49. *Spravka,* July 9, 1953, signed by Ozherelev.

50. *Spravka* on the imprisonment of Etinger, July 30, 1951, 1–2 (KGB archive).

51. Ibid., 3.

52. Ibid.

53. Rubenstein and Naumov, *Stalin's Secret Pogrom,* 60, 50–51.

54. "To the Minister of Internal Affairs of the USSR—Comrade L. P. Beria: Report," March 24, 1953, 1.

55. Sudoplatov, *Special Tasks,* 299.

56. The official designation for this unit was UKR "Smersh" NKO USSR, Directorate of Counterintelligence, Peoples Commissariat of Defense, USSR. This was military counterintelligence responsible for the Special Departments that were taken away from the NKVD by Stalin and given to Abakumov in 1943. However, Abakumov retained offices in the Lubyanka and personnel for SMERSH still came from State Security. See David E. Murphy, Sergei A. Kondrashev, and George Bailey, *Battleground Berlin: CIA vs. KGB in the Cold War* (New Haven, Conn.: Yale University Press, 1997), 11. In 1946 Abakumov was named minister of state security. Ryumin must have been sufficiently well known to Abakumov to be assigned to the Unit for Especially Important Cases. It is unclear whether Abakumov knew that Ryumin would turn against him.

57. Protocol of interrogation of Ryumin, March 30, 1953, 1–4.

58. Interview with Sergei Kondrashev, conducted by Jonathan Brent in January 2000 in Moscow.

59. Ignatiev's letter to Beria, March 27, 1953, 4.

60. Protocol of interrogation of Sheinin, December 15, 1953, 1–2.

61. Excerpt from Ryumin's confession, 37–38.

62. This occurred after Abakumov was removed from office in July 1951. See Rubenstein and Naumov, *Stalin's Secret Pogrom,* xv.

63. Protocol of interrogation of Etinger, November 20, 1950, 1.

64. Protocol of interrogation of Etinger, December 6, 1950, 1–2.

65. Ibid., 3.

66. Protocol of interrogation of Etinger, December 27, 1950, 1–2.

67. *Spravka* on the imprisonment of Etinger, July 30, 1951, 1.

68. Ibid., 1–2.

69. Ibid., 3.

70. Ibid.

71. Ibid., 2–3.

72. Protocol of interrogation of Yanshin, no date, 2.

73. Protocol of interrogation of Likhachev, June 13, 1953, 10–11.

74. A. Malenkov, *O moem otse Georgii Malenkov* (Moscow, 1992), 55.

75. Sudoplatov, *Special Tasks,* 299. See also the account in *Kommersant' Vlast,* April 9, 2002, 70. The version of the story in *Kommersant'* assumes that Ryumin's letter was a provocation by Beria and Malenkov against Abakumov, who had succeeded Beria as minister of the MGB. It assumes that the charges took Stalin by surprise.

76. Sudoplatov, *Special Tasks,* 299.

77. Sudoplatov claimed that Ryumin came from a kulak family (something Ryumin vehemently denied) and that his "brother and sister had been convicted of thievery and that his father-in-law had served as an officer in the White Army of Admiral Aleksandr Kolchak." None of this is mentioned in the extant transcripts of his interrogations or in the 1954 trial documents. Sudoplatov writes that Ryumin's party reprimand had to do with having lost an investigation file on a bus, but the documents at our disposal state that it had to do with the fact that he hadn't fulfilled Abakumov's order and had conducted illegal interrogations. See ibid., 99.

78. Ryumin is referring to the Wismut uranium operation in East Germany. This became the focus of intense American intelligence efforts because Stalin was mining the uranium in violation of Allied agreements. See the account of American intelligence efforts to penetrate the Wismut operation in Murphy, et al., *Battleground Berlin,* 13 & ff. See also p 15 for an account of the Soviet colonel, code-named Icarus, who greatly helped U.S. intelligence in this area. "Icarus" was obviously Salimanov mentioned by Ryumin.

79. "To Comrade Stalin" from Ryumin, July 2, 1951 (KGB archive)

80. *Reabilitatsiya,* 135–36.

81. It is important to see that Malenkov's account refers not to Ryumin's eventual letter to Stalin denouncing Abakumov but to a preliminary note—*zapiska*—in which Ryumin stated that the previous interrogator of Etinger had committed a forgery.

82. "They arrested Abakumov because he allegedly protected doctor Etinger from punishment—the old professor who died in prison." *Reabilitatsiya,* 135.

83. Protocol of interrogation of the arrested Etinger, December 27, 1950, 2.

84. Ibid.

85. In 1951 Likhachev admitted that Ryumin's account of Etinger's "confession" was true and that Abakumov was lying. In 1953, after Stalin's death, Likhachev retracted his earlier confession, which he said was compelled by force, and supported Abakumov's account.

86. Protocol of Interrogation of Likhachev, June 13, 1953, 4–7.

87. Excerpt from the protocol of interrogation of Abakumov by Mokichev, August 8, 1951, 1–2.

88. Ibid., 3.

89. By his own confession, Ryumin had Etinger in his hands for the two months leading up to his death. "From the very beginning and up to the 2nd or the 3rd of March 1951, that is up to the day of Etinger's death, I led the investigation of his case." Protocol of interrogation of Ryumin, May 11, 1953, 1.

90. A favorite term of Stalin's. "Stalin called the Chekists and me hippopotamuses, people not capable of quickly and conscientiously fulfilling the orders of the CC." "To Comrade Beria from Ignatiev, S.," March 27, 1953, 4.

91. Decree concerning the extension of Etinger's custody to 18 February 1951, January 17, 1951, signed by M. D. Ryumin, 1.

92. Excerpt from Ryumin's confession, no date, presumed to be summer 1953, 4–7.

93. Ibid., 4.

94. At his trial, Boris Shimeliovich said that Abakumov, "dissatisfied with my responses . . . said, 'Give him a deadly beating.'" Rubenstein and Naumov, *Stalin's Secret Pogrom,* 288.

95. See Amy Knight, *Beria: Stalin's First Lieutenant* (Princeton, N.J.: Princeton University Press, 1993), 170–73, who suggests that Khrushchev may have had a hand in the affair; Sudoplatov, *Special Tasks,* attributes it to "a vicious power struggle in the Kremlin on the eve of Stalin's death," 298. See also Kostyrchenko, *Out of the Red Shadows,* 286–87, for a somewhat confused picture of the situation.

96. Arkady Vaksberg, *Stalin Against the Jews.* Translated by Antonina W. Bouis (New York: Alfred A. Knopf, 1994), 224.

97. Rubenstein and Naumov, *Stalin's Secret Pogrom,* xii.

98. *Spravka,* July 30, 1951, 1–2.

99. Protocol of interrogation of Durinov, January 13, 1954, 2 (KGB archive).

100. Draft memorandum of information concerning the Abakumov case and the participants in his criminal group, composed by Grishaev, summer 1952, 23 (KGB archive).

101. Report on steps taken by security in accord with the 11 July 1951 decision of the Central Committee on the "Unfavorable situation in the MGB," to Comrade Stalin, November 24, 1952, 15–16.

102. This becomes apparent in the direction the interrogation of Dr. Miron Vovsi took. It will be discussed in later chapters.

103. Kostyrchenko, *Out of the Red Shadows,* 126.

104. Sudoplatov, *Special Tasks,* 301

105. Ibid.

106. Protocol of Interrogation of Shvartsman, November 21, 1952, 14.

107. Excerpts from Ryumin's confession, March 24, 1953, 3.

108. Ignatiev to Beria, March 27, 1953, 6–8.

109. Trotsky had said, "The Party is always right," but Stalin had ultimately transcended the party.

110. Ignatiev to Beria, March 27, 1953, 8. For more on the plot to murder Tito, see Sudoplatov, *Special Tasks*, 335–339. Sudoplatov's account and Ignatiev's are not that dissimilar. Not having come from either the military or intelligence services, Ignatiev was almost totally naïve in matters of intelligence.

111. Although A. S. Shcherbakov was well-known for his anti-Semitic views, Etinger was a frequent and informal guest in his house. According to Likhachev,

"During interrogation, Etinger spoke of the warm feelings he had toward Shcherbakov. ETINGER acknowledged that he experienced enmity toward A. S. SHCHERBAKOV because of his unfriendly relationship to Jews, but that this circumstance played no role whatever in the course of his treatment." Protocol of interrogation of Likhachev, June 13, 1953, 4–5.

112. Ignatiev to Beria, March 27, 1953, 2.

113. Ibid., 4.

114. Ibid., 3.

115. Excerpt from Ryumin's confession, March 24, 1953, 7.

CHAPTER FOUR: THE GRAND PLAN

1. M. F. Shkiryatov (1893–1954), a member of the Presidium of the Central Committee, 1952–1953.

2. *Zakrytoe pismo of the Central Committee,* July 13, 1951, 2.

3. Ibid.

4. Ibid., 2–3.

5. Ibid., 3.

6. Protocol of interrogation of Likhachev, June 13, 1953, 11–12.

7. Excerpts from the protocol of interrogation of Leonov, August 4, 1951, 3 (KGB archive).

8. Ibid., 7.

9. Protocol of interrogation of Likhachev, June 13, 1953, 12.

10. Ibid., 12–13.

11. Ibid., 13.

12. Ibid.

13. Ibid., 14.

14. Protocol of review, July 30, 1951, 4.

15. Ibid., 3.

16. Decree on the termination of the case, March 17, 1956.

17. Protocol of interrogation of Likhachev, June 13, 1953, 11.

18. Draft of a memorandum of information concerning the Abakumov case and the participants in his criminal group, composed by Grishaev, summer 1952, to be sent to the Central Committee, 1.

19. Excerpt from the protocol of interrogation of Abakumov, August 8, 1951, 2.

20. *"To Comrade STALIN"* from S. Ignatiev, April 16, 1952, copies to Malenkov and Beria.

21. Protocol of interrogation of Mironov, December 4, 1953, 3. "Etinger was elderly and had a bad constitution. Despite this, he was systematically summoned to long nightly interrogations without being granted rest in daylight hours. In the end,

the case ended up that once Etinger was not able to hold himself up and, as he stood leaning against the wall, he sat down on the floor. Knowing about this, I went to Etinger's cell and asked what the matter was. He answered me: 'Nothing, apparently I simply got dizzy from lack of sleep.'"

22. Excerpt from the protocol of interrogation of Abakumov, August 8, 1951, 5–6.

23. Shcherbakov died on May 10, 1945.

24. See, for instance, Dimtri Volkogonov, *Stalin: Triumph and Tragedy.* Edited and translated from the Russian by Harold Shukman. (Rocklin, Calif.: Prima Publishing, 1992), 570–571; Knight, *Beria: Stalin's First Lieutenant,* 172–173, attributes the "plot" to Khrushchev; see also Vaksberg's statement in *Stalin Against the Jews* that "It was only later, through the efforts of Mikhail Rumin [sic], whom Stalin came to like . . . that the case was given a Jewish tone, and the iron broom started sweeping up Jews" (243).

25. *Zakrytoe pismo,* 2–3.

26. Ibid., 4.

27. *Spravka* on the statement of Abakumov, October 15, 1952, 1.

28. See Yezhov's statement before the USSR Supreme Court, February 3, 1940, in J. Arch Getty and Oleg V. Naumov, *The Road to Terror: Stalin and the Self-Destruction of the Bolsheviks, 1932–1939.* Translations by Benjamin Sher. (New Haven: Yale University Press, 1999), 560–62.

29. Ibid., 557.

30. K. A. Stolyarov, ed., *General-polkovnik Abakumov, fakti I kommentarii* (Moscow, 1991), 12.

31. David E. Murphy to Brent, November 29, 2001. "It was not so much that American-British intelligence was more effective than Soviet intelligence (I never found it to be the case) but that it was overstaffed and the overage in personnel consisted of people who had been serving under Abakumov (who was well liked) since 1943. Both Stalin and Beria sought ways to reduce this number. It is interesting that the second report bore the same or similar title to the first, causing one to believe it is about the situation in the MGB when actually it dealt with the Doctors' Plot."

32. "Conditions in the MGB and the Need for Reform," to Comrade Stalin, November 30, 1952, 5–9.

33. Report on steps taken by security in accordance with the July 11, 1951, decision of the Central Committee on the "Unsatisfactory Situation in the MGB," to Comrade Stalin, November 24, 1952, 4.

34. Ignatiev formally made this connection in a September 25, 1952, memorandum to Stalin: "According to Ryzhikov's testimony . . . After a four month bed rest regimen, which was prescribed for comrade Shcherbakov, A. S., in connection with his serious illness, the council consisting of BUSALOV, VINOGRADOV, LANG and ETINGER permitted him not only to walk around on the territory

of the dacha, but also to go on automobile trips." To Comrade Stalin from S. Ignatiev, September 25, 1952, 2.

35. Ibid.

36. To Comrade Stalin, November 24, 1952, 10–12.

37. By the time Ogoltsov returned from Uzbekistan on November 20, 1952, Goglidze was on board as a deputy minister and head of the Third Chief Directorate (Military CI). At this crucial time, Beria was kept abreast of internal MGB activity by his close associate Goglidze. Murphy to Brent, November 29, 2001.

38. Kostyrchenko, *Out of the Red Shadows,* 286–87.

39. In fact, it was a meeting of the Presidium of the Central Committee, not the Central Committee as a whole. The Central Committee consisted of hundreds of party members; the Presidium was a much smaller and more select body. Nevertheless, according to V. A. Malishev's notes taken at the time, there may have been more than one hundred members present. Malishev (1902–1957) was a member of the Presidium of the Central Committee during the Nineteenth Congress. He recorded the following comment by Stalin in his notes from this Congress: "Every Jew—is a nationalist, an agent of American intelligence. Jewish nationalists believe that their nation has been saved by the USA (there you can become rich and bourgeois). They believe that they are obligated to the Americans. Among the doctors there were many Jewish nationalists. Unsatisfactory situation in [MGB]. Blunt vigilance. They themselves have confessed that they failed, that they sit in manure. The [MGB] must be cured." *Istochnik,* May 1997, no. 30, 140–41.

40. Ibid. "The dictator was somehow bewildered when he saw on it a statement he himself had written: 'Send to the archives.' However, he quickly recovered himself, ordered his assistant to hold his tongue, and decided that he would carry out his preconceived plan all the same."

41. Decree on the termination of the criminal investigation and the release of the prisoners from custody, March 31, 1953, 11.

42. Decree of the CC KPSS, *On wrecking in medicine,* December 4,1952, 2.

43. Protocol of interrogation of Yegorov, February 7, 1953, 230.

44. Protocol of interrogation of Vlasik, November 21, 1952, 8.

45. Protocol of a face-to-face confrontation between Vinogradov and Karpai, February 18, 1953.

46. Protocol of interrogation of Vinogradov, November 18, 1952, 7–9.

47. Ryumin's confession, March 24, 1953, 3–4.

48. Ibid., 4.

49. This "T" department may well have been the Buro No. 1, created by the Politburo in September 1950 as a supersecret executive action unit. See Teodor Gladkov, *Karol Nelegalov* (Moscow, 2000), 312.

50. Kostyrchenko, *Out of the Red Shadows,* 263.

51. Autograph confession of the witness Maslennikov, November 26, 1952, 2.

52. Ibid., 5.

53. Ibid., 6.

54. Protocol of interrogation of Vlasik, November 21, 1952, 6–7.

55. Confession of Maslennikov, November 26, 1952, 5.

56. Ibid.

57. "In the process of work of the operational unit, of which I was a part, it was known to my deputies RUMYANTSEV, GUZANOV, the head of the 6th department DIVAKOV, that VLASIK was personally close with YEGOROV. He was often with him in his office, at the dacha in Barvikha where they got drunk together. This circumstance significantly complicated the operational department's conduct of normal Chekist work concerning the Lechsanupra of the Kremlin." Ibid., 15.

58. Ibid., 5.

59. Ibid., 6–7.

60. An example of the reporting structure from Maslennikov's confession: "Com. RUMYANTSEV further informed me that the report of 'Yurina' was reported by him to the leader of the Chief Directorate of Security, if I am not mistaken, LYNKO, and the latter reported on this to ABAKUMOV, but no sort of orders concerning the report were received." Ibid., 3.

61. Ibid., 11.

62. Ibid., 13.

63. Ibid., 9–10.

64. Protocol of interrogation of Vlasik, December 4, 1952, 4–5.

65. Protocol of interrogation of Vlasik, January 24, 1953, 2–3.

66. Decree of Criminal Indictment, Moscow, November 25, 1954,1.

67. Protocol of interrogation of Ryumin, June 24, 1953, 1–2.

68. Yefim Smirnov, "Moskovskiye novosti," No. 6, February 7, 1988, 16. Smirnov gives the date of this meeting sometime before January 13, 1953, but it could not have taken place after the fall of 1951 when Stalin ceased taking vacations in the south. It makes sense that it was shortly after Dimitrov had died because it was fresh in Stalin's mind.

69. Protocol of interrogation of the Prisoner Ryumin, June 4, 1953, 4–5; Protocol of Interrogation of the Prisoner Ryumin, June 6, 1953, 2–3.

70. Protocol of Interrogation of Ryumin, June 6, 1953, 3.

71. Excerpt from the protocol of a face-to-face confrontation between Abakumov and Likhachev, July 22, 1952, 1.

72. Ibid., 2.

73. Protocol of interrogation of Ryumin, June 24, 1953, 1–2.

74. Obviously, not all. Certainly not those who were soon to be repatriated to the USSR from the West and faced execution or the gulag.

75. To comrade Stalin from Ryumin, July 2, 1950, 2–3.

CHAPTER FIVE: RECOGNIZING THE ENEMY

1. To Comrade Beria, L. P., from Ignatiev, S. March 27, 1953, 2.

2. To Comrade Beria, L. P., from Goglidze, March 26, 1953, 5.

3. November 7, 1937, entry in *The Diary of Georgi Dimitrov*, ed. by Ivo Barac, forthcoming.

4. *Doklad komissi TsK KPSS*, in *Reabilitatsiya*, 318.

5. *Reabilitatsiya*, 317–18.

6. Published in Vaksberg, *Stalin Against the Jews*, 227–36.

7. Ibid., 235.

8. See Kostyrchenko's account of Cheptsov's handling of the JAC trial, *Out of the Red Shadows*, 130–132.

9. The successor of Dzherzhinsky, founder of the Soviet secret service.

10. Report on steps taken by security in accord and with the July 11, 1951, decision of the Central Committee on the "Unsatisfactory Situation in the MGB," To Comrade Stalin, November 24, 1952, 7.

11. Protocol of a face-to-face confrontation between Ryumin and Kuzmin, February 16, 1954, 8–9.

12. Ibid., 3.

13. Protocol of interrogation of Ryumin, May 22,1953, 1–3.

14. Pitovranov, not a Jew, would play a continuing and important role in Soviet intelligence work. He was arrested "in connection with the 'Abakumov affair' and accused of anti-Soviet activity, sabotage, participation in a 'Zionist conspiracy' in the MGB. He was interrogated until November 1952. He sent a letter to Stalin from his cell with proposals for improving the work of intelligence. In November 1952 upon Stalin's orders he was freed and put at the disposal of the Cadres Directorate of the MGB. On 20 November, he was made a member of the commission for creating a chief intelligence directorate (GRU) of the MGB. On January 5, 1953, he was appointed chief of the foreign intelligence directorate under the GRU." See A. Kolpakidi and D. Prokhorov, *Vneshniaia Razvedka Rossii* (Moscow, 2001), 127.

15. Sudoplatov, *Special Tasks*, 301

16. *Spravka* on the case of Abakumov and the participants in his criminal group, a draft prepared by Grishaev in the summer of 1952 to be sent to the Central Committee, 1–3. See also To Comrade Stalin from S. Ignatiev, September 1952, 1.

17. Protocol of a face-to-face confrontation between Ryumin and Maklyarsky, February 1, 1954, 4.

18. Ibid., 5.

19. See Arch Getty and Oleg V. Naumov, *The Road to Terror: Stalin and the Self-Destruction of the Bolsheviks, 1932–1939,* 248 ff., 494–95.

20. Protocol of a face-to-face confrontation between Ryumin and Maklyarsky, February 1, 1954, 3.

21. *Reabilitatsiya,* 371; this stenographic record was corrected in published accounts. *Reabilitatsiya, politicheskiye protessi, 1930–50* (Moscow, Izdatelstvo politicheskoi literatury, 1991), 53.

22. To L. P Beria from the arrested P. I. Yegorov, March 13, 1953, 27.

23. Hosea, 4.6.

24. Protocol of a face-to-face confrontation between Ryumin and Maklyarsky, February 1, 1954, 6–7.

25. Ibid., 7–8.

26. Ibid., 9.

27. *Pravda,* September 21, 1948, 3. See Joshua Rubenstein's informed account of this essay in *Tangled Loyalties: The Life and Times of Ilya Ehrenburg* (New York: Basic Books, 1996), 258 ff.

28. *Istochnik,* May 1997, no. 30. Stalin had spoken in this fashion since 1947. Ryumin admitted that everyone was guided by this dictum of Stalin. Stalin was said to have repeated this to several people who then repeated it until it became common knowledge of an opinion attributed to Stalin. V. A. Malishev, deputy prime minister of the USSR and a member of the Central Committee, recorded Stalin's comments after the nineteenth Congress that met in October 1952.

29. Protocol of interrogation of Sheinin, December 15, 1953, 2.

30. Ibid.

31. Ibid., 2–3.

32. Ibid., 3.

33. Ibid., 3–4.

34. Ibid., 4.

35. Protocol of a Face-to-Face confrontation between Ryumin and Maklyarsky, February 1, 1954, 6.

36. It may be that Goglidze's close relationship to Beria prevented his arrest at this time.

37. To Beria from Ignatiev, March 27, 1953, 2.

38. Autograph confession of Maslennikov, November 26, 1952, 10. Georgi Dimitrov had been the head of the Comintern and then became the first prime minister of postwar Bulgaria in 1947. A. A. Andreyev was a member of the Central Committee. Andreyev had a Jewish wife named Khazar; he became deaf from illness, was reported to have a cocaine habit, and retired in the Khrushchev era.

39. To Beria from Ignatiev, March 27, 1953, 2.

40. See handwritten confession of Maslennikov, November 26, 1952, 6–10.

41. Ibid., 8–9.

CHAPTER SIX: WAITERS IN WHITE GLOVES

1. David E. Murphy, Sergei A. Kondrashev, and George Bailey, *Battleground Berlin*, 143.

2. *Pravda*, November 21, 1952; quoted in Kostyrchenko, *Out of the Red Shadows*, 278–79.

3. Volkogonov writes that Vinogradov "made his last call on Stalin in 1952 and, finding him in poor condition, advised him to do as little work as possible henceforth. Stalin was furious and Vinogradov was never called on again. Indeed, he was soon arrested." Volkogonov, *Stalin*, 571. No reference for this incident is given by Volkogonov.

4. Protocol of interrogation of Ryzhikov, October 24, 1952, 3.

5. Ibid., 2–3.

6. Lang died of natural causes in 1948.

7. Protocal of interrogation of Ryzhikov, Otober 24, 1952, 4.

8. To Comrade Stalin from S. Ignatiev, October 29, 1952, 1.

9. To the deputy minister of state security of the USSR, Lieutenant Colonel Ryumin, M. D., July 2, 1952, 1.

10. Ibid.

11. Ibid., 3.

12. It may be that Kitaev purposely left Zhdanov's name out of his report for reasons of state security.

13. To the deputy minister of state security of the USSR, lieutenent colonel Ryumin, M.D., July 2, 1952, 2–3.

14. Stenogram, September 6, 1948, 21.

15. Ibid., 4.

16. To Comrade Stalin, November 24, 1952, signed by S. Goglidze, 3–4.

17. Protocol of the interrogation of Ryumin, June 6, 1953, 4.

18. Arrest warrant for Ryumin, March 16, 1953.

19. To Beria from Ignatiev, 2.

20. Yakov Rapoport, *The Doctors' Plot of 1953*. Translated by N. A. Perova and R. S. Bobrova. (Cambridge, Mass.: Harvard University Press, 1991), 77–78.

21. Protocol of interrogation of Sheinin, December 15, 1953, 2.

22. Protocol of the Judicial Session of the Military Collegium of the Supreme Court of the Soviet Union, July 2–7, 1954 (excerpts), 31–32.

23. Protocol of interrogation of Ryumin, June 4, 1953, 2.

24. In protocol of the Judicial Session of the Military Collegium of the Supreme Court of the Soviet Union, July 2–7, 1954, 34–35.

25. Konstantin Simonov, *Glazami cheloveka moego pokoleniya*, in *Znamiya*, no. 4, 1988, 85. Simonov was uncertain about the date of the session but placed it in early to mid-March 1952. The pseudonym, "Robinsky," corresponded to the name of the then editor of *Izvestia*.

26. It is useful to compare Stalin's February 26, 1952, comments about anti-Semitism with his June 14, 1941, TASS announcement refuting rumors of an impending Soviet-German war. Stalin was a master of dissimulation.

27. Excerpt from the statement of Ryzhikov, made to Minister of the MGB Beria, March 26, 1953, 2.

28. *Spravka* on Ryzhikov's arrest and imprisonment, September 30, 1953 (signed by Mishin, deputy head of the internal prison of the MVD USSR)

29. Ibid.

30. Excerpt from Ryumin's confession, no date, 34.

31. Ibid.

32. Protocol of interrogation of Konyakhin, June 6, 1953, 1–2.

33. Protocol of interrogation of Ryumin, June 6, 1953, 1.

34. Ibid., 2.

35. Ibid., 2.

36. Ibid., 3–4.

37. Protocol of the Judicial Session of the military Collegium of the Supreme Court of the Soviet Union, July 2–7, 1954, 30–31.

38. Protocol of interrogation of Ryumin, June 4, 1953, 2.

39. Ibid., 3–4.

40. Excerpt from Ryumin's confession, no date, 33.

41. Dzherzhinsky had been the founder of the Cheka, the original name of the Soviet security service. Over time, the Cheka took on a variety of different names—NKVD, NKGB, MGB, KGB—but the service was often nostalgically remembered as the Cheka long after the name formally went out of use.

42. Ryumin's confession, no date, 3.

43. *Spravka* concerning the Abakumov-Shvartsman case and the participants in his criminal group, draft memorandum, composed by Grishaev, to be sent to the Central Committee, summer 1952.

44. To Comrade Stalin from S. Goglidze, September 24, 1952, 15–16. The way by which Kuznetsov became connected to espionage was the direct result of Stalin's inquiry on this subject to Ignatiev in the fall of 1952. Stalin put the question directly to Ignatiev why the defendants in the Leningrad affair had not been

investigated for espionage. The MGB could uncover nothing to support this suspicion and therefore tied Kuznetsov to the death of Zhdanov.

45. To Comrade Stalin from S. Ignatiev, September 25, 1952, 5–6.

46. Ibid., 6.

47. *Spravka* on RYZHIKOV's arrest and imprisonment, September 30, 1953.

48. To L. P. Beria from the arrested P. I. YEGOROV, March 13, 1953, 8.

49. Protocol of interrogation of Timashuk, August 11, 1952, 1–3. "As in 1948, I still believe that doctors MAIOROV, G. N., KARPAI, S. E., professor-consultant YEGOROV, P. I., VINOGRADOV, V. N. and VASILENKO, V. Kh., who treated comrade ZHDANOV, incorrectly diagnosed the illness of comrade ZHDANOV, A. A., ignored the facts demonstrating that he suffered a recent myocardial infarction, and prescribed a regime for the patient contraindicated by this grave illness."

50. To Beria from Ignatiev, March 27, 1953, 3–4.

51. Ibid., 4.

52. Ibid.

53. To Comrade Stalin, from S. Goglidze, November 24, 1952, 17.

54. Protocol of interrogation of Vovsi, November 25, 1952, 2–3

55. Protocol of interrogation of Vovsi, December 8, 1952, 2–6

56. To Comrade Beria from Goglidze, March 26, 1953, 6.

57. Malishev recorded Stalin in December 1952 as using the phrase, "if I live . . ." in connection with future plans. It was the first time Malishev had ever heard him speak in this fashion. See *Istochnik,* 1997 (already cited).

58. *Molotov Remembers,* 198.

59. To Comrade Stalin from Ignatiev, October 29, 1952, 1.

60. Ibid., 4

61. To Comrade Beria from Goglidze, March 26, 1953, 3.

62. Ibid., 5.

63. To Beria from Ignatiev, March 27, 1953, 5.

64. *Khrushchev Remembers,* 257–258.

65. *Khrushchev Remembers,* 286–287

66. To Beria from Goglidze, 2–3.

67. *Khrushchev Remembers,* 286–87.

68. Rubenstein and Naumov, *Stalin's Secret Pogrom,* xv.

69. To Comrade Beria from Ignatiev, March 27, 1953, 4

70. Excerpt from Ryumin's confession, March 18, 1953, 9. Ignatiev did not testify about this conversation in his report to Beria of March 27, 1953.

71. Ibid.

72. Ibid. Even during the interrogation of prisoners, Ryumin would often say that he was chosen representative of the Central Committee to uncover the Jewish conspiracy.

73. To Comrade Beria from Goglidze, March 26, 1953, 2.

74. See Sudoplatov, *Special Tasks,* 324, for another interpretation of Ryumin's fall having to do with the still little-known Mingrelian affair. It is possible that many factors went into the decision to remove him.

75. Excerpts from Ryumin's confession, no date, 33–38.

76. To the Minister of Internal Affairs of the USSR—Comrade L. P. Beria: Report, from Ignatiev, March 24, 1953, 5–6

77. Ibid., 5.

78. To the Investigative Unit of the MVD USSR from the arrested P. I. Yegorov, March 24, 1953, 2.

79. To L.P Beria from the arrested P. I. Yegorov, March 13, 1953, 2–3.

80. Ibid.

81. Ibid., 2–3. Hanging was considered a more degrading form of execution than shooting.

82. To the Investigative Unit of the MVD USSR From P. I. Yegorov, March 24, 1953, 3.

83. Though investigators hammered away at this accusation for several interrogations, it was eventually dropped. Presumably Stalin gave the order not to pursue it further.

84. To the Investigative Unit of the MVD USSR from Yegorov, March 24, 1953, 17–27.

85. To the Investigative Unit of the MVD USSR from the arrested, P. I. Yegorov, March 29, 1953, 8.

86. Excerpt from the decree on the termination of the criminal prosecution and the freeing of the prisoners from custody, confirmed by the Minister of Internal Affairs of the USSR, L. Beria, March 31, 1953, 9–10.

87. Ibid., 10–13.

88. Ibid., 14–15.

89. To L. P Beria from the arrested P. I. Yegorov, March 13, 1953, 3–4.

90. Ibid., 8.

91. Ibid., 9.

92. Ibid., 8.

93. Excerpt from protocol of interrogation of Vinogradov, December 12, 1952, 1–2.

94. Protocol of interrogation of Vinogradov, November 18, 1952, 1, 7–13. Such "declarations" were made in the late 1930s to major political prisoners in order

to extort confessions. This declaration may have represented Stalin's final resort in the present instance and may be another sign of his preternatural anxiety in prosecuting this case. In fact, none of the promises of the Vozhd were fulfilled in the 1930s, and there is little reason they would have been fulfilled now.

95. Excerpt from the protocol of interrogation of the accused Vinogradov, December 26, 1952, 1.

96. The fact that Shvartsman also admitted to a failed attempt on Malenkov's life suggests that Malenkov was being singled out as a particularly powerful member of the Politburo at the time. See protocol of interrogation of Shvartsman, November 21, 1952, 17–18.

97. To Comrade Stalin, a Report on steps taken by security in accordance with the July 11, 1951, decision of the Central Committee on the "Unsatisfactory Situation in the MGB," November 24, 1952, 1–21.

CHAPTER SEVEN: AN INTELLIGENCE PHANTASMAGORIA

1. Protocol of review of the archival-investigative case No. M-5241, February 24, 1954, 15.

2. Ibid., 1.

3. Protocol of interrogation of Grishaev, September 25, 1953, 5.

4. According to the protocol of review of Varfolomeyev's case (No. M-5241), dated February 24, 1954, six months after his execution, Varfolomeyev's parents immigrated to Japan in 1919. In 1921 they transferred to Manchuria. In 1924 Varfolomeyev entered into an anti-Soviet organization called the "Union of Musketeers." In 1934–1935, he became part of another anti-Soviet organization, the "Union of Anarchists." From 1933 to 1945 Varfolomeyev worked with Japanese intelligence and counterintelligence first in Kharbin and, beginning in 1940, in Tian-Tsin. While a Japanese agent he worked with Simidza, Mori, Nemoto, Yu, and others. From 1935 to 1940, Varfolomeyev was part of the Japanese secret service in the counterintelligence department (third department) of the so-called Main Bureau for Matters of Russian Emigrants. In 1948 Varfolomeyev was recruited for American intelligence in the Far East by an American agent, Jackson, and was transferred to Yu Tsun-Bin. Working for American intelligence, Varfolomeyev collected information on the situation in the Chinese People's Republic, the character of the aid the USSR gave China, and other questions of a political and economic character. He received $400 every month for his services. In 1950 Varfolomeyev was based in Tian-Tsin. It is there that he met Rogalsky.

5. Draft *spravka* on the Varfolomeyev case, prepared probably by Pavel Grishaev, senior investigator, for S. Ignatiev, undated and unsigned, 1–2. The final version of this document was sent to Stalin by Ignatiev in April along with a cover memorandum signed by Ignatiev, with the notation at the top: "Making copies without permission of the Secretariat of the MGB USSR—is forbidden." Ignatiev ended his cover memo to Stalin with the usual comment: "I ask for Your order."

6. Ibid. This is an example of how intelligence reports were being slanted. The United States was not preparing an "armed invasion" of North Korea in 1949. See Cold War International History Project, *Bulletin*, nos. 6–7, Winter 1995/96, 54, 87–89.

7. Draft *spravka* on the Varfolomeyev case, 3

8. There are several fantastic elements in this story. In the first place, U.S. government elements including External Survey Detachment 44 (CIA Shanghai's cover designation) left Shanghai in December 1948 even though the city was not occupied by the Chinese Communists until May 1949. See Arthur B. Darling, *The Central Intelligence Agency: An Instrument of Government* (University Park, Pa.: Pennsylvania State University Press, 1990), 119. Also, American intelligence used South Korea as a base for its most effective intelligence operations against North Korea (see David E. Murphy, Sergei A. Kondrashev, and George Bailey, *Battleground Berlin*, 85–86). As for the purported exploitation of French Catholic missions to fund U.S. operation in China, by 1950 the Chinese had been in control in North China for some time, and the crackdown on foreign missionaries was a high priority. Furthermore, the use of French Catholic rental properties to provide Chinese currency made no sense. Hong Kong money changers were awash with it. Anyone engaged in anti-Communist activities at this late stage of the Chinese Communist occupation would have used U.S. dollars or gold.

9. Draft *spravka* on the Varfolomeyev case, 4.

10. Lammot Dupont died in 1884! His son Pierre restructured General Motors in the 1920s and served on the board until he retired in 1940. This is further evidence of the fantastical nature of this "plot" and the haste with which the MGB concocted it.

11. The idea that the United States was about to declare war on the Soviet Union was prevalent in all Soviet propaganda at the time. For example, the office diary of General Robert Grow, U.S. military attaché in Moscow in 1950–1951 was photographed by MGB agents covering the U.S. embassy and then published in English by East German intelligence. One of the most alarming statements in the book purporting to come from the diary was: "War! As soon as possible. Now!" This phrase was never found in the actual diary.

12. Draft *spravka* on the Varfolomeyev case, 3–8.

13. Protocol of review of the archival-investigative case No. M-5241, February 24, 1954, 5–6.

14. Evidently this was Major General John W. "Iron Mike" O'Daniel, military attaché in Moscow in 1949–1950. O'Daniel had been filmed by the Soviets leaning out a balcony of the U.S. embassy to photograph Soviet aircraft flying overhead in connection with a military parade.

15. Decree concerning the criminal indictment of Ryumin, April 21, 1954, 3.

16. Ibid., 3.

17. Protocol of interrogation of Sheinin, December 15, 1953, 3–4.

18. To the head of the special section for Especially Important Cases of the MVD, USSR, Lieutenant-General Comrade Vlodzimirsky, May 25, 1953, 8.

19. Protocol of interrogation of Grishaev, September 25, 1953, 2–3. Grishaev participated in the Slansky trial and other important cases of the 1950s. He was involved in the falsification of materials but probably did not participate in the torture of prisoners.

20. Ibid., 8.

21. Ibid.

22. To the head of the special section for Especially Important Cases of the MVD, USSR, Lieutenant-General Comrade Vlodzimirsky, May 25, 1953, 8.

23. Protocol of a face-to-face confrontation between Ryumin and Maklyarsky, February 1, 1954, 9. Maklyarsky reports here about his February 1952 interrogation by Ryumin.

24. *Spravka*, February 1952, signed by Grishaev.

25. To the head of the special section for Especially Important Cases of the MVD USSR, Lieutenant-General Comrade Vlodzimirsky, from Grishaev, May 25, 1953, 3.

26. Ibid.

27. Ibid., 2–3.

28. Protocol of the interrogation of Ryumin, May 18, 1953, 1.

29. Protocol of the face-to-face confrontation between Ryumin and Grishaev, February 16, 1954, 5.

30. *Khrushchev Remembers,* 283.

31. David Murphy speculates that military intelligence was hesitant about undertaking a thorough verification because it realized the fantastic nature of the supposed plot and didn't want to risk its sources or become involved in something that was such patent nonsense.

32. Protocol of the face-to-face confrontation between Ryumin and Grishaev, February 16, 1954, 10.

33. To the head of the special section for Especially Important Cases of the MVD, from Grishaev, May 25, 1953, 6.

34. Ibid., 6–7.

35. Ibid., 4. O'Daniel's exact dates in Moscow have not been determined.

36. S. R. Savchenko worked in the MGB foreign intelligence department. It may be that the GRU had already shown itself to be uncooperative, and now Ryumin turned to the MGB for intelligence. Nothing else is known about the Savchenko referred to in the letter.

37. Protocol of the face-to-face confrontation between Ryumin and Grishaev, February 16, 1954, 27.

38. To the head of the special section for Especially Important Cases of the MVD, USSR, from Grishaev, May 25, 1953, 6.

39. Ibid.

40. Ibid., 7.

41. Ibid.

42. Protocol of review of the archival-investigative case No. M-5341, February 24, 1954, 12–15.

43. *Reabilitatsiya,* 19.

44. Ibid., 382.

CHAPTER EIGHT: SPIES AND MURDERERS UNDER THE MASK OF DOCTORS

1. Protocol of the interrogation of Vovsi, November 14, 1952, 1–2.

2. Protocol of interrogation of Yegorov, November 15, 1952, 2–3.

3. Ibid., 4.

4. Ibid.

5. Ibid.

6. Sudoplatov, *Special Tasks,* 302.

7. Protocol of interrogation of Shvartsman, November 21, 1952, 6. It is worth noting that the interrogator of Shvartsman was Grishaev, the same man who was interrogating Varfolomeyev.

8. Ibid., 6–7.

9. Ibid., 8.

10. Ibid., 9.

11. Ibid., 10–14.

12. Protocol of a face-to-face confrontation between Ryumin and Maklyarsky, February 1, 1954, 9.

13. Although Kuznetsov was directly implicated in the investigation, it is clear that Molotov (and possibly Mikoyan) was going to be made part of the conspiracy. Stalin denounced him and Mikoyan at a Central Committee meeting in December 1952.

14. It is thought that this *spravka* was composed in the summer; however, internal evidence suggests that it was composed following Shvartsman's testimony in November 1952 concerning the SDR, the Society for the Struggle Against the Revolution.

15. *Spravka* concerning the Abakumov case and the participants in his criminal group. Handwritten at the top of the document: "Draft memorandum composed by Grishaev on the Abakumov-Shvartsman case," no date, 2.

16. To Comrade Stalin from S. Ignatiev (excerpt), September 1952, 1.

17. Protocol of interrogation of Shvartsman, November 21, 1952, 14–15.

18. It was forbidden to refer to Stalin by name at this time.

19. Protocol of interrogation of Shvartsman, November 21, 1952, 17–18.

20. Ibid., 20.

21. *Reabilitatsiya*, 467.

22. Protocol of Interrogation of Broverman, November 27, 1952, 2–3. The "D" department was the department for diversion and disinformation.

23. Ibid., 3.

24. Ibid., 9–10.

25. Identified as Moisei Elyevich Belkin, who worked in counterintelligence. Ibid., 7–8.

26. Ibid.

27. Protocol of interrogation of Broverman, November 27, 1952 , 7. Broverman is referring to the tumultuous welcome given Golda Meir in Moscow in September 1948 that provoked Stalin's anger and gave him reason to see the strength of Jewish sentiment toward Israel. See discussions in Kostyrchenko, *Out of the Red Shadows*, 104, 122–123, and Rubenstein and Naumov, *Stalin's Secret Pogrom*, 41, 46–47.

28. Report on steps taken by security in accordance with the July 11, 1951, decision of the Central Committee on the "Unfavorable Situation in the MGB," to Comrade Stalin, November 24, 1952, 6–19. This document is signed by Goglidze alone.

29. *Zakrytoe Pismo TsK VKP(b)* [secret letter], 13 July 1951, 2.

30. Ibid., 15–16.

31. Ignatiev signed this November 30 document, probably from his hospital bed.

32. To Comrade Stalin, from Ignatiev, Goglidze, and Ogoltsov, November 30, 1952, 2.

33. Ibid., 2–3.

34. Stalin's speech at the December 1 meeting of the Presidium of the Central Committee, recorded by V. A. Malyshev; quoted by Lev Bezymensky in *Novoye Vremya*, No. 14/98, 38

35. The Presidium was a smaller, more select body of the Central Committee, consisting of perhaps as many as twenty-five to fifty members.

36. Protocol of interrogation of Yegorov, February 7, 1953, 1–2.

37. Protocol of interrogation of Vovsi, December 8, 1952, 2–4.

38. To Comrade Stalin from S. Goglidze, December 16, 1952.

39. *Molotov Remembers,* 222.

40. Protocol of the closed judicial session of the Military Collegium of the Supreme Court of the USSR, January 17, 1955, 23.

41. Ibid.

42. Russian names are declined. Masculine names usually have no ending, feminine names end in "a." Thus Anna Karenina's husband's name is Karenin.

43. Excerpt from protocol of interrogation of Vlasik, February 11, 1953, 1.

44. Ibid., 12.

45. Ibid., 12–14.

46. Protocol of the closed judicial session of the Military Collegium of the Supreme Court of the USSR, January 17, 1955, 25.

47. Ibid., 14.

48. These may still exist but have not been found.

49. Protocol of interrogation of Vlasik, February 5, 1953, 2–3.

50. To the Chairman of the Presidium of the Supreme Soviet of the USSR, Marshal of the Soviet Union, K. Ye. Voroshilov from N. S. Vlasik, April 6, 1955, 1.

51. Ibid., 3.

CHAPTER NINE: THE GREAT STORM

1. Zinovii Sheinis, *Provokatsiya Veka* (Moscow: PIK, 1992), 107.

2. Ibid., 103–105.

3. Rapoport, *The Doctors' Plot of 1953*, 73.

4. Sudoplatov, *Special Tasks*, 320–325

5. *I prmiknuvshii k nim Shepilov*, 13.

6. Lars T. Lih, Oleg V. Naumov, Oleg V. Khlevniuk, eds., *Stalin's Letters to Molotov, 1925–1936*, translated from the Russian by Catherine A. Fitzpatrick (New Haven, Conn.: Yale University Press, 1999), Letter 76, January 1933, 232.

7. Excerpt from protocol of interrogation of Vinogradov, January 6, 1953, 1–2.

8. Kostyrchenko, *Out of the Red Shadows*, 288.

9. Handwritten note from D. Shepilov to Stalin, January 10, 1953.

10. Quoted from Rapoport, *The Doctors' Plot of 1953*, 74.

11. Third from last paragraph in *Pravda*, January 13, 1953.

12. The Joint Distribution Committee was established in America in the 1920s to provide aid to Soviet Jews.

13. Varfolomeyev, a White Russian spy, was caught in Manchuria by the Chinese and handed over to Soviet authorities in 1950. See chapter 7.

14. Ibid., 74.

15. It is astounding that Goglidze repeated, almost word for word, in his March 1953 report to Beria, the language of his November 30 report to Stalin, in which he said, "The entirely correct claim has been made to us that the investigators are working without spirit, that they clumsily use contradiction, slips of the tongue

of the arrested for their unmasking, and do not check credible information even if barely credible." This shows what the feedback loop was in Stalin's bureaucracy. Stalin's own words would be endlessly recycled. The "entirely correct claim" was made by Stalin.

16. *Izvestia*, January 15, 1953, 1, paragraph 5.

17. Rapoport, *The Doctors' Plot of 1953,* 77–78.

18. Ibid., 78.

19. *Pravda,* January 31, 1953, 3.

20. Ibid., 82.

21. *To Com. Malenkov*, January 30, 1953.

22. *To Com. Malenkov,* February 17, 1953.

23. Rubenstein and Naumov, *Stalin's Secret Pogrom.*

24. No mention is made of these especially dangerous state criminals in any of the documents that have come to light after Stalin's death.

25. Protocol of a face-to-face confrontation between the arrested A. A. Vinogradov, V. N., and Karpai, S. Ye., February 18, 1953, 2.

26. Ibid., 2.

27. Ibid., 3.

28. Ibid., 5.

29. Vaksberg, 258 ff. Chesnokov died in the early 1970s before the first published accounts of his alleged participation in day "X." Otherwise it is probable that he would have taken legal steps to deny it. It is entirely possible that the rumor about day "X" was started as a provocation.

30. Ibid., 259.

31. "RYASNOY declared to the late Head of the Soviet government [Stalin] that the confessions of VARFOLOMEYEV were difficult to believe; they did not contain any names and they were unconvincing. To this, the HEAD of the Soviet Government answered that one could expect anything from the Americans, and therefore if the confessions of VARFOLOMEYEV are unconvincing, then you and the MGB must make them convincing for the trial." P. 5 of *Protocol of the face-to-face confrontation between the accused RYUMIN, Mikhail Dimitrievich and the witness GRISHAEV, Pavel Ivanovich,* February 16, 1954.

32. "Ryumin . . . said to me that he intended to put the question about the expulsion of the Jews from Moscow before the government." Protocol of the face-to-face confrontation between Ryumin and Maklyarsky, I. B., February 1, 1954, 9.

33. Rubenstein, *Tangled Loyalties,* 272–274.

34. Sheinis, *Provakatsia Veka,* 107–109. This is the same Ehrenburg whom Ryumin tried to get both Maklyarsky and Sheinin to implicate in the anti-Soviet Jewish nationalism in 1952. See chapter 5, Protocol of Ryumin-Maklyarsky confrontation, February 1, 1954, 6–7.

35. Ibid., 108.

36. Ibid., 109.

37. See *Molotov Remembers,* 222; *Khrushchev Remembers,* 274–275; Volkogonov, *Stalin,* 569. Like Abakumov and Vlasik, Poskrebyshev knew too much about too many matters.

38. Quoted in Rubenstein, *Tangled Loyalties,* 272.

39. *Istochnik,* 1997, no. 1, 143–146.

40. Ibid., 143; citation from the Presidential Archive: "APRF, F. 3. Op 32. D. 17. L. 100–100. True typed copy. Signature by the author." In *Stalin Against the Jews* Vaksberg reproduced a version of this letter, 263–264, with a significant variation: the last paragraph contains the phrase, ". . . inform me about your attitude toward my refusal to sign this document." The letter published in *Istochnik* does not contain the important statement that Ehrenburg had "refused to sign this document."

41. Protocol of interrogation of the arrested VINOGRADOV, February 12, 1953, 7.

42. Rubenstein and Naumov, *Stalin's Secret Pogrom,* 84.

43. Excerpt from the decree (on the termination of criminal prosecution and the release of the prisoners . . .), March 31, 1953, 9.

44. Ibid., 9.

45. To Comrade Stalin, from C. Ignatiev, February 17, 1953 (copy to Malenkov). No. 661/i.

46. See Peter Deriabin, *The Watchdogs of Terror: Russian Body Guards from the Tsars to the Commissars,* 2nd rev. ed. (Frederick, Md., University Publications of America, 1984), 324–327. Deriabin argues that Beria masterminded Stalin's end by "stripping Stalin of all personal security, except for the comparative window dressing of the minor Okhrana officers in his office and household." By February 1953, Deriabin writes, Beria had "completed the emasculation of the Okhrana," which left Stalin vulnerable to Beria's further machinations.

47. Criminal indictment re investigative case No. 5428, signed S. Ignatiev, dated February "__," 1953, p. 8.

CHAPTER TEN: THE END

1. "The History of the Illness of J. V. Stalin (Composed on the Basis of Journal Notes of the Course of the Illness from March 2 to 5, 1953)," 16.

2. *Spravka* on Ryzhikov's arrest and imprisonment, September 20, 1953, 5.

3. Rybin, in fact, had never been to Blizhnyaya until after Stalin's death. His published accounts were in fact composed on the basis of what others told him—twice removed from an actual eyewitness account.

4. The first published reports of Stalin's death in *Pravda* stated that he had suffered his stroke in the Kremlin, suggesting that he had then been removed to

Blizhnyaya. Adam Ulam discounts this and there is little reason to suppose this is what happened. See Ulam, *Stalin: The Man and His Era,* 3–4.

5. Knight, *Beria: Stalin's First Lieutenant,* 179.

6. Peter Deriabin, a former Soviet counterintelligence officer and member of Stalin's bodyguard who defected to the West in 1954, has proposed that Stalin actually suffered a stroke in his Kremlin office, following a supposed "confrontation" between Stalin and the other Politburo members who "knew that for their own survival they had to quash the 'Doctors' Plot.'" Deriabin believes that Stalin was removed to his dacha at a later date. He argues that Beria had been responsible for the "process of stripping Stalin of all personal security" and had been the motive force behind the purge "of Abakumov, the dismissal of Vlasik, the discrediting of Poskrebyshev, the emasculation of the Okhrana [Kremlin Guards]," and several other important developments at the time. Deriabin introduces the interesting point that five days after the death of the Kremlin Guard Kosynkin, the press campaign about the "Doctors' Plot" ended "as suddenly as it had started." Deputy Kremlin commandant General Kosynkin was, according to Deriabin, "the only surviving guard that Stalin could trust." He died of a heart attack on February 17, 1953. On February 22, one week before Stalin's stroke, a lull in press coverage occurred. However, there may have been many reasons for this lull, chief among which was that Stalin was not feeling well. It is known from extant records that the interrogations continued. No evidence exists to support any of Deriabin's conjectures about Stalin's death. See *Watchdogs of Terror,* 325–26.

7. *Khrushchev Remembers,* 340–341.

8. In fact, Stalin drank little except for *Madjoloy,* which is "young wine" with barely any alcoholic content.

9. Volkogonov, *Stalin: Triumph and Tragedy,* 571.

10. Ibid.

11. Svetlana Alliluyeva, *Only One Year,* translated by Paul Chavchavadze (New York: Harper & Row, 1969), 6.

12. *Khrushchev Remembers,* 342.

13. Ibid.

14. A test of reflexes in the foot that can signify abnormalities in the brain or motor functioning.

15. Cheyne-Stokes respiration indicates that periods of fast breathing alternate with periods of slow (or no) breathing.

16. *Molotov Remembers,* 237. A. Avtorkhanov is one of the chief proponents of the idea that Beria murdered Stalin. However, his account in *Zagadka smerti Stalina* [The Mystery of Stalin's death], published in Moscow in 1992, is based entirely on conjecture and circumstantial evidence.

17. *Khrushchev Remembers,* 318.

18. "The History of the Illness of J. V. Stalin," 15 and ff.

19. Ibid., 20.

20. We are indebted to Lawrence S. Cohen, M.D., for first suggesting this possibility, and to Robert Silver, Ph.D., and Phillip Dickey, M.D., for helping to clarify the medical issues involved.

21. V. V. Ivanov (1909–1987) was an investigator in the investigative unit of the MGB. He was fired in 1953 but not imprisoned.

22. L. E. Vlodzimirsky (1903–1953) had been an officer in the Soviet security services since the thirties. He was arrested after Beria in July 1953 and shot in December.

23. To L. P. Beria from the arrested P. I. Yegorov, March 13, 1953, 8.

24. Ibid., 8.

25. Confession of M. D. Ryumin, March 24, 1953, 7.

26. Excerpt from the decree on the termination of the criminal prosecution and freeing of the prisoners from custody, confirmed by the Minister of Internal Affairs of the USSR, L. Beria, March 31, 1953, 18.

27. *Pravda,* April 6, 1953.

28. Beria, *My Father,* 23–24.

29. Ibid., 25.

CONCLUSION: THE CONSPIRATORIAL MIND

1. Rubenstein and Naumov, *Stalin's Secret Pogrom,* 134.

2. *Pravda,* March 11, 1953, 4.

CHRONOLOGY OF THE DOCTORS' PLOT

May 9, 1945	Day of Victory over Nazi Germany.
May 10, 1945	**Shcherbakov dies of heart attack.**
April 1948	Yuri Zhdanov criticizes Lysenko at seminar of scholars.
May 1948	Israel declares statehood.
June 1948	10,000 Jews attend ceremonial service in Moscow Choral Synagogue.
July 1948	A. A. Zhdanov is sent to Valdai.
July 23, 1948	Zhdanov receives call from Shepilov; becomes gravely ill.
July 31–August 7, 1948	Lysenko Conference.
August 7, 1948	Yuri Zhdanov letter to Stalin published in *Pravda*. Karpai delivers opinion on A. A. Zhdanov; goes on vacation.
August 28, 1948	Timashuk flies to Valdai; performs EKG; returns to Moscow.
August 29, 1948	**Zhdanov's second heart attack; Timashuk returns to Valdai in the morning; Timashuk writes letter to Vlasik.**
August 30, 1948	Abakumov puts Timashuk letter and EKG before Stalin; Timashuk conducts second EKG.
August 31, 1948	Vinogradov session with Zelenin, Etinger, and Nezlin in Moscow.
August 31, 1948	**Zhdanov dies; Yegorov returns to Valdai; autopsy performed in Valdai.**

September 4, 1948	Yegorov summons Timashuk to his office; Timashuk writes letter to Suranov, her MGB handler.
September 6, 1948	**Yegorov conducts special session in Kremlin Hospital.**
September 8, 1948	"Yurina" report on "unobjective" character of discussion at the Kremlin Hospital session.
September 8–9, 1948	Timashuk fired and transferred to the second polyclinic of the Kremlin Hospital.
September 15, 1948	Timashuk writes first letter to A. A. Kuznetsov.
September 21, 1948	Ehrenburg's article on Israel in *Pravda*.
January 1949	Timashuk's second letter to Kuznetsov.
January 1949	Arrests of Jewish Antifascist Committee members Lozovsky, Kvitko, Markesh, and others.
January 26, 1949	Zhemchuzhina arrested.
February 16, 1949	Leningrad party attacked as "anti–party."
Spring 1949	Svetlana Alliluyeva marries Yuri Zhdanov.
April 22, 1949	Interrogation of Fefer; description of Etinger's ties to JAC.
August 1949	Kuznetsov, Voznesensky and others arrested in Leningrad affair.
June 1950	Korean War begins.
October 1950	Voznesensky, Kuznetsov, and seven other Leningrad party leaders are shot.
November 22, 1950	**Etinger arrested; interrogated by Abakumov.**
December 1950–January 1951	Varfolomeyev arrested and handed over to Soviets.
March 2, 1951	Etinger dies in Lefortovo.
March 1951	Ryumin formally rebuked by party for handling of Etinger case.
July 2, 1951	**Ryumin to Stalin denouncing Abakumov.**
July 2, 1951	Presidium of Central Committee meets to review Ryumin's letter.
July 4, 1951	Abakumov removed as head of MGB.
July 4, 1951	Abakumov removed as head of MGB.
July 11, 1951	**Memorandum by Central Committee on the**

"Unfavorable Situation in the MGB"; secret letter from Central Committee.

July 12, 1951	Abakumov arrested.
July 16, 1951	Karpai arrested.
July 17, 1951	Report to Vlasik concerning "suspicious agitation of Yegorov."
July 30, 1951	MGB Procurator Kitaev's *spravka* on Etinger case.
August 1951	Maslennikov has personal conversation with Lynko and Vlasik about Timashuk, "Yurina," and Kremlin Hospital session.
October 1951	Ignatiev reports to Stalin on the progress in unmasking the group of terrorist doctors.
October 6, 1951	Interrogation of Belov, A. A. Zhdanov's bodyguard.
October 19, 1951	Arrest of Sheinin.
October 1951	**Ryumin sends letter to Stalin concerning MGB.**
October–November 1951	Ryumin made deputy head of state security.
November 1951	Draft report concerning Timashuk, Yurina, Zhdanov, and Shcherbakov sent to Central Committee.
November 6, 1951	Maklyarsky arrested.
December 1951	Arrest of Jewish "nationalists" in MGB: Raikhman, Navlovsky, Sverdlov, and others.
February 7, 1952	Interrogation of Krongauz.
February 15, 1952	Ryzhikov arrested and accused of murder of Shcherbakov.
February 6, 1952	Awarding of Stalin prizes.
February 26, 1952	Interrogation of Sheinin by Ryumin.
March 1952	Date of alleged American attack on Soviet Union.
April 2, 1952	Ignatiev memo to Stalin concerning Karpai; Stalin rejects her execution.
July 2, 1952	MGB procurator Kitaev's report criticizing "expert examinations" of Shcherbakov's death.
Summer 1952	Grishaev's draft *spravka* on Abakumov-Shvartsman case.
August 12, 1952	**Execution of JAC members.**

mid-September 1952	"Expert examination of Zhdanov's heart" sent to Stalin. Yegorov arrested but falls ill and goes to hospital. Vasilenko arrested.
September 27, 1952	Yegorov's wife arrested.
October 5, 1952	19th Party Congress; Politburo and Orgburo merged to become Presidium of Central Committee, consisting of twenty-five members; Central Committee secretariat expanded from five to ten members; Malenkov gives main address.
October 17, 1952	First formal interrogation of Timashuk.
October 18, 1952	Yegorov rearrested.
November 2, 1952	**Meeting of Ignatiev, Ryumin, Goglidze, and Ryasnoy with Stalin. Stalin orders MGB to beat the doctors with "death blows."**
November 4, 1952	Vinogradov arrested.
November 12, 1952	Malenkov orders Ignatiev to beat the doctors with death blows. Vovsi and Kogan arrested for treatment of Dimitrov.
November 13, 1952	**Ryumin removed from Investigative Unit for Especially Important Cases.** Sokolov is temporarily made head of investigative unit.
November 13, 1952	Ryumin's last letter to Stalin.
November 13 or 14, 1952	Ignatiev meets with Stalin; has heart attack.
November 13, 1952	Vinogradov's first interrogation.
November 15, 1952	Goglidze appointed head of investigation of Doctors' Plot.
November 21, 1952	Interrogation of Shvartsman.
November 21, 1952	Interrogation of Vlasik.
November 24, 1952	**"To Stalin on Steps Taken by Security to Implement Decision of July 11, 1951 concerning MGB."**
November 26, 1952	Interrogation of Maslennikov, head of operational department of the Chief Directorate of Security.
November 27, 1952	Interrogation of Broverman, former deputy head of the MGB.
November 30, 1952	**"To Stalin concerning the need for reform in the MGB."**

December 1, 1952	**Stalin calls meeting of Presidium of Central Committee, denounces Abakumov and Vlasik, holds up the Timashuk letter, calls Politburo members "blind men, kittens."**
December 4, 1952	Decree of the Central Committee on the Doctors' Plot.
December 16, 1952	Arrest of Vlasik.
December 21, 1952	**Stalin's seventy-third birthday; meeting of Politburo at Stalin's dacha; Stalin denounces Molotov and Mikoyan.**
January 9, 1953	Politburo meeting to discuss TASS article on Doctors' Plot.
January 10, 1953	Shepilov draft of the TASS article, "Spies and Murderers Under the Mask of Doctors."
January 13, 1953	**Article on Doctors' Plot published.**
January 20, 1953	Timashuk receives Order of Lenin for unmasking the doctors.
January 30, 1953	Ministry of Internal Affairs recommends construction of special camps for German, Austrian, and other foreign criminals.
February 2, 1953	Yakov Rapoport arrested (released April 3).
February 17, 1953	MGB recommends organization of a special system of camps for foreign prisoners: two camps in Komi ASSR; one camp in Kazakhstan; one camp in Irkutsk.
February 18, 1953	Face-to-face confrontation of Vinogradov and Karpai.
March 1, 1953	Stalin discovered ill in Blizhnyaya dacha.
March 5, 1953	**Stalin dies.**
March 9, 1953	**Stalin's funeral.**
March 17, 1953	Ryumin arrested.
March 27, 1953	"Concerning the History of the Illness of A. A. Zhdanov," by Dr. Vinogradov to Beria.
March 31, 1953	**Decree on termination of criminal prosecution and freeing of prisoners in Doctors' Plot.**
April 2, 1953	To Beria from Vinogradov: Statement on confessions, torture, and falsifications.

April 4, 1953	*Pravda* **announces that doctors had been freed; Timashuk's Order of Lenin revoked.**
April 6, 1953	"Soviet Law is Inviolable" in *Pravda*.
June 18, 1953	Execution of the Rosenbergs in United States.
June 26, 1953	Beria arrested.
September 10, 1953	Varfolomeyev executed.
December 23, 1953	Beria shot.
July 2–7, 1954	Trial and execution of Ryumin.
December 1954	Abakumov shot.

BIBLIOGRAPHY

PUBLISHED BOOKS IN ENGLISH

Akhmatova, Anna. *The Poems of Akhmatova*. Selected, Translated and Introduced by Stanley Kunitz, with Max Hayward. Boston: Little, Brown and Company, 1973.

Alliluyeva, Svetlana. *Only One Year*. Translated by Paul Chavchavadze. New York: Harper & Row Publishers, 1969.

———. *Twenty Letters to a Friend*. Translated by Priscilla Johnson McMillan. New York: Harper & Row, Publishers, 1967.

Beria, Sergo. *Beria, My Father: Inside Stalin's Kremlin*. Edited by Francoise Thom. English translation by Brian Pearce. London: Duckworth, 2001.

Bulgakov, Mikhail. *The Master & Margarita*. Translated from the Russian by Mirra Ginsburg. New York: Grove Press, 1967.

Bullock, Alan. *Hitler and Stalin: Parallel Lives*. New York: Vintage Books, 1993.

Chuev, Felix. *Molotov Remembers: Inside Kremlin Politics, Inside Kremlin Politics*. Conversations with Felix Chuev. Edited and with an introduction and notes by Albert Resis. Chicago: Ivan R. Dee, 1993.

Clark, Katerina. *Petersburg: The Crucible of Cultural Revolution*. Cambridge, Mass.: Harvard University Press, 1995.

Cohen, Stephen F. *Bukharin and the Bolshevik Revolution: A Political Biography, 1888–1938*. With a new introduction by the author. Oxford & New York: Oxford University Press, 1980.

Conquest, Robert. *Stalin: Breaker of Nations*. New York: Penguin Books, 1992.

———. *The Great Terror: A Reassessment*. Revised edition. New York: Oxford University Press, 1990.

Dallin, Alexander, and F. I. Firsov. *Dimitrov and Stalin: 1934–1943, Letters from the Soviet Archives*. Russian documents translated by Vadim A. Staklo. New Haven, Conn.: Yale University Press, 2000.

Deriabin, Peter, S. *Inside Stalin's Kremlin: An Eyewitness Account of Brutality, Duplicity, and Intrigue.* With Joseph Culver Evans. Washington, D.C.: Brassey's, 1998.

Deriabin, Peter. *Watchdogs of Terror: Russian Bodyguards from the Tsars to the Commissars.* 2nd revised edition. University Publications of America, 1984.

Dziewandowski, M. K., *A History of Soviet Russia and its Aftermath.* Fifth edition. Upper Saddle River, N.J.: Prentice Hall, 1997.

Fitzpatrick, Sheila. *The Russian Revolution, 1917–1932.* New York: Oxford University Press, 1982.

Gaddis, John Lewis. *We Now Know: Rethinking Cold War History.* Oxford & New York: Oxford University Press, 1997.

Getty, J. Arch, and Oleg V. Naumov. *The Road to Terror: Stalin and the Self-Destruction of the Bolsheviks, 1932–1939.* Translations by Benjamin Sher. New Haven, Conn.: Yale University Press, 1999.

Joravsky, David. *The Lysenko Affair.* Chicago: The University of Chicago Press, 1986.

Khrushchev, Nikita. *Khrushchev Remembers.* With an Introduction, Commentary and Notes by Edward Crankshaw; Translated and edited by Strobe Talbott. Boston: Little, Brown and Company, 1970.

Knight, Amy. *Beria: Stalin's First Lieutenant.* Princeton, N.J.: Princeton University Press, 1993.

Kostyrchenko, Gennadi. *Out of the Red Shadows: Anti-Semitism in Stalin's Russia.* Amherst, New York: Prometheus Books, 1995.

Laqueur, Walter. *STALIN: The Glasnost Revelations.* New York: Charles Scribner's Sons, 1990.

Leskov, Nikolai. *Selected Tales.* Translated by David Magarshack. With an Introduction by V. S. Pritchard. New York: Farrar, Straus and Cudahy, 1961.

Lih, Lars T.; Oleg V. Naumov; Oleg V. Khlevniuk, editors. *Stalin's Letters to Molotov, 1925–1936.* Translated from the Russian by Catherine A. Fitzpatrick. Foreword by Robert C. Tucker. New Haven, Conn.: Yale University Press, 1995.

Medvedev, Roy. *Let History Judge: The Origins and Consequences of Stalinism.* Revised and expanded edition, edited and translated by George Shriver. New York: Columbia University Press, 1989.

Medvedev, Zhores, A. *The Rise and Fall of T. D. Lysenko.* Translated by I. Michael Lerner, with editorial assistance of Lucy G. Lawrence. Garden City, New York: Doubleday & Co., Inc., 1971.

Murphy, David E.; Kondrashev, Sergei, A.; and Bailey, George. *Battleground Berlin: CIA vs. KGB in the Cold War.* New Haven, Conn.: Yale University Press, 1997.

Pipes, Richard. *A Concise History of the Russian Revolution.* New York: Vintage Books, 1996.

Pipes, Richard, editor. *The Unknown Lenin: From the Secret Archive.* With the Assistance of David Brandenberger. Translations from Russian by Catherine A. Fitzpatrick. New Haven, Conn.: Yale University Press, 1996.

Pryce-Jones, David. *The Strange Death of the Soviet Empire.* New York: Henry Holt & Co., 1995.

Radosh, Ronald, and Joyce Milton. *The Rosenberg File.* Second edition. New Haven, Conn.: Yale University Press, 1997.

Radzinsky, Edvard. *Stalin.* Translated by H. T. Willetts. New York: Doubleday, 1996.

Rapoport, Louis. *Stalin's War Against the Jews: The Doctors' Plot and the Soviet Solution.* New York: The Free Press, 1990.

Rapoport, Yakov. *The Doctors' Plot of 1953.* Translated by N. A. Perova and R. S. Bobrova. Cambridge, Mass.: Harvard University Press, 1991.

Remnick, David. *Lenin's Tomb: The Last Days of the Soviet Empire.* New York: Random House, 1993.

Rubenstein, Joshua and Naumov, Vladimir P., editors. *Stalin's Secret Pogrom: The Postwar Inquisition of the Jewish Anti-Fascist Committee.* Translated by Laura Esther Wolfson. New Haven, Conn.: Yale University Press, 2001.

Rubenstein, Joshua. *Tangled Loyalties: The Life and Times of Ilya Ehrenburg.* New York: BasicBooks, 1996.

Salisbury, Harrison E. *The 900 Days: The Seige of Leningrad.* New York: Da Capo Press, 1985.

Sudoplatov, Pavel, and Anatoli Sudoplatov. *Special Tasks: The Memoirs of an Unwanted Witness—A Soviet Spymaster.* With Jerrold L. and Leona P. Schecter. Foreword by Robert Conquest. Boston: Little, Brown and Company, 1994.

Tucker, Robert C. *Stalin as Revolutionary: A Study in History and Personality.* New York: W. W. Norton & Co., 1973.

———. *Stalin in Power: The Revolution from Above, 1928–1941.* New York: W. W. Norton & Co., 1990.

———, editor. *Stalinism: Essays in Historical Interpretation.* With a new introduction by the editor. New Brunswick, N.J.: Transaction Publishers, 1999.

Ulam, Adam, B. *STALIN: The Man and his Era*. Boston: Beacon Pr., 1989.

Vaksberg, Arkady. *Stalin Against the Jews*. Translated by Antonina W. Bouis. New York: Alfred A. Knopf, 1994.

Volkogonov, Dmitri. *Lenin: A New Biography*. Translated by Harold Shukman. New York: The Free Press, 1994.

————. *STALIN. Triumph and Tragedy*. Edited and translated from the Russian by Harold Shukman. Rocklin, Calif.: Prima Publishing, 1992.

PUBLISHED BOOKS IN RUSSIAN

Gorky, M. *Nesvoevremennye mysli I rassuzhdeniya o revolyutsii I kulture, 1917–1918*. Moscow: MSP "Interkontakt," 1990.

Kaganovich, Lazar. *Pamyatnye zapiska*. Moscow: Vagrius, 1997

Malenkov, A. G. *O moem ottse Georgii Malenkove*. Moscow: NTTs "Tekhnoekos," 1992.

Mikoyan, Anastas Ivanovich. *Tak bylo: razmyshleniya o minuvshem*. Moscow: "Vagrius," 1999.

O polozhenii v biologicheskoy nauke: Stenographicheskii otchet sessii vsesoyuznoy akademii selskokhozyaistvennykh nauk imeni V. I. Lenina, 31 Iyulya–7 Avgusta 1948. Moscow: OGIZ—Selkhozgiz, 1948.

Stolyarov, K. A. *Golgofa: Dokumentalnye povesti*. Vol. I, *General-polkovnik Abakumov. Fakty I kommentarii*. Moscow, 1991.

Tolchanova, Tamara, and Mikhail Lozhnikov, editors. *I primknuvshii k nim SHEPILOV: Pravda o cheloveka, uchenom, voine, politike*. Moscow: "Zvokhitsa-MG," 1998.

Yakovlev, A. N., editor; A. Artizov; Yu. Sigachev; V. Shevchuk; V. Khlopov, compilers. *Reabilitatsiya: kak eto bylo. Dokumenty prezidiuma TsK KPSS i drugie materially, Mart 1953–Fevral 1956*. Moscow: Mezhdunarodnyi Fond "Demokratiya," 2000.

Yakovlev, A. N., editor; Andrei, Artizov, and Oleg Naumov, compilers. *Vlast I khudozhestvennaya intelligentsiya: Dokumenty TsK RKP(b), VChK-OGPU-NKVD o kulturnoy politike. 1917–1953*. Moscow: Mezhdunarodnyi Fond "Demokratiya," 1999.

Yakovlev, A. N., editor; V. Naumov, and Yu. Sigachev, compilers. *Lavrentii Beria. 1953: Stenogramma iyulskogo plenuma TsK KPSS I drugie dokumenty*. Moscow: Mezhdunarodnyi Fond "Demokratiya," 1999.

Sheinis, Zinovii. *Provokatsiya veka*. Moscow: PIK, 1992.

INDEX